T0145354

Studies in Fuzziness and Soft Computing

Volume 423

Series Editor

Janusz Kacprzyk, Systems Research Institute, Polish Academy of Sciences, Warsaw, Poland

The series "Studies in Fuzziness and Soft Computing" contains publications on various topics in the area of soft computing, which include fuzzy sets, rough sets, neural networks, evolutionary computation, probabilistic and evidential reasoning, multi-valued logic, and related fields. The publications within "Studies in Fuzziness and Soft Computing" are primarily monographs and edited volumes. They cover significant recent developments in the field, both of a foundational and applicable character. An important feature of the series is its short publication time and world-wide distribution. This permits a rapid and broad dissemination of research results.

Indexed by SCOPUS, DBLP, WTI Frankfurt eG, zbMATH, SCImago.

All books published in the series are submitted for consideration in Web of Science.

Shahnaz N. Shahbazova · Ali M. Abbasov ·
Vladik Kreinovich · Janusz Kacprzyk ·
Ildar Z. Batyrshin
Editors

Recent Developments and the New Directions of Research, Foundations, and Applications

Selected Papers of the 8th World Conference on Soft Computing, February 03–05, 2022, Baku, Azerbaijan, Vol. II

 Springer

Editors
Shahnaz N. Shahbazova ⓘD
Institute of Control Systems, Ministry
of Science and Education of the Republic
of Azerbaijan
Baku, Azerbaijan

Azerbaijan State University of Economics,
UNEC
Baku, Azerbaijan

Vladik Kreinovich
Department of Computer Science
University of Texas at El Paso
El Paso, TX, USA

Ildar Z. Batyrshin
Centro de Investigacion en Computacion
Instituto Politecnico Nacional
Ciudad de México, Mexico

Ali M. Abbasov
Institute of Control Systems, Ministry
of Science and Education of the Republic
of Azerbaijan
Baku, Azerbaijan

Azerbaijan State University of Economics,
UNEC
Baku, Azerbaijan

Janusz Kacprzyk
Systems Research Institute
Polish Academy of Science
Warsaw, Poland

ISSN 1434-9922   ISSN 1860-0808 (electronic)
Studies in Fuzziness and Soft Computing
ISBN 978-3-031-23478-1   ISBN 978-3-031-23476-7 (eBook)
https://doi.org/10.1007/978-3-031-23476-7

This Springer imprint is published by the registered company Springer Nature Switzerland AG
The registered company address is: Gewerbestrasse 11, 6330 Cham, Switzerland

Preface

These two volumes constitute the selected papers from of the 8th World Conference on Soft Computing or WConSC 2022, held on February 03–05, 2022, in Baku, Azerbaijan. The conference was organized by the Institute of Control System, Ministry of Science and Education of the Republic of Azerbaijan, University of California, Berkeley, California, USA, and the Azerbaijan State University of Economics UNEC, Azebaijan.

In 1991, Zadeh introduced the concept of Soft Computing—a consortium of methodologies that collectively provide a foundation for the conception, design, and utilization of intelligent systems. One of the principal components of Soft Computing is Fuzzy Logic. Today, the concept of Soft Computing is growing rapidly in visibility and importance. Many papers written in the eighties and early nineties were concerned, for the most part, with applications of fuzzy logic to knowledge representation and common sense reasoning. Soft Computing facilitates the use of fuzzy logic, neurocomputing, evolutionary computing, and probabilistic computing in combination, leading to the concept of hybrid intelligent systems.

Combining these intelligent systems tools has already led to a large number of applications, which shows the great potential of Soft Computing in many application domains.

The volumes cover a broad spectrum of Soft Computing techniques, including theoretical results and practical applications that help find solutions for industrial, economic, medical, and other problems.

The conference papers included in these proceedings were grouped into the following research areas:

- Soft Computing and Fuzzy Logic
- Fuzzy Applications and Fuzzy Control
- Fuzzy Logic and Its Applications
- Soft Computing and Logic Aggregation
- Fuzzy Model and Z-Information
- Fuzzy Neural Networks
- Artificial Intelligence and Expert Systems

- Probabilistic Uncertainty
- Intelligent Methods and Fuzzy Approach
- Fuzzy Methods and Applications in Medicine
- Fuzzy Control Systems
- Fuzzy Logic and Fuzzy Sets
- Fuzzy Simulation
- Fuzzy Recognition
- Fuzzy Applications and Measurement Systems

During the WConSC 2022 conference, we had two panels. The first panel titled **Lotfi A. Zadeh Legacy: Inspirations for Mathematics, Systems Theory, Decision, Control** featured presentations by Prof. Janusz Kacprzyk, Prof. Oscar Castillo, Prof. Ildar Z. Batyrshin, Prof. Nadezhda Yarushkina, Prof. Valentina E. Balas, Prof. Vadim Stefanuk, Prof. Jozo Dujmovic, and Prof. Shahnaz N. Shahbazova. The second panel titled **Fuzzy Logic and Soft Computing, Computational Intelligence, and Intelligent Technologies** featured Prof. Vladik Kreinovich, Prof. Oscar Castillo, Prof. Takahiro Yamanoi, Prof. Aminaga Sadiqov, Prof. Averkin Alexey, Prof. Ildar Z. Batyrshin, Prof. Marius Balas, Prof. Nishchal Kumar Verma, and Prof. Shahnaz N. Shahbazova. Many thanks to all the panel speakers for their very valuable memories of Prof. Lotfi A. Zadeh and for their interesting talks.

At the WConSC 2022 conference, we had sixteen keynote speakers: Prof. Janusz Kacprzyk (Poland), Prof. Vladik Kreinovich (Texas, USA), Prof. Oscar Castillo (Mexico), Prof. Ildar Z. Batyrshin (Mexico), Prof. Nadezhda Yarushkina (Russia), Prof. Valentina E. Balas (Romania), Prof. Vadim Stefanuk (Russia), Prof. Alexey Averkin (Russia), Prof. Marius Balas (Romania), Prof. Nishchal Kumar Verma (India), Prof. Imre Rudas (Hungary), Prof. Elizabet Chang (Austria), Prof. Asaf Hajiyev (Azerbaijan), Prof. Praveen K. Khosla (India), Prof. Takahiro Yamnoi (Japan), and Prof. Shahnaz N. Shahbazova (Azerbaijan). Summaries of their talks are included in this book.

The editors of this book would like to acknowledge all the authors for their contributions that kept the quality of the WConSC 2022 conference at a high level. We have received invaluable help from the members of the International Program Committee and from the conference. Thanks to all of them, we had remarkably interesting presentations and stimulating discussions.

We express our gratitude to the Institute of Control Systems of the Ministry of Science and Education of the Republic of Azerbaijan, the Director General of the Institute, Prof. Ali M. Abbasov, for assistance in organizing the conference and publications of the 8th World Conference on Soft Computing.

Our special thanks go to Janusz Kacprzyk, Editor-in-Chief of the Springer book series *Studies in Fuzziness and Soft Computing*, for the opportunity to organize this guest-edited volume.

We are grateful to the Springer staff, especially to Dr. Thomas Ditzinger (Senior Editor, *Applied Sciences and Engineering*, Springer-Verlag), for their excellent collaboration, patience, and help during the preparation of these volumes.

We hope that the volumes will provide useful information to professors, researchers, and graduated students in the area of Soft Computing techniques and applications and that all of them will find this collection of papers inspiring, informative, and useful. We also hope to see you all at future World Conferences on Soft Computing.

Baku, Azerbaijan   Prof. Shahnaz N. Shahbazova
Baku, Azerbaijan   Prof. Ali M. Abbasov
El Paso, USA   Prof. Vladik Kreinovich
Warsaw, Poland   Prof. Janusz Kacprzyk
Mexico City, Mexico   Prof. Ildar Z. Batyrshin

Contents

Fuzzy Recognition

Problems Difficulties in Recognizing Flat Figures

Shahnaz N. Shahbazova and **Ali M. Abbasov**

Abstract The paper explores classes of visual objects that have not yet received a stable solution, and private methods need artificial adjustments with subsequent difficulties with integration into the general and completely harmonious mode of the system. Flat figures act as one of these classes, and a model of particular approaches to problem classes of objects is built to the solutions of problems associated with them.

Keywords Artificial adjustments · Recognition stability · Artificial personality

1 Introduction

This paper explores the problem of recognition of visual and audio information in the natural environment, building an algorithmic model that imitates the human abilities of motor (unconscious or non-intelligent) pattern recognition.

Although the vast majority of modern developments in the field of information recognition are closed commercial projects and their theoretical achievements are practically not disclosed, the entire direction of research is in decline, and this, given that the last decades have been characterized by the rapid development of information technology in general, speaks of its theoretical and practical impasse [1–5].

Almost all more or less successful projects only confirmed the finite area of applicability and limited reliability of recognition results.

Sh. N. Shahbazova (✉) · A. M. Abbasov
Ministry of Sciences and Education of the Republic of Azerbaijan, Institute of Control Systems, 68 Bakhtiyar Vahabzadeh Street, Baku AZ1141, Azerbaijan
e-mail: shahbazova@gmail.com; shahnaz_shahbazova@unec.edu.az; shahbazova@cyber.az

Department of Digital Technologies and Applied Informatics, Azerbaijan State University of Economics, UNEC, Baku AZ1001, Azerbaijan

A.M. Abbasov
e-mail: pr.dr.abbasov@gmail.com

3

The research, of which this work is a part, is generally devoted to the development of a machine intelligence system with a gradual increase in the complexity of the behavioral model of an artificial personality (AP) in order to experimentally study the issues of artificial intelligence [6–10].

This research is aim at obtain an algorithmic model capable of extracting "meaningful" objects from the surrounding world for linking with the corresponding vocabulary concepts, which are atomic (indivisible) building blocks of intelligence. The question of the base language for describing vocabulary concepts is not fundamental, it is enough that in the target language it would be possible to explain the concepts: object, property, space, time, and action, i.e. almost every language in the world [11–15].

The aim of the work is to build a system for identifying objects of visual and sound information with decision-making based on the theory of fuzzy sets and neural networks.

2   Problems Without Sustainable Solutions

In this paper, problems are collected, in principle, related to the recognition problems being solved in work, but still not amenable to solutions by the developed identification methods [16].

These problems include:

- Recognition of a mirror image—any reflections;
- Recognition of secondary objects—the impossibility of binding and, accordingly, recognition of individual parts of the object (for example, by the hand, that this is a human hand, special training is required with the dictionary concept of "human hand");
- Recognition of flat figures—this type of visual image includes symbols, images and figures drawn on the surface (with imitation of the depth of the scene and, accordingly, the use of parsing objects by standard techniques).

The relationship of these problems to the "motor" human abilities of recognition, on the one hand, is not in doubt, since a person almost instantly orients himself in all of the situations listed, but on the other hand, to ask the question at what age a person masters these abilities, it turns out that only with the development of the imagination [17]. If, in general, imagination is considered an intermediary between the intellect and the natural environment, then with the help of willpower, a person is quite capable of imagining a reality that differs from the surrounding reality or finding hidden relationships and patterns that distinguish a person with a developed intellect. Thus, although the listed problems are of great research importance, most likely they cannot be fully resolved within the framework of this work due to their close relationship with the functioning of highly complex levels of intelligence [18].

3 Difficulties in Recognizing Flat Figures

In the course of experimental work with various types of visual information, various drawn geometric shapes turned out to be special in terms of recognition stability.

Despite the simple visual structure, they were extremely difficult to identify, since, being representatives of one class, they represent an extremely diverse set. It is possible that only highly intelligent animals (for example, primates, and dolphins) are capable of identifying the geometric shapes of humans [19–23].

Classically, the presentation of the sheet and the geometric figures drawn on it are flat objects (see Fig. 1a) and, accordingly, their focal lengths are equal (in fact, points A and B only indicate the approximate center of the processed area), the original approach is based on changing this factor.

If we do not take into account intermediate solutions that give almost 100% results, but only for particular experimental cases, then the only training method that gave quite satisfactory results is the method of converting the flat form of geometric shapes into volumetric ones, assuming that the sheet (on which the figure is drawn) is a distant background (see Fig. 1b, the focal length of point A is 0, i.e. as far as possible) [11, 24].

This approach makes it possible to make a clear parsing of visual objects, but a stable recognition method has not yet been found (for example, about 65% accuracy in letter recognition and about 75% in word recognition).

A strict limitation is accepted in the work that the color of the sheet is always monochromatic, otherwise, it is practically impossible to conduct experiments, since the system begins to carefully examine the sheet itself and the time of the experiment is greatly extended. But in general, this limitation is also characteristic of a person,

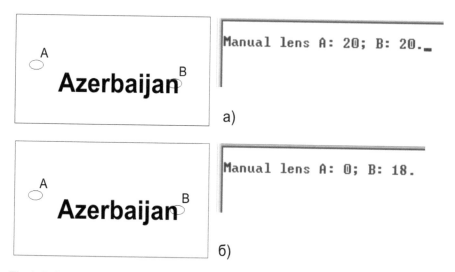

Fig. 1 Imitation of the volume of flat geometric shapes

everyone probably noticed how tiring it can be to read text against a non-uniform background and that the most interesting thing is that reading speed drops from the motor (0.1–0.5 s) to almost intellectual time (1–2 s), i.e. in fact, a person begins to recognize consciousness [12, 25–28].

As another complex experimental case, we can give an example of two cubes placed next to each other, which in 90% of cases will be recognized as one object, since they are characterized by one focal length, and it will be possible to separate them as two only if the Mach effect in preprocessing will give a good definition of the boundary (which at best will make up the remaining 10% of cases) (Fig. 2).

The figure below (see Fig. 3) shows another example, in which the visual object depicts the well-known psychological effect of jumping focus, which, simulated in the VS, also does not have stable recognition even for the case of one solid side of the cube (see Fig. 4 a and b), and their recognition is stable only as a flat figure when the background of the cube is represented by zero focus, and the edges have a certain focal length and, therefore, recognition is carried out by comparing the "relief" of the surface of the flat figure [29–31].

The main difference with natural intelligence is most likely due to the fact that a person gets used to the study of three-dimensional objects and therefore, diagonal gaze movements in any direction are associated with movement deep into the image and, accordingly, gives volume to the figure, which is not observed in a computer system, since the focal length does not change, and an artificial decrease in the focal length away from the main geometric center of mass leads to, although much better recognition, at the same time reduces the overall stability of the decisions made.

Many experiments on the recognition of various drawn geometric shapes (including points and lines) gave rather weak results compared to the general background, and most of the successful and sustainable solutions were mainly associated with the use of pre-processing, which helps to highlight the contrast of visual borders and then give them an artificial "height". compared to the background, in the focal length function.

Fig. 2 Illustration of the complexity of recognizing combined objects

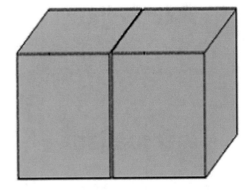

Fig. 3 Illustration of the
psychological effect of focus
instability

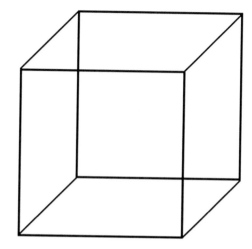

Fig. 4 The effect of
artificial 3-dimensionality of
a flat figures, a and b

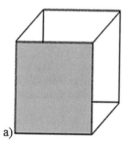 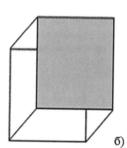

a)   б)

4 Conclusion

In this paper as a whole researches classes of visual objects that have not yet received
a stable solution, and particular methods need artificial adjustments with subsequent
difficulties in integrating into a general and completely harmonious mode of operation of the system. Flat figures act as one of these classes, and a model of particular approaches to problem classes of objects is built on the solutions to problems
associated with them.

References

1. Anisimov, B.V., Kurganov, V.D., Zlobin, V.K.: Recognition and Digital Image Processing. -
   M. Higher School (1984)
2. Bakut, P.A., Kolmogorov G.S., Vornovitsky, I.E.: Image segmentation: thresholding methods.
   Zarubezh Radioelectron 10(S), 4–24 (1987)
3. Nesteruk, V.F., Porfiryeva, N.N.: Contrast law of light perception. Opt. Spectro. T. XXIX(6),
   1138–1143 (1970)

4. Golovko, V.A.: Neurointelligence: Theory and Applications. Book 1. Organization and Training of Neural Networks with Direct and Feedback—Brest: BPI, p. 260 (1999)
5. Rabiner, L., Gould, B.: Theory and application of digital signal processing. M.: Mir, (1978)
6. Multidimensional signals and images under the action of interference: Diss. doctor of tech. Sciences. Novosibirs (1997)
7. Sotnik, S.L.: Lecture notes on the course "Fundamentals of designing artificial intelligence systems", 1997–1998. http://pmn.narod.ru/ii/ii_kat.htm, http://www.uic.nnov.ru/~nik/docs/ai/lecthtml/htm/lectures.htm
8. Kalintsev, Y.K.: Speech intelligibility in digital vocoders. - M.: Radio Commun. (1991)
9. Matrosov, V.L.: Theory of algorithms: Proc. for mat. specialist. ped. universities. – M.: Prometheus (1989)
10. Jacobsen, X., Zscherpel, U., Perner, P.: A Comparison Between Neural Networks and Decision Trees. Lecture Notes in Artificial Intelligence—Machine Learning and Data Mining in Pattern Recognition, pp. 144–158 (1999)
11. Pavlidis, T.: Algorithms for Computer Graphics and Image Processing. Radio Commun. 1986
12. Zhuravel, I.M.: A Short Course in the Theory of Image Processing. MAT-LAB Consulting Center of SoftLine Company. http://www.matlab.ru/imageprocess/book2/index.asp
13. Shahbazova, S.N., Suleymanov, M.R.: Stages of recognition probability images. In: Proceedings of AzTU (2004)
14. Shahbazova, S.N., Bonfig, K.V., Suleymanov, M.R.: Questions of construction of system of recognition of non-standard image and speech in the architecture of a hybrid vehicle based on neuro-fuzzy logic. In: 2nd International Conference on Application of Fuzzy Systems and Software Computing and Analysis System, Decision and Control, Barcelona, Spain (2004)
15. Shahbazova, S., Grauer, M., Suleymanov, M.: The development of an algorithmic model of object recognition using visual and sound information based on neuro-fuzzy logic. In: Annual Conference of the North American Fuzzy Information Processing Society-NAFIRS, 19 Mar 2011
16. Andrews, G.: Two-dimensional transformation. In: Huanga, T. (ed.) Image Processing and Digital Filtering. – M.: Mir. pp. 36–52 (1979)
17. Digital processing of television and computer images: Textbook. In: Zubarev, Y.B., Dvorkovich, V.P. (eds.) - M.: International Center for Technical Information (1997)
18. Findeisen, W., Malinowski, K.: Two-level control and coordination for dinamisal systems. Archiwum automatiki i telemechaniki. T. XX1V, N1, 3–27
19. Fu, K., Gonzalez, R., Lee, K.: Robotics: Translated from English. In: Gradetsky, V.G (ed.) – M.: Mir (1989)
20. Mariton, M., Drouin, M., Abou-Kandil, H., Duc, G.: Une nouvelle methode de decomposition-coordination. 3 e partie: Application a la commande coordonnees-hierarchisee des procesus complexes, N3, 243–259 (1985)
21. Petrou, M.: Learning in Pattern Recognition. Lecture Notes in Artificial Intelligence—Machine Learning and Data Mining in Pattern Recognition, pp. 1–12 (1999)
22. Ranganath, S., Arun, K.: Face recognition using transform features and neural networks. Pattern Recogn. 30, 1615–1622 (1997)
23. Sabah, G.: Knowledge representation and natural language understanding. AI Commun. 6(3–4), 155–186 (1993)
24. Bakhvalov, N.S., Kobel'kov, G.M., Zhidkov, N.P.: Numerical Methods. – M.: Nauka (1987)
25. Dunin-Barkovsky, I.V., Smirnov, N.V.: Probability Theory and Mathematical Statistics in Engineering. M.: GITTL (1955)
26. Vasil'ev, K.K., Krasheninnikov, V.R.: Filtering Methods for Multidimensional Random Fields. Sarat Publishing House, Saratov (1990)
27. Vitkus, R.Y., Yaroslavsky, L.P.: Adaptive linear filters for image processing. Adapt. Methods Image proc. – M.: Nauka (1988)
28. Duda, R., Hart, P.: Pattern recognition and scene analysis. – M.: Mir (1976)

29. Zagoruiko, N.G.: Recognition methods and their application. M. Sov. Radio (1972)
30. Gruzman, I.S., Kirichuk, V.S., Kosykh, V.P., Peretyagin, G.I., Spektor, A.A.: Digital Image Processing in Information Systems: Textbook. Publishing House of NSTU, Novosibirsk (2000)
31. Ivanov, L.D. :Variations of Sets and Functions. – M.: Nauka (1975)

The Problem of Parsing and Recognizing Individual Objects of Information on Base of Fuzzy Logic

Zinyet R. Huseynova

Abstract This paper presents a method for solving the algorithm of the model of recognition of visual and audio information of the environment by constructing an algorithmic model that simulates the human abilities of the motor (unconscious or non-intellectual) behavior. The paper is devoted to the development of a machine intelligence system with a gradual increase in the complexity of the behavioral model of an artificial personality (AP) to experimentally study the issues of artificial intelligence.

Keywords Recognition of Visual and Sound Objects · Modulated Speech Information · Perception of Information · Parsing Visual Information

1 Introduction

One of the most critical from the point of view of the success of recognition of visual and sound objects is the problem of qualitative analysis of incoming information. Although there are few works trying to solve this problem, some aspects of this work and other related problems can be found in the works [1–8].

The main distinguishing feature of the research was the task of finding the most natural (i.e., capable of arising evolutionarily with the development of biological species) ways of perceiving visual and sound objects of information and functioning on the basis of simple physical laws.

The goal set in the work does not allow the use of any technical means that have facilitated at least some stages of recognition, for example, special light sources, sound location techniques, specially modulated speech information, etc.

The problems of parsing and recognition are not accidentally combined into one section, the relationship between the allocation of a potential object of information and its immediate recognition attempt is very high, and any attempts to separate

Z. R. Huseynova (✉)
1Nakhchivan State University, University Campus, 7012 Nakhchivan, Azerbaijan
e-mail: zhuseynova123@gmail.com

© The Author(s), under exclusive license to Springer Nature Switzerland AG 2023
Sh. N. Shahbazova et al. (eds.), *Recent Developments and the New Directions of Research, Foundations, and Applications*, Studies in Fuzziness and Soft Computing 423,
https://doi.org/10.1007/978-3-031-23476-7_2

them (to optimize calculations) led to almost complete loss of stability of the results. However, the course of experiments has shown that sequential recognition does not always lead to a successful solution of the problem of identifying information objects.

As you can see, if any theoretical and practical difficulties arise in the course of work, then attempts are made in research to simulate human behavior in a similar situation, this philosophy in the work often helped to get out of deadlocks, albeit at the cost of more computationally wasteful algorithms. Perhaps in the future, serious research on human behavioral patterns will allow us to get closer to the issues of artificial intelligence [9–11].

At the heart of the problem of parsing information is the inability of the computing system to distinguish between visual and sound objects among noise, which it is impossible to get rid of:

For visual objects, noise is:

- background,
- light distortion,
- own non-uniform color,
- own and extraneous shadow overlays,
- overlapping objects with each other,
- etc.

For sound objects, due to less informative content, noise is:

- background accompaniment,
- mutual distortion of sound information, etc.

At the same time, not only speech information is accepted as sound information, but also other sounds that can be described by vocabulary concepts, and those bearing aesthetic or psychological significance will be inaccessible to computing systems (at least without the use of intelligence).

2 Recognition of All Available Objects

Strengthening the framing of potential visual and sound objects at the stage of preliminary information processing significantly improves the characteristics that contribute to the correct analysis of information objects, but still not enough to move to the recognition stage.

With detailed recognition of all available objects, the quality of the selection of information objects from the environment comes to the fore. The peculiarity of humans and animals is the simplicity and naturalness with which this task is solved by the intellect, although there is a possibility that most of the load lies on the subconscious [12].

The method of local foci is difficult to describe mathematically, since it is more of a programmed algorithm than any function, and this algorithm consumes more than 80% of computing resources and 95% of the experiment time.

Despite the fact that there are no serious differences in the algorithm of the program processing visual and audio information, there is a difference in the stages of processing and in the way, information is presented, therefore, in those areas that have their own characteristics will be considered in more detail.

The difference in the nature of visual and sound information is expressed in the fact that flat perception of information is sufficient for the study of sound information, i.e. the front of the sound wave and time (see Fig. 1) represent the necessary and sufficient amount of information to recognize the information contained in it (the location of the sources of sound information can be neglected due to the imperfection of the receiving devices), while the recognition of visual information requires taking into account the depth of the scene (see Fig. 2) including a virtual one, which is a 3-dimensional image in photographs and in televisions, which also have depth, despite their flat way of displaying. Suffice it to note that when developing graphical program interfaces, designers specifically frame active components (menus, buttons, text fields, etc.) with an artificial shadow, as this gives a natural volumetric image and helps to reduce the strain on users' eyes.

Before proceeding to the presentation of the algorithm for parsing information objects [13], we determine that:

- the concept of focus in the method of local foci and the criterion of focal length refers to the usual optical effect of the basic course of physics, the phenomenon of focusing on certain areas of the scene by analogy with devices for automatic adjustment of image sharpness and simulating the change in refraction of the lens of the human eye when studying a visual object;
- the concept of frequency density is understood as a raster area on which the values of frequencies (amplitudes or other indicators) of a sound fragment are marked with dots (or circles, dashes, etc.) in order to obtain a density map (the number of raster areas may vary);

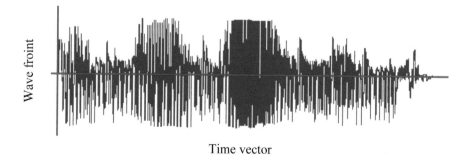

Time vector

Fig. 1 Audio information. The plane of the sound wave and the time vector

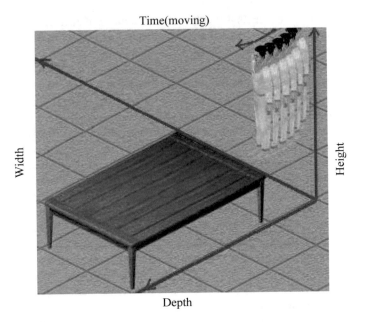

Time(moving)

Width

Height

Depth

Fig. 2 Visual information. Scene volume and time vector

- the concept of a scene means a closed space (room) with objects placed in it, in general, a real room with real objects is meant, but for the reasons discussed in the previous sections of this chapter, in our case it is a virtual model (i.e. computer simulation) of a room with 3-dimensional models of various real objects [14, 15].

3   Visual Information

Studies of the possibilities of parsing visual information have shown, and experimental data have confirmed, that the depth parameter of the scene is extremely important for parsing the scene into visual objects, which consists in the fact that, relative to some observer, each object located in some area of the studied scene corresponds to its own focal length [16, 17].

Thus, the method is a series of multiple measurements of focal lengths, the levels of which show an approximate map of the surface of a visual object, and if the focal length increases sharply, then either this area is outside the boundaries of the object or there is a through hole in the object, in any case by after some time, the system has at its disposal both a map of focal lengths and a real image of this (focal) area. Such two-part information about the image under study, called a trial image-contour, is the most informative model of all the tested methods, despite some informative redundancy, which in this case can be neglected [14, 15].

Of course, not in all cases it is possible to obtain the correct focal length (in a virtual environment, unlike a real one, determining the distances between two points is solved trivially—$\sqrt{(x_1 - x_2) + (y_1 - y_2) + (z_1 - z_2)}$), this factor will introduce a certain statistical error associated with random selections of measured coordinates to build a focal length map (image-contour). It is quite possible that there are (or will be created) optical devices (simulating the human eye) capable of focusing on individual sections of a visual object with measuring the appropriate distance, without which the transition from a virtual environment to a real one will be practically impossible.

The stage of visual information recognition is a function of calculating the similarity of objects that the system has already trained and the received image-contour of the target object. The algorithm of the similarity function is based on comparison with a small level of permissible error of the focal maps of the objects being compared [18].

Perhaps this factor is intensively used in the evolution of animals (both the devices of the visual system and the appearance), if you do not take into account the usual means of disguising under the natural environment, which are also difficult for humans, then, for example, drawings on the body of animals easily identified by humans may make predators hunting them a kind of blurred spots that require additional focusing time or identification, at the same time, a similar technique takes place with drawings on predators (for example, a cobra pattern).

In favor of the applicability of the local focus method, it says that this is practically the only way to naturally determine the integrity of an object, since there are no ways to determine whether two adjacent (visually) pieces of the scene belong to the same object of information, while neither enlarging nor reducing the trial pieces improves the characteristic, and any solution without taking into account the change in focal length will inevitably have low stability.

An additional consequence of the use of focal length in the study is the possibility of correctly taking into account the volume of objects and their orientation in space, and, accordingly, to make a comparison not blindly by image points, but taking into account the bulges and concavities mainly characterizing the objects of visual information.

These conclusions are not any principle or law, it is possible to create such visual scenes in which its consequences can be questioned.

There is no evidence that there is no other way of parsing the scene into its constituent objects and subsequent recognition, however, no other has been found in the course of research a method that would combine such naturalness and universality of the application of the basic physical properties of the natural environment.

4 Audio Information

One of the main methods of speech recognition is the processing and analysis of speech phonemes. And although the idea of creating a dictionary of phonemes will allow you to store words in the form of a sequence of corresponding indexes, and recognition will reduce to the selection of phonemes from the sound stream with a parallel search for the resulting sequence in the dictionary, it looks very logical and effective—the stability of the decisions it makes remains at a fairly low level. In practice, this idea works with very serious limitations solved by various research groups, and in general, by far the most successful this method solves the problem of identifying a person by his voice [19].

With the second method of recognizing audio information adopted in the work, everything is somewhat simpler. The minimum element of speech is the word and, accordingly, there is no need for precise adjustment of any coefficients of characteristics of allophones for different speakers. The comparison takes place according to the main characteristics of the sound series, although the method used in the work gives quite satisfactory results, a relatively small number of sound objects still participate in experiments, and therefore other researchers may well use completely different methods of comparing sound information up to the use of relatively new methods of wavelet analysis [20].

The redundancy of this method and the main reason for the rejection of this method in the development of speech recognition systems is the need to store whole dictionary instances, this becomes burdensome with a dictionary of 500 words and above. In this paper, this property is not only not a disadvantage, but can also be regarded as an advantage of the naturalness of the learning process (by analogy with a child who can speak before he begins to study individual letters or phonemes).

Therefore, in the case of sound information, it is mainly necessary only to select the target frequency domain belonging to the sound object under study from the general sound stream [21].

The analysis and recognition of sound objects is based on another property of information, different from that used for the analysis and recognition of visual objects, which consists in the fact that, relative to some listener, each sound information in a certain time period corresponds to its own frequency domain. This definition contains some error, since several sound objects are quite capable of overlapping each other, but in general this property has shown quite stable results in experiments.

As such, there is no mathematical proof of this definition, but in practice this property can be easily verified by trying to listen to the simultaneous speech of at least two interlocutors with the same type of voice (not necessarily similar), very soon it turns out that this is extremely tedious (most likely due to participation in the recognition of intelligence). At the same time, it is much easier to listen to different types of voices at the same time (for example, male and female voices), although it is also somewhat tedious [22].

Figure 3 shows the frequency density of a sound fragment with the word "circle" (circled with a red oval) pronounced by a female voice against the background of interference, if this word was pronounced by a male (ordinary) voice, then the main frequency density would be in the area circled with a blue square. There are much more variations with children's and immature voices, but in general they are more subtle (the frequency is higher) and the frequency spot area is much smaller, i.e. the range involved is much narrower than that of a mature male or female voice (it is impossible to confuse the concept of range with loudness).

If there are several density clusters, the recognition procedure is carried out for each of them, regardless of their type.

The longer the audio information, the greater the data density, which in general negatively affects the quality of parsing, therefore, after reaching 25–30% of the filling (derived experimentally), it becomes necessary to add a new bitmap image [23].

Analysis of the frequency densities of artificial sounds have a characteristic unnatural appearance (in the form of dense dots, lines or waves), at the same time, live sounds have serious differences from the characteristic type of human speech Fig. 4.

The interference accompanying in one volume or another all the sounds of the natural environment are predominantly chaotic in nature and can be filtered out by simply ignoring the sound information that is not included in the target area [24].

The following Fig. 5 shows the characteristic of the frequency densities of non-speech information. This is a short fragment of a symphonic work, transferred to a bitmap image, in itself may well represent some abstract value (artistic). Several violins form the upper strokes, the wind instruments split the lower ones, and the percussion point in the lower right corner.

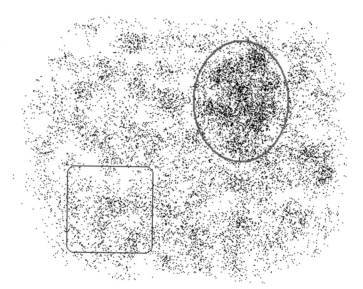

Fig. 3 Frequency density of the word "circle" (circled in red oval) against a background with strong interference

Fig. 4 Example of using
multiple frequency densities

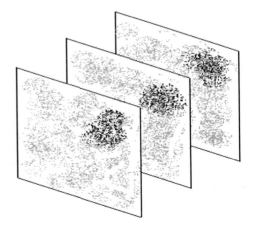

Fig. 5 Illustration of
frequency densities of
non-speech information
(fragment of a symphony)

5   Conclusion

This paper provides a way to solve the problem of isolating information objects from
the general picture and relative to each other, using the method of local foci for visual
information and the method of frequency densities for sound information. The given
justification boils down to the fact that the applied one is quite capable of disassem-
bling a visual scene and a sound fragment of high complexity, as well as recognizing
their constituent objects. When using hardware with sufficient computing power, the
method of algorithmic parallelization will allow you to bring real computing time to
virtual.

References

1. Aizenberg, I.N., Aizenberg, N.N., Krivosheev, G.A.: Multi-valued and Universal binary neurons: learning algorithms. Applications to Image Processing and Recognition. Lecture Notes in Artificial Intelligence—Machine Learning and Data Mining in Pattern Recognition, pp. 21–35 (1999).
2. Astova J., Haavisto P.: Vector median filter. Proceedings of IEEE, Vol. 78. №. 4, p. 678 (1990)
3. Belhumeur, P.N., Hespanha, J.P., Kriegman, D.J.: Eigenfaces versus Fisherfaces: recognition using class specific linear projection. IEEE Trans. Pattern Anal. Mach. Intell. **19**, 711–720 (1997)
4. Kaiser, G., Kaplan, S.: Parallel and distributed incremental attribute evaluation algorithms for multiuser software development environments. ACM Trans. Softw. Eng. Methodol. **2**(1), 47–92 (1993)
5. Knerr, S., Personnaz, L., Dreyfus, G.: Handwritten digit recognition by neural networks with single-layer training. IEEE Trans. Neural Netw. **3**, 962–968 (1992)
6. Krzyzak, A., Dai, W., Suen, C.Y.: Unconstrained handwritten character classification using modified backpropagation model. Proceedings of 1st International Workshop on Frontiers in Handwriting Recognition. Montreal, Canada, CTP, pp. 155–166 (1990)
7. Martin, G.L., Pitman, J.A.: Recognizing hand-printed letters and digits using backpropagation learning. Neural Comput. **3**, 258–267 (1991)
8. Wilson, I.D.: Foundations of hierarhical control. Int. J. Control **29**(6), p. 899–933 (1979)
9. Michalska H., Ellis J.E., Roberts P.D.: Joint coordination method for the steady-state control of large-scale systems. Int. J. Syst. Sci. (5): 605–618 (1985)
10. Shahbazova, S., Grauer, M., Suleymanov, M.: The development of an algorithmic model of object recognition using visual and sound information based on neuro-fuzzy logic. In: Annual Conference of the North American Fuzzy Information Processing Society-NAFIRS, March 19 (2011)
11. Vetter, T., Poggio, T.: Linear object classes and image synthesis from a single example image. IEEE Trans. Pattern Anal. Mach. Intell. **19**, 733–742 (1997)
12. Findeisen, W., Malinowski, K.: Two-level control and coordination for dinamisal systems. Archiwum automatiki i telemechaniki. T. XX1V, N1, p.3–27
13. Foltyniewicz, R.: Efficient high order neural network for rotation, translation and distance invariant recognition of gray scale images. Lecture Notes in Computer Science—Computer Analysis of Images and Patterns, pp. 424–431 (1995)
14. Ketcham, D.J.: Real time image enhancement technique. In: Proceedings of SPIE/OSA Conference on Image Processing. Pacific Grove, California, vol. 74, pp. 120–125 (1976)
15. Kralik, J., Stiegler, P., Vostry, Z., Zavorka, J.: Modelovani dynamiky rozsahlych siti. Praha, Akademia, p. 364 (1984)
16. Jacobsen, X., Zscherpel, U., Perner, P.: A comparison between neural networks and decision trees. Lecture Notes in Artificial Intelligence—Machine Learning and Data Mining in Pattern Recognition, pp. 144–158 (1999)
17. LeCun, Y., Matan, O., Boser, B., Denker, J.S., Henderson, D., Howard, R.E., Hubbard, W., Jackel, L.D., Baird, H.S.: Handwritten zipcode recognition with multilayer networks. In: Proceedings of International Conference on Pattern Recognition, Atlantic City (1990)
18. Milanova M., Almeida P. E. M., Okamoto J. and Simoes M. G.: Applications of cellular neural networks for shape from shading problem. lecture notes in artificial intelligence. Machine Learning and Data Mining in Pattern Recognition, pp. 51–63 (1999)
19. Shahbazova S.N., Suleymanov, M.R.: Stages of recognition probability images. In: Proceedings of AzTU (2004)
20. Shahbazova, S.N., Bonfig, K.V., Suleymanov, M.R.: Questions of construction of system of recognition of non-standard image and speech in the architecture of a hybrid vehicle based on neuro-fuzzy logic. In: 2nd International Conference on Application of Fuzzy systems and software computing and analysis system, Decision and Control, Barcelona, Spain (2004)

21. Tzafestas, S.G.: Large-scale systems modeling in distributed-parameter control and estimation. In: Modeling Simul. Eng. 10th IMACS World Congr. Syst. Simul. and Sci. Comput., Montreal, 8–13 Aug. 1982, v.3" Amsterdam e.a., p.69–77 (1983)
22. Translation, W.W., Report, T.: Reprinted in machine translation of languages, 1955, pp. 15–23. MIT Press, Cambridge, MA (1949)
23. Winograd, T.: Language as Cognitive Process, vol. 1. Addison-Wesley, Syntax (1983)
24. Yoon, K.S., Ham, Y.K., Park, R.-H.: Hybrid approaches to frontal view face recognition using the hidden Markov model and neural network. Pattern Recogn. **31**, 283–293 (1998)

Call Center and Robot Call Center How to Reduce the Cost

Rauf Fatullayev, Orkhan Fatullayev, and Abulfat Fatullayev

Abstract This article explores the application of artificial intelligence in call centers. It is shown how an artificial intelligence-based robot operator can stop the constant increase in call center costs and significantly reduce call center costs. It also describes the process of developing robot operators and their work in real time.

Keywords Call center · Artificial Intelligence · Azerbaijani

1 Introduction

Call center. Wikipedia defines a call center as follows: A call center (British spelling: call centre) is a centralised office used for receiving or transmitting a large volume of enquiries by telephone. An inbound call centre is operated by a company to administer incoming product or service support or information enquiries from consumers.

A modern call center is sometimes referred to as a contact center. A contact center, a further extension to call centers, administers centralized handling of individual communications, including letters, faxes, live support software, social media, instant messages, and email.[1]

The Oxford English Dictionary defines the term "call center" as follows: "a call center is an office staffed and equipped to handle large numbers of telephone calls, using computer technology to assist in the management of calls, supply of information, etc.; especially, such an office providing the centralized customer contact and customer service functions of a large organization.[2]

---

[1] https://en.wikipedia.org/wiki/Call_centre#cite_note-2.

[2] https://www.oxfordlearnersdictionaries.com/definition/american_english/call-center.

R. Fatullayev · A. Fatullayev (✉)
Ministry of Science and Education of the Republic of Azerbaijan, Institute of Control Systems, B.Vahabzadeh Str, 68, Az 1141, Baku, Azerbaijan
e-mail: fabodilmanc@gmail.com

O. Fatullayev
Nizami Dist, Robotroniks LTD, M. Sherifli Street 19, AZ1002 Baku, Azerbaijan

© The Author(s), under exclusive license to Springer Nature Switzerland AG 2023
Sh. N. Shahbazova et al. (eds.), *Recent Developments and the New Directions of Research, Foundations, and Applications*, Studies in Fuzziness and Soft Computing 423,
https://doi.org/10.1007/978-3-031-23476-7_3

From an organizational point of view, the call center is organized as follows: it consists of a team of operators who conduct dialogues with customers, one or more supervisors who control the work of operators and redistribute calls to them, and a technical support team.

The history of call centers. The 20 s of the XX century can be considered the beginning of the emergence of call centers. Before telephones, correspondence with clients was conducted mainly by mail. Telephones also allowed "live" communication with customers, through telephone conversations with them.

Starting with the installation of telephones, most of the state bodies in Azerbaijan also created separate telephone rooms, which, in addition to answering incoming calls, also provided them with outgoing information.

These telephone rooms can be considered the "embryo" of call centers.

Call centers appeared in Azerbaijan in the 70 s. One of the first call centers was the "09" inquiry service, "say the address and get the phone number for this address" or vice versa.

2 Modern Call Centers and Their Problems

Modern call centers originated in the USA. In the United States, call centers emerged as a specific area of the market and technology at a time when it became possible to conduct mass advertising campaigns through radio and TV, which caused a very large volume of requests from potential consumers interested in the advertised product [1].

Thus, a need arose for a means of effectively processing telephone calls as a result of advertising campaigns.

This is how these call centers developed as marketing advertising tools, and it was only then that they came up with the idea that call centers could not only receive a huge volume of incoming calls but also conduct marketing campaigns themselves.

Therefore, the word combination "call center" was associated with a room, in which several tens or hundreds of operators sat with headsets and other necessary "ammunition." But the call center paradigm has evolved significantly over the past few years [2, 3].

The main reason for this is the ever-increasing number of calls to call centers.

The increase in requests makes it impossible for a small number of operators to respond to all inquiries. Therefore, on the one hand, a potential buyer is lost, on the other hand, a negative opinion about the organization is formed.

A "rough" solution to this problem may be to increase the number of operators. This leads to an increase in the number of call center staff and thus a decrease in the organization's revenue.

Thus, as it is said in chess, the situation of "zugzwang[3]" arises.

---

[3] A position in checkers and chess in which any move by a player leads to a deterioration in his position.

3 Determination of the Optimal Number of Operators

Let us try to answer such a question:

1. The customer does not wait in line for more than 20 s,
2. At least 80% of calls received during the working day are answered.

How many operators are required for the call center under these conditions?

This problem was solved by the Danish mathematician A.K. Erlang at the beginning of the last century [4, 5].

Let's admit that most of the costs of any call center are the salaries of operators, the purchase of furniture and equipment to create working conditions for them, training, etc. costs.

If we want to significantly reduce the cost of any call center, we have to cut staff. No other action will have such an impact on cost reduction.

For example, let's look at the 2018 report of the State Agency for Public Services to Citizens (ASAN service) under the President of the Republic of Azerbaijan [6]:

"Since August 2018, the call center has been operating in a new office with a total area of about 2000 square meters. "The center is equipped with tables, computers, the necessary software ..., headphones with a silencer function, a microphone, and other modern technical and other equipment, ... The Call Center has launched the Cisco UCCE (Cisco Unified Contact Center Enterprise) IP telephony system, which allows applying the latest innovations in the organization of services in this area. In total, 90 employees and 10 team leaders will work in the office, as well as working groups of 20 employees, which are expected to participate in the organization of the work." Can you imagine how much this organization requires? Hundreds of thousands of dollars are needed per year to maintain such a call center.

We will now answer the question of the "optimal number of operators" mentioned above in the example of the Call Center No. 195–1 of the Ministry of Economy.

First, it should be noted that the Cabinet of Ministers of Azerbaijan adopted Resolution No. 50, dated February 25, 2015, on the rules for establishing call centers and regulating their work.

According to the decree, the rules for the organization of call centers were approved in order to (a) timely meet the requests of citizens and (b) ensure transparency and accessibility of government agencies.

According to this document, periodic monitoring should be performed in call centers in order to measure and improve service quality and to identify and prevent recurrence of shortcomings in the future [7].

At present, call centers operate in most ministries and large government agencies, businesses, banks, and other organizations. All these organizations are interested in reducing the cost of maintaining call centers.

It should be noted that the large number of staff is also characteristic of some other call centers [8].

For example, one such call center is the Call Center No. 195–1 of the Ministry of Economy [8].

Table 1 Call center of the ministry of economy (195–1)

Period	Calls (×1000)	Answered calls (×1000)	Unanswered calls (×1000)	Unanswered calls (%)
2020	1047.2	463.2	584.1	55.8
2019	674.2	377.0	297.2	44.1
2018	354.3	256.3	98.0	27.7
2017	414.6	312.5	102.1	24.6
2016	316.3	220.4	95.9	30.3

Table 2 Calculation of the umber of operators (195–1)

Period	Average call duration (Sec.)	Average post processing time (Sec.)	Number of calls per 8 h	Customer waiting time on the line (Sec.)	Number of operators typical/Max
2020	242	30	2705	20	44 / 52
2019	272	30	1976	20	37 / 42
2018	215	30	1104	20	18 / 20
2017	216	30	1296	20	21 / 23
2016	248	30	872	20	17 / 19

Table 1 shows the number of calls received by the call center of this ministry in the last 5 years (2016–2020) [9].

The A.K. Erlang formulas and the data in the table above (Table 1) can be used to calculate the optimal number of operators [10]. But the easiest way is to use a calculator based on these formulas (Erlang calculator[4]).

Table 2's column 6 contains the results of computing the number of operators for each row.

How should the table's results be interpreted?

Operators, for example, spent an average of 242 s (4 min and 2 s) on each phone call in 2020, with a 30-s pause before answering the next call. It is also known that during the working day, 2705 calls were received.

With a 20-s wait time, it takes 44 operators during regular hours and 52 operators during peak hours to answer at least 80% of incoming calls.

As a result, it can be argued that the "Achilles heel" of contact centers is the rising expense of their upkeep as the quantity of calls increases over time.

Thus, two options are possible:

1. Either reduce the number of operators while increasing user wait times, and conceqvently reduce the quality of service,
2. Or increase the number of operators while ignoring rising call center costs. In both cases, the profit of the call center is reduced.

---

[4] https://www.callcentrehelper.com/tools/erlang-calculator/day-planner.php.

The next section provides information on the new Robot Call Center and its structure, which will allow us to solve this problem.

4 Robot Call Center[5]

Hereinafter, we will name the call center, which operates on the basis of artificial intelligence (AI), a "Robot call center" (hereinafter RCC).

The idea of applying AI to call centers is not new, but it does not have a long history.

AI technologies are used in call centers for the following tasks: intelligent voice menus (IVR) for handling incoming calls, autodialing of clients, converting telephone conversations to text for further processing, evaluation of employee performance and customer satisfaction, solving the staffing problem of large call centers, etc. [11].

The development of such applications facilitates the work of operators, improves the quality of customer service, and allows call center management to evaluate the work of operators. But it does not replace operators and does not allow a reduction in the number of hired workers.[6]

However, it should be emphasized that the use of AI in terms of replacing operators in call centers is a completely new issue, and there is no noticeable software in this area.

The RCC provides for a more radical application of AI in call centers—the replacement of human operators by robot operators.

If we discard all the "additional details", we will get a "clean" call center, the scheme of which is shown in Fig. 1.

In the simplest version, the call center consists of an operator and a supervisor who oversees its work. The client calls, and the operator raises his phone handset, listens to his question, asks clarifying questions, if necessary. He/She voices the answer and ends the conversation.

During the conversation, the supervisor monitors the process and intervenes when necessary (for example, when the operator has difficulty answering the client's question) (the client does not feel this interference).

Now imagine that instead of an operator, AI (we will call it a robot-operator) responds to customers. AI does not mean "global" AI that is omniscient and can answer all customer questions. The robot-operator is trained in the services that are part of the "authority" of the call center and can only answer questions about these services.

---

[5] The project "Development of the Robot Call Center" supports by National Academy of Sciences of Azerbaijan (ANAS).

[6] https://www.liveintellect.ru/tags/ii_dlja_call-centrov/.

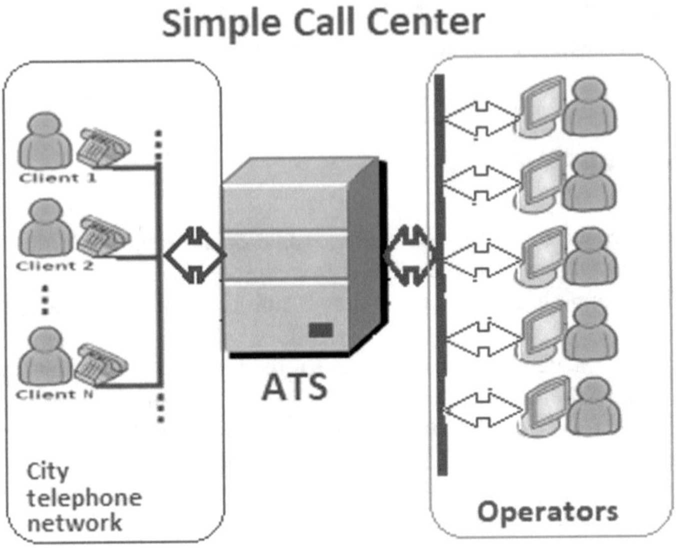

Simple Call Center

Fig. 1 Scheme of the simplest call center

This also applies to human-operators. The operator of the call center, for example, of the Ministry of Health answers only questions related to the field of health, does not answer questions related to the Ministry of Economy, and at best advises the client to contact the appropriate call center.

In the same way, the robot operator is trained according to certain rules and databases. If the question is not within the scope of the knowledge taught to him, then the robot directs it to the human operator or supervisor.

Thus, in order to get out of a difficult situation, it is necessary to keep a certain number of operators in RCCs. It should be noted that a similar situation is observed in traditional call centers. In this case, the operator stops the conversation and informs the client that he will be called in the next few hours. For example, "in 2020, 8417 application forms were compiled in the category of complex questions…".

At the end of this section, we note that, unlike other call centers, the RCC is designed as a self-learning system.

Thus, when the robot-operator cannot understand the question, he directs it to the operator. In this case, the operator's answer is added to the database of "incomprehensible questions" together with the client's question, and the database is processed over time and added to the robot operator's knowledge base. Thus, the next time such questions are asked, the robot "understands" and answers them. There is no need to direct the question to the operator.

5   Robot-Operator

Undoubtedly, the most interesting element of the RCC is the robot operator.

To understand how it works, let's look at a fragment that describes how a robot-operator works in real-time question-and-answer mode (Fig. 2).

As shown in Fig. 2, the robot operator, like the human operator, gradually approaches the response expected by the client and finally sounds the required response.

The translation of this dialogue is given below:

User: I want to get a passport; how much do I have to pay?
System: Please tell, is this your first-time getting a
passport or changing your passport?
User: For the first time
System: How old are you?
User: 45 years old
> > db_search(domain:'passport',intent:'say_price',slots:
(renew_type:'new',age:45)
System: Amount of state duty: 60 manats for 10 days, 120
manats for 5 days, 210 manats for 1 day

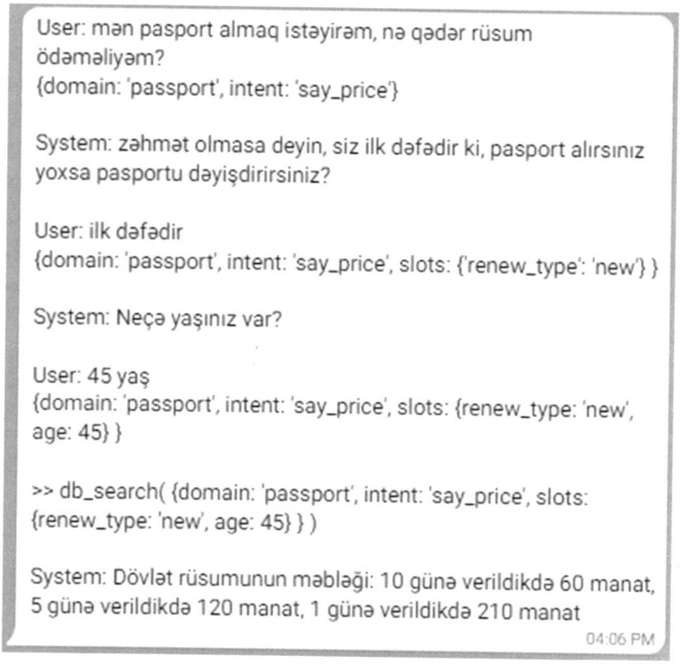

```
User: mən pasport almaq istəyirəm, nə qədər rüsum
ödəməliyəm?
{domain: 'passport', intent: 'say_price'}

System: zəhmət olmasa deyin, siz ilk dəfədir ki, pasport alırsınız
yoxsa pasportu dəyişdirirsiniz?

User: ilk dəfədir
{domain: 'passport', intent: 'say_price', slots: {'renew_type': 'new'} }

System: Neçə yaşınız var?

User: 45 yaş
{domain: 'passport', intent: 'say_price', slots: {renew_type: 'new',
age: 45} }

>> db_search( {domain: 'passport', intent: 'say_price', slots:
{renew_type: 'new', age: 45} } )

System: Dövlət rüsumunun məbləği: 10 günə verildikdə 60 manat,
5 günə verildikdə 120 manat, 1 günə verildikdə 210 manat
                                                                      04:06 PM
```

Fig. 2 Fragment of real-time work of the robot-operator

```
[
    {
        "domain": "passport",
        "renew_types": [
            {
                "renew_type": "new",
                "case_description": "yeni pasport aldıqda",
                "required_docs": [
                    {
                        "age_range": "0-17",
                        "required_docs": "şəxsiyyət vəsiqəsi; dövlət rüsı
                    },
                    {
                        "age_range": "18-35",
                        "required_docs": "şəxsiyyət vəsiqəsi; dövlət rüsı
```

Fig. 3 The robot-operator knowledge database

In order to obtain such an accurate dialogue, a database has been created that contains "Domain", "Intent", and "Slots". The services provided by the call center (Domain names), the purpose of the client (Intents) and the possible answers (Slots) are included in a special order - in the database in JSON[7] format. A small fragment of this base is shown in Fig. 3.

Natural Language Processing (NLP) technologies were used to create the robot operator. These are ASR (Automatic Speech Recognition), NLU (Natural Language Understanding), TOD (Task-Oriented Dialogue), and TTS (Text To Speech).

For example, in Fig. 3, the customer asks, "I want to get a passport, how much do I have to pay?" Let's follow the work of the robot operator to answer the question.

1. As soon as the question is asked, the ASR converts the sound signal to text;
2. The NLU system analyzes the question and determines to which domain the question belongs and what is its purpose [12, 13]: {domain: "passport", intent: "say_price"};
3. The TOD system ensures dialogue with the client [14, 15];
4. The TTS system allows the response in text format to be delivered to the client in human voice.

After developing a robot operator, you can begin to create a RCC where the main load is borne by robot operators.

Because while AI lags far behind human intelligence, in such call centers, in addition to robot operators, there will also be human operators.

But since the robot operator is a self-learning system, over time, the proportion of robot operators in the operators will gradually increase, and the number of human

---

[7] JSON-Java Script Object Notation.

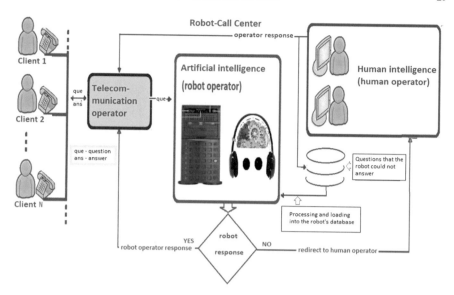

Fig. 4 Robot-call center

operators will decrease, until eventually their (human operators) number will reach the "optimal" minimum.

And fears that robotic operators will not be able to do their job are unfounded.

Why? Since a call center robot does not need to have all human knowledge, it only needs to know the answers to questions about its organization.

The Robot-Call Center offered by us is designed as a self-learning system (Fig. 4).

Figure 3 shows that if the robot operator does not understand the client's question, he sends it to the human operator. In this case, the answer of the human operator is added to the database of "unrecognized questions" together with the client's question, and this database, after some processing, is added to the main knowledge base of the robot. Thus, the next time such questions are asked, the robot will already understand and answer them.

6   Discussion

In the previous sections, during the analysis of costs in call centers, we showed that the costliest item in call centers is related to the number of operators.

After the creation of robot operators, it will be possible to reduce the number of human operators several times.

In our opinion, this will not happen soon, but the ability of robot operators to self-learn will allow them to completely (or in large numbers) replace human operators in the near future.

Because AI is constantly evolving, and it will undoubtedly replace natural intelligence in various local areas of human activity.

The implementation of the RCC project is an example of support for this work in Azerbaijan.

Acknowledgements The authors of this work express their gratitude to academician Ali Abbasov, director of the National Academy of Sciences' Institute of Control Systems, for his unwavering support for the development of new scientific fields in Azerbaijan, particularly AI technology. This project would not have been possible without his enthusiastic support and active participation.

References

1. Gans, N., Koole, G., Mandelbaum, A.: Telephone Call Centers: Tutorial, Review, and Research Prospects. Manuf. Serv. Oper. Manag. **5**(2), 79–141 (2003). https://doi.org/10.1287/msom.5.2.79.16071
2. Helper Magazine: http://www.callcentrehelper.com/the-history-of-the-call-centre-15085.html
3. L.H. Kahn. et al "Modelling Hybrid Human-Artificial Intelligence Cooperation: A Call Center Customer Service Case Study". 2020 IEEE International Conference on Big Data (Big Data). 2020, Pages: 3072–3075. DOI Bookmark: https://doi.org/10.1109/BigData50022.2020.9377747
4. Erlang, Agner K. (1909), [Probability Calculation and Telephone Conversations], "Sandsynlighedsregning og Telefonsamtaler" Nyt Tidsskrift for Matematik (in Danish), 20 (B): 33–39, JSTOR 24528622
5. Principle works of A.K. Erlang. The theory of probabilities and telephone conversations. First published in "Nyt Tidsskrift for Matematik" B, Vol. 20 (1909), p. 33. (https://web.archive.org/web/20110719123704/http://oldwww.com.dtu.dk/teletraffic/erlangbook/pps131-137.pdf)
6. Azərbaycan Respublikası Prezidenti Yanında Vətəndaşlara Xidmət və Sosial İnnovasiyalar Agentliyinin (ASAN xidmət) 2018-ci il hesabatı. p. 7, pp 41. (http://www.vxsida.gov.az/media/shares/12compressed.pdf)
7. "Dövlət orqanlarında çağrı mərkəzlərinin fəaliyyətinin təşkili Qaydaları"nın təsdiq edilməsi haqqında. Azərbaycan Respublikası Nazirlər Kabinetinin Qərarı. (http://www.e-qanun.az/framework/29632)
8. A.Abbasov, R.Fatullayev, A.Fatullayev. Speech technologies market ın Azerbaijan. American journal of Manegement. Vol 21(3), 2021, 95-101. https://doi.org/10.33423/ajm.v2Ii3
9. İqtisadiyyat Nazirliyi yanında Dövlət Vergi Xidmətinin Çağrı Mərkəzi. 2016–2020-ci illər üzrə hesabatlar (https://www.taxes.gov.az/az/page/hesabat)
10. J.Gao, M.Galley and L.Li (2019), "Neural Approaches to Conversational AI", Foundations and Trends® in Information Retrieval: Vol. 13: No. 2–3, pp 127–298.
11. Brown, L., Gans, N., Mandelbaum, A., Sakov, A., Shen, H., Zeltyn, S., Zhao, L.: Statistical Analysis of a Telephone Call Center. J. Am. Stat. Assoc. **100**(469), 36–50 (2005). https://doi.org/10.1198/016214504000001808
12. L.Archawaporn, W.Wongseree. "Erlang C model for evaluate incoming call uncertainty in automotive call centers", 2013 International Computer Science and Engineering Conference (ICSEC), 4–6 Sept. 2013, Pages:109 – 113, DOI: https://doi.org/10.1109/ICSEC30767.2013

13. E. Hosseini-Asl, B. McCann, C.S. Wu, S. Yavuz, R. Socher. "A Simple Language Model for Task-Oriented Dialogue". arXiv:2005.00796 [cs.CL]
14. P.Budzianowski, T.H. Wen, B.H.Tseng, I.Casanueva, U. Stefan, R. Osman. M. Gac. MultiWOZ - A Large-Scale Multi-Domain Wizard-of-Oz Dataset for Task-Oriented Dialogue Modelling. EMNLP, (2018): pp. 5016–5026
15. J.Devlin, M.W. Chang, K. Lee, K. Toutanova. "BERT: Pre-training of Deep Bidirectional Transformers for Language Understanding".

Learning and Self-education of Complex Types of Objects

Shahnaz N. Shahbazova⓪ **and Zinyet R. Huseynova**

Abstract This paper presents information that, on the one hand, provides additional experimental, algorithmic and functional information, but on the other hand, reveals the essence of the problems of the upcoming research studies. In parallel, a mechanism is presented that allows the system to move to a new level of learning—self-learning. The developed self-learning method is a delayed learning mechanism, when the system learns to recognize new objects without a special command and when information characterizing a new object can enter the system as needed (depends on the system operator).

Keywords Recognition · Visual information · Sound information · Artificial personality · Self-learning mode

1 Introduction

The problem of learning any system that will make certain decisions presupposes the existence of mechanisms for the formation of a base of reference objects with certain rules for analyzing the goals that are achieved during learning. In the paper, when determining the quality of learning, first of all, the results of experiments and the average stability of the results of recognition reliability are analyzed, which is why the material devoted to learning the system and the study of complex types of

Sh. N. Shahbazova (✉)
Department of Digital Technologies and Applied Informatics, Azerbaijan State University of Economics, UNEC, Baku AZ1001, Azerbaijan
e-mail: shahbazova@gmail.com; shahnaz_shahbazova@unec.edu.az; shahbazova@cyber.az

Ministry of Science and Education of the Republic of Azerbaijan, Institute of Control Systems, 68 Bakhtiyar Vahabzadeh Street, Baku AZ1141, Azerbaijan

Z. R. Huseynova
Nakhchivan State University, University Campus, 7012 Nakhchivan, Azerbaijan
e-mail: zhuseynova123@gmail.com

visual information is presented at the end of the work, after highlighting all the main research results [1–3].

This research is not limited only to the study of learning issues, but also a problem is posed, and a solution is proposed for implementing the system's self-learning mode, which will allow us to consider artificial personality (AP) as a self-learning (in a limited sense) hardware and software complex.

Purely technical problems associated with the course of the learning process are also investigated—this is the overflow of available disk capacities due to the recording of images of information objects. The proposed recording modes have their own characteristics and require careful study, the implementation features of which are discussed below [4–6].

In conclusion, special types of visual information are studied, which have only particular special solutions, and which can be conditionally inscribed in a natural model of information recognition. With sound information, there is no such problem due to the fact that any "special" information, one way or another, necessarily associated with the intellectual or psychological levels of the nervous system, is not considered in this paper.

2 Learning the System for New Information Objects

Learning the system with new objects and, accordingly, replenishment of the information library at the current stage of system development is feasible only thanks to the complete control of the virtual experimental environment by the researcher [7].

Learning consists of several stages, 4 of which are fundamental:

- object preparation—creation of a two-dimensional model for a photograph and a three-dimensional model for a visual object (digitization of a fragment for a sound object);
- selection of a dictionary concept is an important intellectual stage, the researcher needs to make a list (perhaps with one element) of dictionary concepts that, in his opinion, correspond and are quite associated with a new object;
- learning—training is a special experiment in which the system is asked to recognize a new object of information (usually only one new object is placed in a room at a time), during the experiment, after the object is isolated from the environment and at the end of the recognition stage, the system receives a list corresponding to the new object vocabulary concepts;
- memorization of an object is a critically important stage, the qualitative implementation of which is decisive for the stable functioning of the system as a whole, it is a function that creates about 20 copies of a memorized object, each of which contains small (within 15%) distortions in the geometric dimensions of objects and their orientation (time and volume for a sound object), as well as color distortions, including a black and white image (tone distortion for a sound object), then the resulting modified copies and the original are randomly distributed among

approximately 1% of the available system (but not less than the number of modified copies) modular processes by writing to the appropriate databases (see Fig. 1).

The learning process is quite laborious and lengthy [8–11]. The stage of object preparation requires significant efforts to convert the information object into the appropriate digital form (see Fig. 2), and training takes a lot of computer time (see Table 1).

As can be seen from the summary report (see Table 2), during recognition, a new object ("bedside table") was erroneously recognized ("table"), and then recorded under the dictionary concept attached to it ("bedside table").

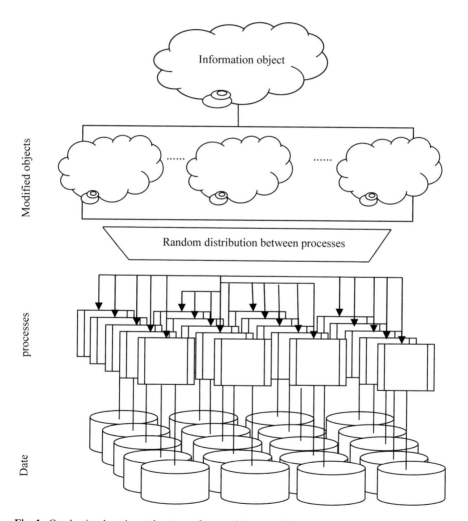

Fig. 1 Conducting learning—the stage of memorizing an object

Fig. 2 Special learning
experiment

Table 1 Report of a special experiment on the conduct of learning

Real time	Virtual time	Coordinates AP (X, Y)	Angle and focal length AP $(\alpha, \underline{R})$	Recognized visual or sound object
0:00:00	0:00:00.00	5000, 5000	1.5708, 0	Stage—Start
0:11:27	0:00:00.08	5000, 5000	1.6692, 0	Stage—Find
0:25:12	0:00:00.08	5000, 5000	1.7144, 0	Stage—Find
...				
5:37:24	0:00:01.77	5000, 5000	3.3638, 0	Stage—Find
5:55:36	0:00:01.78	5000, 5000	3.4423, 0	Stage—Find
6:01:49	0:00:01.89	4871, 4958	3.4423, 2088	Stage—Move
6:20:25	0:00:01.98	4801, 4954	3.4423, 1972	Stage—Move
...				
8:49:10	0:00:02.83	3479, 4507	3.4423, 116	Stage—Move
9:18:52	0:00:03.03	3479, 4507	3.4344, 125	Stage—Research
10:00:45	0:00:03.13	3479, 4507	3.4222, 115	Stage—Research
...				
84:31:11	0:00:26.45	3479, 4507	3.3729, 114	Stage—Research
84:56:59	0:00:26.60	3479, 4507	3.4246, 118	Stage—Research
86:45:24	0:00:27.33	3479, 4507	3.4246, 118	Stage—Make decision—Possible visual-object "table"
86:45:24	0:00:27.33	3479, 4507	3.4246, 118	Stage—Finish

Table 2 Summary report of the training experiment

Real time	Virtual time	Virtual object	Sound object	Reliability
86:45:24	0:00:27.33	Possible visual-object "table"		58.84%
86:57:48	0:00:27.33	Save visual-object {"bedside table"}		

As can be seen from the general report (see Table 1), the learning process is not reflected in it in any way, this is due to the fact that the learning process suspends the operation of the virtual room in order to avoid information collisions that will lead to a situation where the information used by the system is outdated due to the update of the source.

3 System Learning. Self-learning Mode

The self-learning mode is currently an experiment within an experiment, the positive result of which will make it possible to give naturalness to the process of teaching the system new visual and sound objects associated with vocabulary concepts [12–15].

The self-learning mode is based on the idea of self-study of the available visual and sound information FIs in order to parse and recognize objects as nameless images.

In a procedural form, self-learning is carried out in the following steps:

1. An object unfamiliar to the system is placed in the virtual room and the recognition procedure is started;
2. If in the course of recognition, a target object that gives a recognition accuracy of less than 50% or creates an indefinite disputable situation with a reliability of less than 70%, then the object is stored under an unnamed marker;
3. The same object is again placed in the virtual room and the recognition procedure is started again;
4. If in the course of recognition the target object gives the reliability of recognition, as a corresponding nameless marker, comparable to 70% and above and at the same time does not create disputes with other objects, then the object is considered identified;
5. Otherwise, you must return to the second stage.

Upon receiving a successful identification of the target item in 8–12 cases for the same unnamed marker, the system is allowed to ask the question "What is this?" and the operator either introduces a dictionary concept, and it becomes a full-fledged marker, or remains for later (people who ask a question do not always receive an answer). If the marker is already present in the system, then a set of visual or audio instances of the object will be added to the existing set [16–19].

With sound objects, there is an additional uncertainty that requires the artificial intervention of the researcher in the process of self-learning, associated with the uncertainty of the main group of frequencies that carry the information load.

The fact is that, unlike the interfering details of visual information (which can be ignored by reducing the detail, resolution and some smoothing filters), the details of audio information are of a completely different nature and there is no other way to select information objects (at least in the course of experiments this cannot be done) succeeded, except to recognize it in the stream of sound information, and any attempt to clean or filter the sound range only introduces additional distortions.

To recognize sound information, the indication of the frequency range is mainly required for the first 5–8 information objects, which lay the library of groups of sound frequencies, which are subsequently used for "homing" to speech information.

In the process of growth, a person learns to understand which sound information carries a verbal load for a person, which is informational, and which is psychological. Most likely, in the first year of life, a person learns to distinguish between the types of these three groups.

Basically, only verbal information is available to us, which has a fairly clear system of formation and is quite well identified.

In the course of the experiment, there is no need to consistently train the system with new information objects, although the number of experiments conducted does not yet give exact grounds for such an assertion (it is possible that one fuzzy recognized object will be similar to another object of a completely different type).

Simple numeric codes are used as unnamed dictionary concepts. This is quite feasible, since the dictionary concept plays a secondary role, playing the role of an identifier (marker) in the recognition process, which allows the researcher to control the correct functioning of the system [20, 21].

Of course, this mechanism cannot be compared with the human capabilities of self-learning, but it is quite consistent with it, if we do not take into account the ability of the intellect to search for patterns, which allows a minimum number of questions to cover the entire spectrum of object varieties, but in general, this procedure can be taken similar to a situation where a child asks about the same question when meeting an unfamiliar object, animal or word.

In this type of experiment, there are not enough opportunities to directly control the correct functioning of the system at each step, and after only a few stages of self-learning, it is possible to check the quality of the correctness of the associative binding.

4 Conclusion

This paper presents information provides additional experimental, algorithmic and functional information. Also reveals the essence of the problems of the upcoming research studies. In parallel, a mechanism is presented that allows the system to move to a new level of learning—self-learning. The developed self-learning method is a delayed learning mechanism, when the system learns to recognize new objects without a special command and when information characterizing a new object can enter the system as needed (depends on the system operator).

References

1. Altunin A.E., Semukhin M.V. Models and algorithms for making decisions in fuzzy conditions: Monograph. Tyumen: Publishing House of the Tyumen State University, 2000. 352 p. http://www.plink.ru/tnm/gl51.htm

2. Anisimov B. V., Kurganov V. D., Zlobin V. K. Recognition and digital image processing—M .: Higher School (1984)

3. Shahnaz Shahbazova, Manfred Grauer, Musa Suleymanov, The development of an algorithmic model of object recognition using visual and sound information based on neuro-fuzzy logic, Annual Conference of the North American Fuzzy Information Processing Society-NAFIRS, March 19, 2011.

4. Kaiser, G., Kaplan, S.: Parallel and distributed Incremental Attribute Evaluation Algorithms for Multiuser Software development Environments. ACM Trans. Softw. Eng. Methodol. 2(1), 47–92 (1993)

5. Lee, S-W., Kim, Y.J.: Off-line recognition of totally unconstrained handwritten numerals using multilayer cluster neural network. In: Proceedings of the 12th IAPR International Conference on Pattern Recognition. Jerusalem, Israel. (1994). Стр. 507–509.

6. Shahbazova, S.N., Suleymanov, M.R.: Stages of recognition probability images. In: Proceedings of AzTU (2004)

7. Andrey A., Demo.design 3D programming FAQ, release 2, 2:5036/5.47@fidonet, e-mail: shodan@chat.ru

8. Digital processing of television and computer images: Textbook. In: Zubarev Yu. B., Dvorkovich, V.P., (ed.) M.: International Center for Technical Information (1997)

9. Forsyth J., Malcolm M., Mowler K.: Machine methods of mathematical computations—M.: Mir (1980)

10. Fu K., Gonzalez R., Lee K. Robotics: Translated from English. In: Gradetsky, V.G. (ed.)—M.: Mir (1989)

11. Fu K.S.: Sequential methods in pattern recognition and machine learning—M.: Nauka (1971)

12. Fu, K.S.: Structural methods in pattern recognition–M.: Mir (1977)

13. Huang S., Schreiber V., Tretiak O.: Image processing/Sat. scientific tr. Image Process. In: Andrews H.-M (ed.), Using Digital Comput., S. 38–39 (1973)

14. Gonzalez, R.C., Wintz P.: Digital image processing. Addison–Wesley. Reading. Massachusetts, p. 505 (1987)

15. Shahbazova Sh.N., Bonfig K.V., Suleymanov M.R.: Questions of construction of system of recognition of non-standard image and speech in the architecture of a hybrid vehicle based on neuro-fuzzy logic. In: 2nd International Conference on Application of Fuzzy systems and software computing and analysis system, Decision and Control, Barcelona, Spain (2004)

16. Gutta, S., Wechsler, H.: Face recognition using hybrid classifiers. Pattern Recogn. 30, 539–553 (1997)

17. Tzafestas S.G. Large-scale systems modeling in distributed-parameter control and estimation. In: Modeling and Simul. Eng. 10th IMACS World Congr. Syst. Simul. and Sci. Comput., Montreal, pp. 8–13 (1982), v.3" Amsterdam e.a., p.69–77 (1983)

18. Valentin, D., Abdi, H., O'Toole, A.J,. Cottrell, G.W.: Connectionist models of face processing: a survey. In: Pattern Recognition vol. 27, pp. 1209–1230 (1994)

19. Vetter, T., Poggio, T.: Linear object classes and image synthesis from a single example image. IEEE Trans. Pattern Anal. Mach. Intell. 19, 733–742 (1997)

20. Hall, E.L.: Almost uniform distribution for computer image enhancement. IEEE Trans. Comput. 23(2), pp. 207–208 (1974)

21. Petrou, M.: Learning in pattern recognition. Lecture Notes in Artificial Intelligence—Machine Learning and Data Mining in Pattern Recognition, pp. 1–12, (1999)

22. Ketcham, D.J.: Real time image enhancement technique. In: Proceedings SPIE/OSA Conference on Image Processing—Pacific Grove, Vol. 74, pp. 120–125, California (1976)
23. Ranganath, S., Arun, K.: Face recognition using transform features and neural networks. Pattern Recogn. **30**, 1615–1622 (1997)
24. Sabah, G.: Knowledge Representation and Natural Language Understanding. AI Commun, **6** (3–4), pp. 155–186 (1993)

Probabilistic Uncertainty

Why Moving Fast and Breaking Things Makes Sense?

Francisco Zapata, Eric Smith, and Vladik Kreinovich

Abstract In the traditional approach to engineering system design, engineers usually come up with several possible designs, each improving on the previous ones. In coming up with these designs, they try their best to make sure that their designs stay within the safety and other constraints, to avoid potential catastrophic crashes. The need for these safety constraints makes this design process reasonably slow. Software engineering at first followed the same pattern, but then realized that since in most cases, failure of a software test does not lead to a catastrophe, it is much faster to first ignore constraints and then adjust the resulting non-compliant designs so that the constrains will be satisfied. Lately, a similar "move fast and break things" approach was applied to engineering design as well, especially when designing autonomous systems whose failure-when-testing is not catastrophic. In this paper, we provide a simple mathematical model explaining, in quantitative terms, why moving fast and breaking things makes sense.

Keywords Engineering design · Software engineering · Move fast and break things

1 Formulation of the Problem

General engineering problems: reminder. Whatever we design, we have an objective function that we want to optimize:

- From the business viewpoint, we want to maximize profit.
- When we design computers, we want computations to be as fast as possible.

F. Zapata (✉) · E. Smith · V. Kreinovich
University of Texas at El Paso, El Paso, TX 79968, USA
e-mail: fcozpt@outlook.com

E. Smith
e-mail: esmith2@utep.edu

V. Kreinovich
e-mail: vladik@utep.edu

© The Author(s), under exclusive license to Springer Nature Switzerland AG 2023
Sh. N. Shahbazova et al. (eds.), *Recent Developments and the New Directions of Research, Foundations, and Applications*, Studies in Fuzziness and Soft Computing 423,
https://doi.org/10.1007/978-3-031-23476-7_5

- When we are interested in saving environment, we want to make sure that the corresponding chemical process produces as little pollution as possible, etc.

What is also important is that there are always constraints, limitations. Here are some examples:

- A construction company contracted to build a federal office building may want to maximize its profit and use the cheapest materials possible—but in this desire, the company is restricted by the contract, and it is also restricted by the building regulations: according to these regulations, the building must be able to withstand strong winds, floods, and/or earthquakes that can occur in this area.
- A company that sets up a chemical plant may want to save money on filtering—which is often very expensive—but it has to make sure that the resulting pollution does not exceed the thresholds set by national and local regulations.
- Designers of a fast small plane aimed at customers interested in speed may sacrifice a lot of weight to gain more speed—but they still need to make sure that the plane is sufficiently safe.

How engineering problems used to be solved. To find the optimal solution, a company—or an individual investor—go from one design to another, trying to come up with the best possible design. In this process, maximum efforts were undertaken to make sure that at all the stages, the corresponding design satisfies all the needed constraints. These efforts make sense: whether we tests a car or a plane or a chemical process, a violation of safety and health constraints may be disastrous and may even lead to loss of lives.

Sometimes, an experimental plane would crash, often killing a pilot. Sometimes, a bridge would collapse—all these catastrophes would remind other designers of the importance to stay within the constraints.

These constraints provided sufficient safety, but they also have a negative effect: the need to make sure that each new design is within the constraints drastically slows down the progress.

Software design first followed the same pattern. At first, folks who designed software followed the same general patterns: whenever they changed a piece of a big software package to a more efficient (and/or more effective) one, they first patiently made sure that this change will not cause any malfunctions. This "making sure" slowed down the progress.

New idea: move fast and break things. With time, software designers realized that they do not necessarily need to be so cautious. Unless the software they design is intended for life-critical systems like nuclear power stations or airplane control, a minor fault is tolerable and usually does not lead to catastrophic consequences. They realized that they can move much faster if they come up with designs that may not necessarily satisfy all the constraints at first—corresponding corrections can be done later, and that even with some time spent on these corrections, still this new paradigm led to faster design.

This new practice was explicitly formulated as "move fast and break things" by Mark Zuckerberg, then CEO of Facebook. This phrase even became (for several years) the motto of the Facebook company—and, informally, of the whole Silicon Valley; see, e.g., [1, 2].

This idea moved to engineering design. Interestingly, this idea—first originated in software design and first intended for software design only—eventually moved to general engineering as well. The reason for this transition is that with the increased automation, most test crashes stopped being dangerous to humans: if a self-driving car or a pilotless plane crashes, there is no immediate danger to people—unless they accidentally happen to be nearby.

The main pioneer of using this idea in engineering was Elon Musk, who used this successfully, in particular, in his space exploration efforts.

How can we explain the success of this idea? The idea of moving fast and breaking things seem to work for many projects. (Sometimes, it does not work—but, on the other hand, sometimes projects based on the traditional engineering design techniques do not work either.)

But why does this moving-fast idea work? Understanding why it works for many projects is important: this way, we will be able:

- To better understand when it works and when it does not, and
- In situations where this idea works, come up with the best possible way of using this idea.

This is what we do in this paper: we provide a natural simple quantitative model explaining why this idea, in general, works.

2 Description of the Model

What we want to optimize. As we have mentioned, in engineering design, we need to select some quantities $x_1, \ldots, x_n$— parameters of the design—for which a given objective function $f(x_1, \ldots, x_n)$ attains its largest possible value among all the tuples $x = (x_1, \ldots, x_n)$ that satisfy all the given constraints.

How to describe constraints. In general, constraints have the form of inequalities $\ell_i(x_1, \ldots, x_n) \le r_i(x_1, \ldots, x_n)$ for some quantities ℓ_i and r_i. Each such constraint can be equivalently reformulated as $g_i(x_1, \ldots, x_n) \ge 0$, where we denoted

$$g_i(x_1, \ldots, x_n) \overset{\text{def}}{=} r_i(x_1, \ldots, x_n) - \ell_i(x_1, \ldots, x_n).$$

So, we have a finite numbers of constraints that the desired design must satisfy:

$$g_1(x_1, \ldots, x_n) \ge 0, \ldots, g_m(x_1, \ldots, x_n) \ge 0.$$

Several numbers are non-negative if and only if the smallest of these numbers is non-negative. Thus, satisfying the above m constraints is equivalent to satisfying a single constraint

$$g(x_1, \ldots, x_n) \geq 0 \tag{1}$$

where we denoted

$$g(x_1, \ldots, x_n) \overset{\text{def}}{=} \min(g_1(x_1, \ldots, x_n), \ldots, g_m(x_1, \ldots, x_n)).$$

Resulting formulation of the general problem. We want to maximize a function $f(x_1, \ldots, x_n)$ under the constraint (1).

Additional complexity: need to take uncertainty into account. At first glance, this sounds like a usual optimization problem, for which many algorithms are known. However, in many practical situations, there is an additional complexity—that both the objective function $f(x_1, \ldots, x_n)$ and the function $g(x_1, \ldots, x_n)$ that describes the constraints are only approximately known.

Indeed, if both these functions were exactly known, there would be no need for testing different designs—we would just be able to solve the corresponding constrained optimization problem and implement it.

Mathematical fact: the solution is usually on the edge of constraints. In general, the largest value of the objective function is attained:

- Either inside the area where the constrains are satisfied, i.e., where $g_i(x_1, \ldots, x_n) > 0$ for all i,
- Or at the border of this area, i.e., when $g_i(x_1, \ldots, x_n) = 0$ for some i, and thus,

$$g(x_1, \ldots, x_n) = 0. \tag{2}$$

In the first case, the solution is a local maximum. So, in looking for this maximum, we can simply ignore all the constraints, they are automatically satisfied.

An important case where we do need to take constraints into account is the second case, when the maximum is attained at the border. This is the case that we will consider in this paper.

How the corresponding problem is solved under uncertainty: possible preliminary stage. We start with the first design $x^{(1)} = \left(x_1^{(1)}, \ldots, x_n^{(1)}\right)$.

In some cases, this first design is already on the border (2)—or at least close to this border.

In many other cases, however, this design is usually rather far from the area where the constraints are not satisfied. So at first, we can kind-of ignore the constraints and try to modify the values x_i so as to increase the value of the objective function $f(x_1, \ldots, x_n)$. After several consequent improvements, we get closer and closer to the optimal design—and thus, closer and closer to the border (2). At the end of this possible preliminary stage, we get so close to the border that the constraints can no longer be ignored.

Main stage of the design process. Once we have reached a point close to the border, for which the value $\varepsilon \overset{\text{def}}{=} g(x_1, \ldots, x_n) > 0$ is small, the main stage of the design process starts: finding the optimal solution while taking constraints into account.

Let us analyze how this main stage is performed in general, and what is different when we perform this stage in the traditional engineering methodology and in the moving-fast software-motivated methodology.

3 The Main Stage of Optimization: General Analysis

What we know and what we want. At the beginning of this main stage, we have a design $(x_1, \ldots, x_n)$ which is close to the border, i.e., for which—at least approximately—the condition (2) is satisfied. We know that this design is not optimal, so we want to find a modified design

$$(x_1 + \Delta x_1, \ldots, x_n + \Delta x_n)$$

for which the value of the objective function is larger.

Ideal case, when we have the exact knowledge of the objective function and of the constraints. Let us first consider the ideal case, when we know the exact expressions both:

- for the objective function $f(x_1, \ldots, x_n)$ and
- for the function $g(x_1, \ldots, x_n)$ that describes the constraints.

It is known that, in general, a constraint optimization problem is equivalent to the unconstrained optimization problem of maximizing the expression

$$f(x_1, \ldots, x_n) + \lambda \cdot g(x_1, \ldots, x_n) \tag{3}$$

for an appropriate value λ; this value is known as the *Lagrange multiplier*.

For unconstrained optimization, one of the most natural optimization techniques is *gradient method*, when, by choosing the values Δx_i, we follow the direction in which the objective function increases the most—i.e., the direction of its gradient. For the equivalent objective function (3), this means that we select

$$\Delta x_i = \alpha \cdot \frac{\partial}{\partial x_i}(f(x_1, \ldots, x_n) + \lambda \cdot g(x_1, \ldots, x_n))$$

for an appropriate value $\alpha > 0$, i.e.,

$$\Delta x_i = \alpha \cdot (f_i + \lambda \cdot g_i), \tag{4}$$

where we denoted

$$f_i \stackrel{\text{def}}{=} \frac{\partial f}{\partial x_i} \text{ and } g_i \stackrel{\text{def}}{=} \frac{\partial g}{\partial x_i}.$$

The value of the Lagrange multiplier λ must be determined from the condition that we will be moving along the border, i.e., that the original zero value of the function $g(x_1, \ldots, x_n)$—that describes this border as the set of all the tuples x for which $g(x_1, \ldots, x_n) = 0$—should not change. In precise terms, we should have

$$g(x_1 + \Delta x_1, \ldots, x_n + \Delta x_n) = 0,$$

i.e.,

$$0 = g(x_1 + \Delta x_1, \ldots, x_n + \Delta x_n) \approx g(x_1, \ldots, x_n) + \sum_{i=1}^{n} g_i \cdot \Delta x_i = \sum_{i=1}^{n} g_i \cdot \Delta x_i.$$

Substituting the expression (4) into this formula, we get

$$\sum_{i=1}^{n} f_i \cdot g_i + \lambda \cdot \sum_{i=1}^{n} g_i^2 = 0,$$

hence

$$\lambda = -\frac{\sum_{i=1}^{n} f_i \cdot g_i}{\sum_{i=1}^{n} g_i^2}. \tag{5}$$

Realistic case, when we take uncertainty into account. As we have mentioned, in practice, we only know the objective function $f(x_1, \ldots, x_n)$ and the function $g(x_1, \ldots, x_n)$ (that describes the constraints) only approximately. As a result, we only know the approximate values of the corresponding derivatives, i.e., we know the values $\widetilde{f}_i \approx f_i$ and $\widetilde{g}_i \approx g_i$ for which

$$\widetilde{f}_i = f_i + \Delta f_i \text{ and } \widetilde{g}_i = g_i + \Delta g_i$$

for some reasonably small values Δf_i and Δg_i.

Let $\delta > 0$ be the accuracy with which we know these derivatives. This means that for each i, we have $|\Delta f_i| \leq \delta$ and $|\Delta g_i| \leq \delta$.

Of course, since these approximate values are the only information we have, we use these approximate values when we decide on which parameters to use for the next design. In other words, we take

$$\Delta x_i = \alpha \cdot (\widetilde{f}_i + \widetilde{\lambda} \cdot \widetilde{g}_i), \tag{6}$$

where

$$\widetilde{\lambda} = -\frac{\sum\limits_{i=1}^{n} \widetilde{f}_i \cdot \widetilde{g}_i}{\sum\limits_{i=1}^{n} (\widetilde{g}_i)^2}. \tag{7}$$

The only remaining question is what value α we should use. Let us show how the choice of α depends on which of the two methodologies we use: the traditional engineering methodology or the moving-fast software motivated methodology.

4 The Main Stage of Optimization: Case of the Traditional Engineering Methodology

Idea. In the traditional engineering methodology, we select the value α so as to make sure that, no matter what the actual values f_i and g_i are, we remain within the safe domain, i.e., that we still have $g(x_1 + \Delta x_1, \ldots, x_n + \Delta x_n) \geq 0$.

Based on this idea, what value α do we choose. For relatively small changes Δx_i we have

$$g(x_1 + \Delta x_1, \ldots, x_n + \Delta x_n) \approx g(x_1, \ldots, x_n) + \sum_{i=1}^{n} g_i \cdot \Delta x_i. \tag{8}$$

Since $\widetilde{g}_i = g_i + \Delta g_i$, we have $g_i = \widetilde{g}_i - \Delta g_i$, therefore

$$g(x_1 + \Delta x_1, \ldots, x_n + \Delta x_n) \approx g(x_1, \ldots, x_n) + \sum_{i=1}^{n} \widetilde{g}_i \cdot \Delta x_i - \sum_{i=1}^{n} \Delta g_i \cdot \Delta x_i. \tag{9}$$

Because of our selection (6) of the values Δx_i, we have

$$\sum_{i=1}^{n} \widetilde{g}_i \cdot \Delta x_i = 0,$$

thus

$$g(x_1 + \Delta x_1, \ldots, x_n + \Delta x_n) \approx g(x_1, \ldots, x_n) - \sum_{i=1}^{n} \Delta g_i \cdot \Delta x_i. \tag{10}$$

We have denoted the current value of $g(x_1, \ldots, x_n)$ by ε, so

$$g(x_1 + \Delta x_1, \ldots, x_n + \Delta x_n) \approx \varepsilon - \sum_{i=1}^{n} \Delta g_i \cdot \Delta x_i. \tag{11}$$

We want to make sure that this value remains non-negative for all possible combinations of values Δg_i for which $|\Delta g_i| \leq \delta$, i.e., we want to make sure that for all these combinations, the sum

$$\sum_{i=1}^{n} \Delta g_i \cdot \Delta x_i \qquad (12)$$

remains smaller than or equal to ε. In other words, we want to make sure that the largest possible value of the sum (12) is smaller than or equal to ε.

The sum (12) attains its largest possible value when each term $\Delta g_i \cdot \Delta x_i$ in this sum is the largest possible. Each of these terms is a linear function of Δg_i. So:

- when $\Delta x_i \geq 0$, this term is an increasing function of Δg_i and therefore, its largest value is attained when the variable Δg_i attains its largest possible value $\Delta g_i = \delta$; the corresponding largest value of this term is therefore equal to $\delta \cdot \Delta x_i$;
- when $\Delta x_i \leq 0$, this term is a decreasing function of Δg_i and therefore, its largest value is attained when the variable Δg_i attains its smallest possible value $\Delta g_i = -\delta$; the corresponding largest value of this term is therefore equal to $-\delta \cdot \Delta x_i$.

Both cases can be describe by a single formula $\delta \cdot |\Delta x_i|$. Thus, the largest value of the sum (12) is equal to

$$\sum_{i=1}^{n} \delta \cdot |\Delta x_i| = \delta \cdot \sum_{i=1}^{n} |\Delta x_i|.$$

Substituting the expression (6) into this formula, we conclude that this largest value is equal to

$$\delta \cdot \alpha \cdot \sum_{i=1}^{n} \left| \tilde{f}_i + \tilde{\lambda} \cdot \tilde{g}_i \right|.$$

Thus, the condition that this largest value is smaller than or equal to ε takes the form

$$\delta \cdot \alpha \cdot \sum_{i=1}^{n} \left| \tilde{f}_i + \tilde{\lambda} \cdot \tilde{g}_i \right| \leq \varepsilon,$$

i.e., equivalently, that

$$\alpha \leq C \cdot \varepsilon, \qquad (13)$$

where we denoted

$$C \stackrel{\text{def}}{=} \frac{\varepsilon}{\delta \cdot \sum_{i=1}^{n} \left| \tilde{f}_i + \tilde{\lambda} \cdot \tilde{g}_i \right|}.$$

Conclusion of this section. The desired optimal solution is located on the border. In the process of optimization, we get closer and closer to the optimal solution—and thus, the closer and closer to the border.

According to the formula (13), in the traditional engineering approach, the smaller the distance ε from the current solution to the border, the smaller our next modification can be—and thus, the slower our progress towards the optimal design.

5   The Main Stage of Optimization: Case of the Moving-Fast Software-Motivated Methodology

In the case of moving-fast methodology, we are not limiting our next design by any constraints, so we can make a big step—probably violating the constraints, but then moving them back. Here, we do not having any slowing-down inequality like (13), so we get to the optimal solution much faster.

Thus, we indeed explained, in quantitative terms, why the moving-fast methodology is much faster than the traditional engineering one.

Acknowledgments This work was supported in part by the National Science Foundation grants 1623190 (A Model of Change for Preparing a New Generation for Professional Practice in Computer Science), and HRD-1834620 and HRD-2034030 (CAHSI Includes), and by the AT&T Fellowship in Information Technology.

It was also supported by the program of the development of the Scientific-Educational Mathematical Center of Volga Federal District No. 075-02-2020-1478, and by a grant from the Hungarian National Research, Development and Innovation Office (NRDI).

References

1. Parzych, D.: The fallacy of move fast and break things (2020). https://devops.com/the-fallacy-of-move-fast-and-break-things/
2. Tuplin, J.: Move Fast and Break Things: How Facebook, Google, and Amazon Cornered Culture and Undermined Democracy. Brown, and Company, New York, Little (2017)

Shall We Be Foxes or Hedgehogs: What Is the Best Balance for Research?

Miroslav Svítek, Olga Kosheleva, Shahnaz N. Shahbazova⑩, and Vladik Kreinovich

Abstract Some researchers have few main ideas that they apply to many different problems—they are called hedgehogs. Other researchers have many ideas but apply them to fewer problems—they are called foxes. Both approaches have their advantages and disadvantages. What is the best balance between these two approaches? In this paper, we provide general recommendations about this balance. Specifically, we conclude that the optimal productivity is when the time spent on generating new ideas is equal to the time spent on understanding new applications. So, if for a researcher, understanding a new problem is much easier than generating a new idea, this researcher should generate fewer ideas—i.e., be a hedgehog. Vice versa, if for a researcher, generating a new idea is easier than understanding a new problem, it is more productive for this person to generate many ideas—i.e., to be a fox. For researchers for whom these times are of the same order, we provide explicit formulas for the optimal research strategy.

M. Svítek
Faculty of Transportation Sciences, Czech Technical University in Prague, Konviktska 20, 110 00 Prague 1, Czech Republic
e-mail: svitek@fd.cvut.cz

O. Kosheleva · V. Kreinovich (✉)
University of Texas at El Paso, 500 W. University, El Paso, TX 79968, USA
e-mail: vladik@utep.edu

O. Kosheleva
e-mail: olgak@utep.edu

Sh. N. Shahbazova
Department of Digital Technologies and Applied Informatics, Azerbaijan State University of Economics, UNEC, Baku AZ1001, Azerbaijan
e-mail: shahbazova@gmail.com; shahnaz_shahbazova@unec.edu.az; shahbazova@cyber.az

Ministry of Science and Education of the Republic of Azerbaijan, Institute of Control Systems, 68 Bakhtiyar Bahabzadeh Street, Baku AZ1141, Azerbaijan

© The Author(s), under exclusive license to Springer Nature Switzerland AG 2023
Sh. N. Shahbazova et al. (eds.), *Recent Developments and the New Directions of Research, Foundations, and Applications*, Studies in Fuzziness and Soft Computing 423, https://doi.org/10.1007/978-3-031-23476-7_6

1 Foxes and Hedgehogs

Foxes and hedgehogs: a positive viewpoint. In his famous essay [1], Isaiah Berlin, an American philosopher, divided all the thinkers into:

- *hedgehogs*, who have one main idea (or a few main ideas) and apply it (them) to several problems, and
- *foxes*, who have many different ideas.

Some great thinkers were hedgehogs (Freud and Zadeh come to mind right away), some—like Aristotle—were foxes. At first glance, it looks like both types of thinkers could reach great results. But each of these two types has its limitations.

Foxes: a negative viewpoint. At first glance, what can be wrong with having many interesting ideas, with always learning many interesting ideas? Well, the problem is that you may spread yourself too thin.

For example, in mathematical logic, Georg Kreisel was one of the most productive authors, publishing many papers with interesting ideas; see, e.g., [3, 4]. This did not bother hedgehogs, but several foxes—eagerly interested in learning new ideas—complained that they have no time to do their own research: they have to read all new papers by Kreisel.

Hedgehog: a negative viewpoint. Lotfi Zadeh, himself clearly a hedgehog, liked to emphasize what can go wrong with this approach, by reminding us of the saying that if all you have is a hammer, then everything starts looking like a nail.

We have seen many examples of this in politics, when an originally successful idea gets used everywhere; in popular medicine, where successful medicines like antibiotics gets too overused, etc. In Russia, where several of us are from, we had a silly joke showing this problem. A young man wants to become a writer, so he is taking an entrance exam to the writer's program.

- What can you say about Tolstoy's War and Peace?
- Never read it.
- ??? Did not you say that you want to become a writer?
- Yes, but I want to be a writer, not a reader.

In science, some hedgehogs become such writers-not-readers: they may have had a great idea, but later on, their reluctance to adopt new ideas makes them not very productive. This even happened to great Einstein, who started as a fox—e.g., his Nobel prize was for photo-effect, not for relativity—but who spent several not-very-productive last decades on a single not-very-successful idea of a unified field theory.

There should be a balance. Since both extremes can be counterproductive, there should be a balance between these two extremes, a balance that leads to the maximal possible productivity.

What is this balance? In this paper, we provide a simple model of the situation, and we use this model to provide recommendations on the best balance.

2 Let Us Model the Situation

We need to generate new ideas. The whole idea of research is to solve problems that no one was able to solve before. This means that the existing ideas are not enough to solve the corresponding problem—you need to have a new idea, or at least a new twist on an existing idea.

Generating ideas: notations. Let us assume that a researcher spend time t_I on developing a new idea (or a new twist on a new idea). Then, if during a certain period of time T_0, the researcher comes up with I ideas, then overall, during this period, this researcher spends time $T_I = t_i \cdot I$ on coming up with new ideas.

Fox and hedgehog. For a hedgehog, $I \approx 1$, while for a fox, the number of new ideas I is much larger than 1: $I \gg 1$.

Understanding problems: notations. To be able to solve a problem, it is important to spend some time understanding this problem. This is not easy—especially if this problem is from an area which is different from the researcher's main area of expertise. Let us denote the average time needed to understand a problem by t_P, and the number of different problems the researcher learns during the period T_0 by P. Then overall, during this period, the researcher spends time $T_P = t_P \cdot P$ on learning new problems.

We need to apply these ideas. The whole purpose of coming up with new ideas is to solve problems—and the whole purpose of learning a problem is to try to solve it. If one idea is not working on a problem, a reasonable approach is to apply a different idea. Some problems are solved, most are not—unless we are dealing with a genius who solves all the problems, and such geniuses are rare. In general, a researcher applies all his/her ideas to all the problems that he/she tries to solve—otherwise, what is the purpose of learning a new problem if you do not try to solve it by using all ideas you have?

Let t_0 denote the time that it takes, on average, to try one idea on one problem. Then, to try each of I ideas on each of P problems, we need time $t_0 \cdot I \cdot P$.

Resulting constraint. The overall time that a researcher spends on inventing ideas, learning the problems, and trying ideas on problems cannot exceed T_0. Thus, we have the following constraint:

$$t_I \cdot I + t_P \cdot P + t_0 \cdot I \cdot P \le T_0. \tag{1}$$

What do we want? The main objective of research is to solve problems. The more problems we solve altogether, the more successful we are in our research efforts. From this viewpoint, we should therefore aim for maximizing the number of solved problems.

How many problems can we solve this way? A priori, we do not know which idea will work on which problem. So, it is natural to assume that for each pairs of an idea and a problem, there is the same probability that this particular idea will solve this particular problem. This assumption is known as Laplace Indeterminacy Principle; see, e.g., [2]. Let p_0 denote this joint probability. This probability means that out of all $I \cdot P$ pairs, the proportion of those that lead to solution is equal to p_0. Thus, the overall number of problems solved by a researcher is equal to

$$p_0 \cdot I \cdot P. \tag{2}$$

So, we arrive at the following optimization problem.

Resulting optimization problem. Let us assume that we are planning for time period T_0. For a given researcher, we know:

- the average time t_I that it takes this researcher to come up with a new idea or a new twist on an idea;
- the average time t_P that it takes this researcher to understand a new problem;
- the average time t_0 that it takes this researcher to apply an idea to a problem; and
- the probability p_0 that a randomly selected idea will solve a randomly selected problem.

We want to find the number of ideas I and the number of problems P that maximize the expected number (2) of solved problems under constraint (1).

Let us now solve this problem.

3 Let Us Solve the Resulting Optimization Problem and Thus Find the Optimal Balance Between Fox and Hedgehog Strategies

First simplification. If in the constraint (1), we have a strict inequality, this would mean that we can increase either I or P (or both) without violating the constraint and thus, increase the value of the objective function (2). Thus, the maximum of the objective function is attained when in the constraint (1), we have equality, i.e., when

$$t_I \cdot I + t_P \cdot P + t_0 \cdot I \cdot P = T_0. \tag{3}$$

So, we have a problem of optimizing the objective function (2) under the constraint (3).

Second simplification. In terms of T_I and T_P, we have

$$I = \frac{T_I}{t_I}, \quad P = \frac{T_P}{t_P}, \text{ and thus, } t_0 \cdot I \cdot P = c \cdot T_I \cdot T_P, \tag{4}$$

where we denoted

$$c \stackrel{\text{def}}{=} \frac{t_0}{t_I \cdot t_P}. \tag{5}$$

In these terms, the constraint (3) takes the form

$$T_I + T_P + c \cdot T_I \cdot T_P = T_0, \tag{6}$$

and the objective function (2) takes the form

$$p_0 \cdot I \cdot P = c_0 \cdot T_I \cdot T_P, \text{ where } c_0 \stackrel{\text{def}}{=} \frac{p_0}{t_I \cdot t_P}. \tag{7}$$

So, the problem becomes: to maximize the expression (7) under the constraint (6).

Let us use Lagrange multiplier method. Since the constraint has the form of equality, we can use the Lagrange multiplier method to solve the corresponding constrained optimization problem. Namely, for some λ, the original constrained optimization problem is equivalent to the unconstrained problem of optimizing the expression

$$c_0 \cdot T_I \cdot T_P + \lambda \cdot (T_I + T_P + c \cdot T_I \cdot T_P - T_0). \tag{8}$$

For an unconstrained optimization problem, maximum is attained when all the partial derivatives are equal to 0.

Differentiation the expression (8) with respect to T_I and equating the derivative to 0, we conclude that

$$c_0 \cdot T_P + \lambda + \lambda \cdot c \cdot T_P = 0,$$

hence

$$T_P \cdot (c_0 + \lambda \cdot c) = -\lambda,$$

and

$$T_P = -\frac{\lambda}{c_0 + \lambda \cdot c}. \tag{9}$$

Similarly, differentiation the expression (8) with respect to T_P and equating the derivative to 0, we conclude that

$$c_0 \cdot T_I + \lambda + \lambda \cdot c \cdot T_I = 0,$$

hence

$$T_I \cdot (c_0 + \lambda \cdot c) = -\lambda,$$

and

$$T_I = -\frac{\lambda}{c_0 + \lambda \cdot c}. \tag{10}$$

First conclusion. By comparing the expressions (9) and (10), we conclude that we have

$$T_I = T_P, \tag{11}$$

i.e., that *the time spent on inventing new ideas should be equal to the time spent on learning new problems.*

So fox or hedgehog? From (11), we conclude that

$$I = \frac{t_P}{t_I} \cdot P. \tag{12}$$

So:

- For researchers for whom $t_P \ll t_I$, i.e., for whom it is much easier to understand a new problem than to come up with a new idea, it is better to generate fewer ideas but apply them to many problems—in other words, to be a hedgehog.
- On the other hand, for researchers for whom $t_I \ll t_P$, i.e., for whom it is much easier to come up with a new idea that to understand a new problem, it is better to generate many ideas but apply them to fewer problems—in other words, to be a fox.

For the cases when the times t_I and t_P are of the same order, the formula (12) provides the desired optimal balance.

So what are the optimal values of P and I? In the optimal case, when $T_I = T_P$, the constraint (6) takes the form

$$2T_I + c \cdot T_I^2 = T_0. \tag{13}$$

By solving this quadratic equation, we get

$$T_I = T_P = \frac{\sqrt{1 + c \cdot T_0} - 1}{c}, \tag{14}$$

where c is determined by the formula (5). Thus,

$$I = \frac{T_I}{t_I} = \frac{t_P}{t_0} \cdot \left(\sqrt{1 + \frac{t_0}{t_I \cdot t_P} \cdot T_0} - 1 \right) \tag{15}$$

and

$$P = \frac{T_P}{t_P} = \frac{t_I}{t_0} \cdot \left(\sqrt{1 + \frac{t_0}{t_I \cdot t_P} \cdot T_0} - 1 \right). \tag{16}$$

Acknowledgments This work was supported in part by the National Science Foundation grants 1623190 (A Model of Change for Preparing a New Generation for Professional Practice in Computer Science), and HRD-1834620 and HRD-2034030 (CAHSI Includes), and by the AT&T Fellowship in Information Technology.

It was also supported by the program of the development of the Scientific-Educational Mathematical Center of Volga Federal District No. 075-02-2020-1478, and by a grant from the Hungarian National Research, Development and Innovation Office (NRDI).

References

1. Berlin, I.: The Hedgehog and the Fox: An Essay on Tolstoy's View of History. Princeton University Press, Princeton, New Jersey (2013)
2. Jaynes, E.T., Bretthorst, G.L.: Probability Theory: The Logic of Science. Cambridge University Press, Cambridge (2003)
3. Odifreddi, P. (ed.): Kreiseliana: About and Around Georg Kreisel. A. K. Peters/CRC Press, Natick, Massachusetts (1996)
4. Weingartner, P., Leeb, H.-P.: Kreisel's Interests: On the Foundations of Logic and Mathematics. College Publications, Rickmansworth (2020)

How to Gauge Students' Ability to Collaborate?

Christian Servin, Olga Kosheleva, Shahnaz N. Shahbazova ⓘ**,
and Vladik Kreinovich**

Abstract Usually, we mostly gauge individual students' skills. However, in the modern world, problems are rarely solved by individuals, it is usually a group effort. So, to make sure that students are successful, we also need to gauge their ability to collaborate. In this paper, we describe when it is possible to gauge the students' ability to collaborate; in situations when such a determination is possible, we explain how exactly we can estimate these abilities.

Keywords Education · Student collaboration · Group projects · Gauging collaboration skills

1 Formulation of the Problem

Gauging ability to collaborate is important. In most classes, we test the students' individual knowledge and the individual ability to apply this knowledge. However,

C. Servin
Information Technology Systems Department, El Paso Community College (EPCC), 919 Hunter Dr., El Paso, TX 79915 -1908, USA
e-mail: cservin1@epcc.edu

O. Kosheleva · V. Kreinovich (✉)
University of Texas at El Paso, El Paso, TX 79968, USA
e-mail: vladik@utep.edu

O. Kosheleva
e-mail: olgak@utep.edu

Sh. N. Shahbazova
Department of Digital Technologies and Applied Informatics, Azerbaijan State University of Economics, UNEC, Baku AZ1001, Azerbaijan
e-mail: shahbazova@gmail.com; shahnaz_shahbazova@unec.edu.az; shahbazova@cyber.az

Ministry of Science and Education of the Republic of Azerbaijan, Institute of Control Systems, 68 Bakhtiyar Bahabzadeh Street, Baku AZ1141, Azerbaijan

Sh. N. Shahbazova et al. (eds.), *Recent Developments and the New Directions of Research, Foundations, and Applications*, Studies in Fuzziness and Soft Computing 423, https://doi.org/10.1007/978-3-031-23476-7_7

in the modern world, most problems are solved by collaboration, not individually. While the need for collaboration seems to have increased, collaboration itself is not a new phenomenon: many historians believe that the ability to successfully collaborate was the main factor that made our species dominant; see, e.g., [1].

So, to gauge the students' readiness to solve real-life problems, it is important to gauge not only their individual abilities, but also their ability to collaborate, to solve the problems in collaboration with others.

Gauging ability to collaborate is not easy. A natural way to gauge the ability to collaborate is to combine students into groups, and to assigns tasks to these groups. This way, by grading the result, we can gauge the ability of the group to collaborate. The problem is that it is not easy to translate this information into individual grades:

- If a group has been successful, this does not necessarily mean that all members of this group mastered the art of collaboration. So, if we give everyone from a successful group a very good grade, for some students who have not yet mastered this skill very well, the resulting grade will be undeserved.
- Similarly, if a group has not been very successful, this does not necessarily mean that all members of this group deserve a bad grade on collaboration abilities, a few of them may be better—and so for them, the bad grade based on the project as a whole would also be undeserved.

Remaining problem and what we do in this paper. So, a fair estimation of the students' ability to collaborate is still an important challenge. In this paper, we provide a possible way to solve this challenging problem.

2   How Group Productivity Depends on the Ability to Collaborate

What is given. In other to gauge the students' ability to collaborate, it is important to understand how the group's productivity depends on the students' ability to collaborate. For this purpose, let us introduce natural notations.

For each student i, we will denote:

- this student's individual skills by s_i,
- this student's ability to collaborate by c_i, and
- the amount of effort that the student applied by e_i.

Based on this data, we want to describe the productivity p. In other words, we want to come up with a formula that describes productivity of a group of n people as a function of these inputs:

$$p = p(s_1, \ldots, s_n, c_1, \ldots, c_n, e_1, \ldots, e_n).$$

How to come up with a model: main idea and the resulting formula. To come up with a simple model, we will use only the smallest terms in the Taylor expansion which are consistent the commonsense understanding of the situation.

In general, the first terms in the Taylor expansion are linear terms, so, from the purely mathematical viewpoint, it may seem reasonable to use these terms here as well, i.e., to take

$$p = p_0 + \sum_{i=1}^{n} p_{si} \cdot s_i + \sum_{i=1}^{n} p_{ci} \cdot c_i + \sum_{i=1}^{n} p_{ei} \cdot e_i.$$

However, from the commonsense viewpoint, this formula makes no sense.

First, if no one has any skills, individual or collective, there is no productivity. So, when $s_i = c_i = 0$, we should have $p = 0$. This implies that for all possible values of e_i, we should have $p_0 + \sum_{i=1}^{n} p_{ei} \cdot e_i = 0$. This means that $p_0 = 0$ and $p_{ei} = 0$ for all i.

Similarly, if none of the students applies any effort, there will be no productivity. This implies that $p_{si} = p_{ci} = 0$, so all linear terms should be 0s.

From the commonsense viewpoint, the only possibility to get some productivity is:

- either when at least one student has non-zero individual skills s_i and non-zero effort e_i; the simplest term with this property is the product term $e_i \cdot s_i$;
- or at least two students $i \neq j$ have non-zero ability to collaborate and apply non-zero efforts; the simplest term with this property is $e_i \cdot c_i \cdot e_j \cdot c_j$.

Since we decided to limit ourselves to the smallest non-zero terms—which is usually called the first approximation—we thus conclude that the desired expression for p should be a linear combination of terms $e_i \cdot s_i$ and $e_i \cdot c_i \cdot e_j \cdot c_j$, i.e., we should have

$$p = \sum_{i=1}^{n} a_i \cdot e_i \cdot s_i + \sum_{i<j} b_{ij} \cdot e_i \cdot c_i \cdot e_j \cdot c_j, \tag{1}$$

for some coefficients $a_i > 0$ and $b_{ij} > 0$.

A priori, we have no reasons to believe that some student's skills affect the resulting productivity in different ways. Thus, all the coefficients a_i should be equal to each other: $a_1 = \cdots = a_n$. Let us denote the common value of a_i by a. Similarly, all the coefficients b_{ij} corresponding to different pairs (i, j) should be equal to each other. Let us denote their common value by b. Then, formula (1) takes the following simplified form

$$p = a \cdot \sum_{i=1}^{n} e_i \cdot s_i + b \cdot \sum_{i<j} e_i \cdot c_i \cdot e_j \cdot c_j. \tag{2}$$

Let us simplify this formula. According to formula (2), the only way the value s_i enters the formula is via the product $e_i \cdot s_i$. There is no way to separate these two quantities—and this makes sense: if a student does not even try, how can we determine whether this student has the skills? So, the only thing that we can observe are not "hidden" skills s_i, but the actually applied skills $\widetilde{s}_i \overset{\text{def}}{=} e_i \cdot s_i$. Similarly, we cannot observe the hidden ability to collaborate, we can only observe the product $\widetilde{c}_i \overset{\text{def}}{=} e_i \cdot c_i$. In terms of these "actual" variables, formula (2) takes the following simplified form

$$p = a \cdot \sum_{i=1}^{n} \widetilde{s}_i + b \cdot \sum_{i<j} \widetilde{c}_i \cdot \widetilde{c}_j. \tag{3}$$

Finally, to make this formula even simpler, we can re-scale the student-characterizing parameters $\widetilde{s}_i$ and $\widetilde{c}_i$ into $S_i \overset{\text{def}}{=} a \cdot \widetilde{s}_i$ and $C_i \overset{\text{def}}{=} \sqrt{b} \cdot \widetilde{c}_i$. In terms of these re-scaled values, formula (3) gets the following form:

Final description of our model. The productivity of a group has the form

$$p = \sum_{i=1}^{n} S_i + \sum_{i<j} C_i \cdot C_j, \tag{4}$$

where

- the value S_i describes the individual skills of the ith student, and
- the value C_i describe the ability of the ith student to collaborate.

What we want. Based on the observed productivity values p corresponding to different groups—including "groups" consisting of only one student—we want to reconstruct the values C_i (and, of course, the values S_i as well).

3 Analysis of the Problem: When We Can Determine the Values C_i (and How) and When We Cannot

Simplest case: two students. Let us start with the simplest case of two students. In this case, we do not have much of a choice:

- we can give both students individual assignments, and thus, by observing the resulting productivity $p_i = S_i$, find their individual skills S_i, and
- we can also give them a joint assignment, and observe the joint productivity

$$p_{12} = S_1 + S_2 + C_1 \cdot C_2.$$

Based on the result of the joint assignment, we get the value $S_1 + S_2 + C_1 \cdot C_2$. Once we know S_1 and S_2, we can therefore determine the product $C_1 \cdot C_2$. However, based only on the product, we cannot determine individual numbers C_1 and C_2.

This impossibility makes perfect mathematical sense: we only have three possible measurement results p_1, p_2, and p_{12}, so we only have three equations for four unknowns S_1, S_2, C_1, and C_2—not enough to uniquely determine all the desired quantities S_i and C_i.

Next simplest case—three students: analysis. In the case when we have three students:

- we can give all students individual assignments, and thus, by observing the resulting productivity $p_i = S_i$, find their individual skills S_i, and
- we can group them into pairs $\{1, 2\}$, $\{2, 3\}$, and $\{1, 3\}$, and observe the joint productivities

$$p_{12} = S_1 + S_2 + C_1 \cdot C_2, \ p_{23} = S_2 + S_3 + C_2 \cdot C_3, \text{ and}$$
$$p_{13} = S_1 + S_3 + C_1 \cdot C_3.$$

Based on the results of these assignments, we can find the products

$$P_{12} \overset{\text{def}}{=} C_1 \cdot C_2 = p_{12} - p_1 - p_2,$$

$$P_{23} \overset{\text{def}}{=} C_2 \cdot C_3 = p_{23} - p_2 - p_3, \text{ and}$$

$$P_{13} \overset{\text{def}}{=} C_1 \cdot C_3 = p_{13} - p_1 - p_3.$$

The product $P_{12} \cdot P_{23} \cdot P_{13}$ of all three products is equal to $(C_1 \cdot C_2 \cdot C_3)^2$, thus $C_1 \cdot C_2 \cdot C_3 = \sqrt{P_{12} \cdot P_{23} \cdot P_{13}}$. By dividing this product by the known expression for $C_2 \cdot C_3 = P_{23}$, we conclude that

$$C_1 = \frac{C_1 \cdot C_2 \cdot C_3}{C_2 \cdot C_3} = \frac{\sqrt{P_{12} \cdot P_{23} \cdot P_{13}}}{P_{23}} = \sqrt{\frac{P_{12} \cdot P_{13}}{P_{23}}}.$$

Similarly, we can determine all three values C_i. Thus, we arrive at the following method.

Case of three students: how to determine the values C_i describing the students' ability to collaborate. We give each student an individual assignment, and observe the resulting productivity $p_i = S_i$. This way, we determine the values S_i.

We then give each pair of students a group assignment and thus determine the corresponding group productivities p_{12}, p_{23}, and p_{13}. Based on these values, we compute $P_{ij} = p_{ij} - p_i - p_j$, and then compute

$$C_1 = \sqrt{\frac{P_{12} \cdot P_{13}}{P_{23}}}; \quad C_2 = \sqrt{\frac{P_{12} \cdot P_{23}}{P_{13}}}; \quad C_3 = \sqrt{\frac{P_{13} \cdot P_{23}}{P_{12}}}.$$

Case of three students: possible alternative methods. For each student i, we need to determine 2 values S_i and C_i. So, for 3 students, we need to determine $3 \cdot 2 = 6$

parameters. For this, we need to perform 6 experiments—which is exactly what the above method does.

In addition to these 6 experiments, we could also make a group of all 3 students, so overall, we have 7 possible experiments, corresponding to groups

$$\{1\}, \{2\}, \{3\}, \{1, 2\}, \{2, 3\}, \{1, 3\}, \text{ and } \{1, 2, 3\}.$$

Let us show that any 6 of these experiments enable us to uniquely determine all the desired values S_i and C_i.

Indeed, in the above method, we omitted the $\{1, 2, 3\}$ experiment. What if we omit one the individual-measuring experiments? Without losing generality, let us assume that we miss experiment $\{1\}$. In this case, we get

$$S_2 = p_2, S_3 = p_3, \text{ and } C_2 \cdot C_3 = p_{23} - p_2 - p_3.$$

We also know the values

$$p_{12} = S_1 + S_2 + C_1 \cdot C_2,$$

$$p_{13} = S_1 + S_3 + C_1 \cdot C_3, \text{ and}$$

$$p_{123} = S_1 + S_2 + S_3 + C_1 \cdot C_2 + C_2 \cdot C_3 + C_1 \cdot C_3.$$

In this case,

$$p_{12} + p_{23} + p_{13} = 2 \cdot (S_1 + S_2 + S_3) + C_1 \cdot C_2 + C_2 \cdot C_3 + C_1 \cdot C_3$$

and thus,

$$p_{12} + p_{23} + p_{13} - p_{123} = S_1 + S_2 + S_3.$$

Since we know S_2 and S_3, we can therefore determine S_1 as the difference

$$S_1 = (p_{12} + p_{23} + p_{13} - p_{123}) - p_2 - p_3.$$

Once we know S_1, we can determine all the values C_i as above.

What if we omit one of the paired experiment? Without losing generality, let us assume that we miss experiment $\{2, 3\}$. In this case, we have all the values $S_i = p_i$, and we also have

$$p_{12} = S_1 + S_2 + C_1 \cdot C_2,$$

$$p_{13} = S_1 + S_3 + C_1 \cdot C_3, \text{ and}$$

$$p_{123} = S_1 + S_2 + S_3 + C_1 \cdot C_2 + C_2 \cdot C_3 + C_1 \cdot C_3.$$

Thus, we can determine $C_1 \cdot C_2 = p_{12} - p_1 - p_2$, $C_1 \cdot C_3 = p_{13} - p_1 - p_3$, and

$$C_1 \cdot C_2 + C_2 \cdot C_3 + C_1 \cdot C_3 = p_{123} - p_1 - p_2 - p_3.$$

Thus, we can find the remaining value $C_2 \cdot C_3$ as

$$
\begin{aligned}
C_2 \cdot C_3 &= (C_1 \cdot C_2 + C_2 \cdot C_3 + C_1 \cdot C_3) - C_1 \cdot C_2 - C_1 \cdot C_3 \\
&= (p_{123} - p_1 - p_2 - p_3) - (p_{12} - p_1 - p_2) - (p_{13} - p_1 - p_3) \\
&= p_{123} - p_{12} - p_{13} + p_1.
\end{aligned}
$$

Once we know $C_2 \cdot C_3$, we can determine the values C_i as above.

General case. In the general case, we can divide students into groups of 3 and follow one of the above procedures for each triple.

Caution. As we have mentioned, to determine $2n$ unknowns S_i and C_i, we need to have at least $2n$ results—i.e., we need to perform at least $2n$ measurements. It is important to notice that the very fact that we have performed $2n$ measurements does not necessarily mean that we can uniquely determine all $2n$ values.

An important counterexample is when all the groups have the same size k. Let us show that in this case, the unique determination is not possible. Indeed, let us show that in this case, the same observations p_g corresponding to different k-element groups $g \subset \{1, \ldots, n\}$ are consistent not only with the actual values C_i but also with modified values $C_i' = C_i + \delta$. Indeed, for each i and j, we have

$$
C_i' \cdot C_j' = (C_i + \delta) \cdot (C_j + \delta) = C_i \cdot C_j + \delta \cdot C_i + \delta \cdot C_j + \delta^2.
$$

Thus, if we add up these products for all $(k - 1) \cdot k / 2$ pairs $i, j \in g$, we get

$$
\sum_{i,j\in g,\ i<j} C_i' \cdot C_j' = \sum_{i,j\in g,\ i<j} C_i \cdot C_j + (k - 1) \cdot \delta \cdot \sum_{i\in g} C_i + \frac{(k - 1) \cdot k}{2} \cdot \delta^2.
$$

Thus, we have

$$
\sum_{i,j\in g,\ i<j} C_i' \cdot C_j' = \sum_{i,j\in g,\ i<j} C_i \cdot C_j + \sum_{i\in g} \delta_i,
$$

where we denoted

$$
\delta_i \stackrel{\text{def}}{=} (k - 1) \cdot \delta \cdot C_i + \frac{k - 1}{2} \cdot \delta^2.
$$

Therefore,

$$
\sum_{i,j\in g,\ i<j} C_i \cdot C_j = \sum_{i,j\in g,\ i<j} C_i' \cdot C_j' - \sum_{i\in g} \delta_i,
$$

and thus, for each k-element group g, we have

$$
p_g = \sum_{i\in g} S_i + \sum_{i,j\in g,\ i<j} C_i \cdot C_j = \sum_{i\in g} S_i + \sum_{i,j\in g,\ i<j} C_i' \cdot C_j' - \sum_{i\in g} \delta_i,
$$

i.e.,

$$p_g = \sum_{i \in g} S'_i + \sum_{i,j \in g,\ i<j} C'_i \cdot C'_j,$$

where we denoted $S'_i \overset{\text{def}}{=} S_i - \delta_i$.

Thus, indeed, the same observations p_g are consistent not only with the actual values S_i and C_i, but also with different values S'_i and $C'_i = C_i + \delta$. Thus, *to uniquely determine the values S_i and C_i, we need to have groups of different sizes.*

Acknowledgements This work was supported in part by the National Science Foundation grants 1623190 (A Model of Change for Preparing a New Generation for Professional Practice in Computer Science), and HRD-1834620 and HRD-2034030 (CAHSI Includes), and by the AT&T Fellowship in Information Technology.

It was also supported by the program of the development of the Scientific-Educational Mathematical Center of Volga Federal District No. 075-02-2020-1478, and by a grant from the Hungarian National Research, Development and Innovation Office (NRDI).

Reference

1. Harary, Y.N.: Homo Deus: A Brief History of Tomorrow. HarperCollins, New York (2017)

Need to Combine Interval and Probabilistic Uncertainty: What Needs to Be Computed, What Can Be Computed, What Can Be Feasibly Computed, and How Physics Can Help

Julio Urenda, Vladik Kreinovich, and Olga Kosheleva

Abstract In many practical situations, the quantity of interest is difficult to measure directly. In such situations, to estimate this quantity, we measure easier-to-measure quantities which are related to the desired one by a known relation, and we use the results of these measurement to estimate the desired quantity. How accurate is this estimate? Traditional engineering approach assumes that we know the probability distributions of measurement errors; however, in practice, we often only have partial information about these distributions. In some cases, we only know the upper bounds on the measurement errors; in such cases, the only thing we know about the actual value of each measured quantity is that it is somewhere in the corresponding interval. Interval computation estimates the range of possible values of the desired quantity under such interval uncertainty. In other situations, in addition to the intervals, we also have partial information about the probabilities. In this paper, we describe how to solve this problem in the linearized case, what is computable and what is feasibly computable in the general case, and, somewhat surprisingly, how physics ideas – that initial conditions are not abnormal, that every theory is only approximate – can help with the corresponding computations.

Keywords Interval uncertainty · Probabilistic uncertainty · Feasible algorithms · Physics helps computing

---

J. Urenda
Departments of Mathematical Sciences and Computer Science,
University of Texas at El Paso, 500 W. University, El Paso, TX 79968, USA
e-mail: jcurenda@utep.edu

V. Kreinovich (✉)
Department of Computer Science, University of Texas at El Paso,
500 W. University, El Paso, TX 79968, USA
e-mail: vladik@utep.edu

O. Kosheleva
Department of Teacher Education, University of Texas at El Paso,
500 W. University, El Paso, TX 79968, USA
e-mail: olgak@utep.edu

© The Author(s), under exclusive license to Springer Nature Switzerland AG 2023
Sh. N. Shahbazova et al. (eds.), *Recent Developments and the New Directions of Research, Foundations, and Applications*, Studies in Fuzziness and Soft Computing 423,
https://doi.org/10.1007/978-3-031-23476-7_8

1 Need to Combine Interval and Probabilistic Uncertainty: Linearized Case

Need to take uncertainty into account when processing data. In practice, we are often interested in a quantity y which is difficult to measure directly. Examples are distance to a star, amount of oil in a well, tomorrow's weather.

A solution to this problem is to find easier-to-measure quantities $x_1, \ldots, x_n$ related to y by a known dependence $y = f(x_1, \ldots, x_n)$. Then, we measure x_i and use measurement results $\widetilde{x}_i$ to compute an estimate $\widetilde{y} = f(\widetilde{x}_1, \ldots, \widetilde{x}_n)$ for the desired quantity y. Such computations are usually called *data processing*.

Measurements are never absolutely accurate, so even if the model f is exact, $\widetilde{x}_i \neq x_i$ leads to $\Delta y \overset{\text{def}}{=} \widetilde{y} - y \neq 0$. It is important to use information about measurement errors $\Delta x_i \overset{\text{def}}{=} \widetilde{x}_i - x_i$ to estimate the accuracy Δy; see, e.g., [23].

We often have imprecise probabilities. The usual assumption is that we know the probabilities of different values of measurement errors Δx_i. How can we find these probabilities?

To find them, we measure the same quantities:

- with our measuring instrument (MI) and
- with a much more accurate MI, with $\widetilde{x}_i^{\text{st}} \approx x_i$.

However, in two important cases, this does not work: in the case of state-of-the art-measurements, and in the case of measurements on the shop floor. In the first case, we use state-of-the-art measuring instruments, so more accurate instruments are available. In the second case, it is, in principle, possible to accurately calibrate each sensor, but that would cost too much.

In both cases, we have partial information about probabilities. Often, all we know is an upper bound $|\Delta x_i| \leq \Delta_i$. Then, the only thing that we know about the actual (unknown) values x_i of the measured quantities is that $x_i \in [\widetilde{x}_i - \Delta_i, \widetilde{x}_i + \Delta_i]$. Then, the only thing that we know about $y = f(x_1, \ldots, x_n)$ is that

$$y \in [\underline{y}, \overline{y}] \overset{\text{def}}{=} \{f(x_1, \ldots, x_n) : x_i \in [\widetilde{x}_i - \Delta_i, \widetilde{x}_i + \Delta_i]\}.$$

Computing this interval $[\underline{y}, \overline{y}]$ is known as *interval computation*; see, e.g., [4, 17, 19].

Data processing: example. Let us provide an example of data processing. Suppose that we want to measure coordinates X_j of an object. To find these coordinates, we measure the distance Y_i between this object and objects with accurately known coordinates $X_j^{(i)}$:

$$Y_i = \sqrt{\sum_{j=1}^{3} \left(X_j - X_j^{(i)}\right)^2}.$$

After the measurements, we know the results $\widetilde{Y}_i$ of measuring Y_i. We want to estimate the desired quantities X_j.

Usually linearization is possible. In most practical situations, we know the approximate values $X_j^{(0)}$ of the desired quantities X_j. These approximation are usually reasonably good, in the sense that the difference $x_j \overset{\text{def}}{=} X_j - X_j^{(0)}$ are small.

In terms of x_j, we have $Y_i = f(X_1^{(0)} + x_1, \ldots, X_n^{(0)} + x_n)$. When the differences x_i are small, we can safely ignore terms quadratic in x_j. Indeed, even if the estimation accuracy is 10% (0.1), its square is 1% $\ll$ 10%. We can thus expand the dependence of Y_i on x_j in Taylor series and keep only linear terms:

$$Y_i = Y_i^{(0)} + \sum_{j=1}^{n} a_{ij} \cdot x_j, \quad Y_i^{(0)} \overset{\text{def}}{=} f_i(X_1^{(0)}, \ldots, X_n^{(0)}), \quad a_{ij} \overset{\text{def}}{=} \frac{\partial f_i}{\partial X_j}.$$

Least squares. Thus, to find the unknowns x_j, we need to solve a system of approximate linear equations $\sum_{j=1}^{n} a_{ij} \cdot x_i \approx y_i$, where $y_i \overset{\text{def}}{=} \tilde{Y}_i - Y_i^{(0)}$. Usually, it is assumed that each measurement error is normally distributed with 0 mean and known standard deviation σ_i.

The distribution is indeed often normal: the measurement error is a joint result of many independent factors, and the distribution of the sum of many small independent errors is close to Gaussian; this result is known as the Central Limit Theorem; see, e.g., [24].

0 mean also makes sense: we calibrate the measuring instrument by comparing it with a more accurate one, so if there was a bias (non-zero mean), we delete it by re-calibrating the scale.

It is also usually assumed that measurement errors of different measurements are independent. In this case, for each possible combination $x = (x_1, \ldots, x_n)$, the probability of observing $y_1, \ldots, y_m$ is equal to the product of the corresponding probabilities:

$$\prod_{i=1}^{m} \left(\frac{1}{\sqrt{2\pi} \cdot \sigma_i} \cdot \exp\left(-\frac{\left(y_i - \sum_{j=1}^{n} a_{ij} \cdot x_j\right)^2}{2\sigma_i^2} \right) \right).$$

It is reasonable to select x_j for which this probability is the largest, i.e., equivalently, for which

$$\sum_{i=1}^{n} \frac{\left(y_i - \sum_{j=1}^{n} a_{ij} \cdot x_j\right)^2}{\sigma_i^2} \to \min.$$

For every confidence level γ, the confidence set S_γ, i.e., the set off all combinations x which are possible with this degree of confidence, can be determined by the formula

$$S_\gamma = \left\{ x : \sum_{i=1}^{n} \frac{\left(y_i - \sum_{j=1}^{n} a_{ij} \cdot x_j\right)^2}{\sigma_i^2} \le \chi^2_{m-n,\gamma} \right\}.$$

Sometimes this set is empty; this means that some measurements are outliers.

Need to take into account systematic error. In the traditional approach, we assume that $y_i = \sum_{j=1}^{n} a_{ij} \cdot x_j + e_i$, where the measurement error e_i has 0 mean. However, sometimes, in addition to the random error $e_i^r \overset{\text{def}}{=} e_i - E[e_i]$ with 0 mean, we also have a systematic error $e_i^s \overset{\text{def}}{=} E[e_i]$:

$$y_i = \sum_{j=1}^{n} a_{ij} \cdot x_j + e_i^r + e_i^s.$$

Sometimes, we know the upper bound Δ_i: $|e_i^s| \le \Delta_i$.

What can we then say about x_j?

Comment. In other cases, we have different bounds $\Delta_i(p)$ corresponding to different degree of confidence p; this is known as the *fuzzy case*; see, e.g., see, e.g., [1, 5, 18, 20, 21, 29].

Combining probabilistic and interval uncertainty: main idea. If we knew the values e_i^s, then we would conclude that for $e_i^r = y_i - \sum_{j=1}^{n} a_{ij} \cdot x_j - e_i^s$, we have

$$\sum_{i=1}^{m} \frac{(e_i^r)^2}{\sigma_i^2} = \sum_{i=1}^{m} \frac{\left(y_i - \sum_{j=1}^{n} a_{ij} \cdot x_j - e_i^s\right)^2}{\sigma_i^2} \le \chi^2_{m-n,\gamma}.$$

In practice, we do not know the values e_i^s, we only know that these values are in the interval $[-\Delta_i, \Delta_i]$. Thus, we know that the above inequality holds for some $e_i^s \in [-\Delta_i, \Delta_i]$.

The above condition is equivalent to $v(x) \le \chi^2_{m-n,\gamma}$, where

$$v(x) \overset{\text{def}}{=} \min_{e_i^s \in [-\Delta_i, \Delta_i]} \sum_{i=1}^{m} \frac{\left(y_i - \sum_{j=1}^{n} a_{ij} \cdot x_j - e_i^s\right)^2}{\sigma_i^2}.$$

So, the set S_γ of all combinations $X = (x_1, \ldots, x_n)$ which are possible with confidence $1 - \gamma$ is: $S_\gamma = \{x : v(x) \le \chi^2_{m-n,\gamma}\}$. The range of possible values of x_j can be obtained by maximizing and minimizing x_j under the constraint $v(x) \le \chi^2_{m-n,\gamma}$.

Comment. In the fuzzy case, we have to repeat the computations for every p.

How to check consistency. We want to make sure that the measurements are consistent – i.e., that there are no outliers. This means that we want to check that there exists some $x = (x_1, \ldots, x_n)$ for which $v(x) \leq \chi^2_{m-n,\gamma}$. This condition is equivalent to ([25]):

$$v \overset{\text{def}}{=} \min_{x} v(x) = \min_{x} \min_{e^s_i \in [-\Delta_i, \Delta_i]} \sum_{i=1}^{m} \frac{\left(y_i - \sum_{j=1}^{n} a_{ij} \cdot x_j - e^s_i\right)^2}{\sigma_i^2} \leq \chi^2_{m-n,\gamma}.$$

This is indeed a generalization of probabilistic and interval approaches. In the case when $\Delta_i = 0$ for all i, i.e., when there is no interval uncertainty, we get the usual Least Squares.

Vice versa, for very small σ_i, we get the case of pure interval uncertainty. In this case, the above formulas tend to the set of all the values for which $\left| y_i - \sum_{j=1}^{n} a_{ij} \cdot x_j \right| \leq \Delta_i$. For example, for m repeated measurements of the same quantity, we get the intersection of the corresponding intervals.

So, the new idea is indeed a generalization of the known probabilistic and interval approaches.

From formulas to computations. The expression $\left(y_i - \sum_{j=1}^{n} a_{ij} \cdot x_j - e^s_i\right)^2$ is a convex function of x_j. The domain of possible values of $e^s = (e^s_1, \ldots, e^s_m)$ is also convex: it is a box $[-\Delta_1, \Delta_1] \times \ldots \times [-\Delta_m, \Delta_m]$. There exist efficient algorithms for computing minima of convex functions over convex domains; these algorithms also compute locations where these minima are attained; see, e.g., [13] and references therein. Thus, for every x, we can efficiently compute $v(x)$ and thus, efficiently check whether $v(x) \leq \chi^2_{m-n,\gamma}$.

Similarly, we can efficiently compute v and thus, check whether $v \leq \chi^2_{m-n,\gamma}$ — i.e., whether we have outliers.

The set S_γ is convex. We can approximate the set S_γ by

- taking a grid G,
- checking, for each $x \in G$, whether $v(x) \leq \chi^2_{m-n,\gamma}$, and
- taking the convex hull of "possible" points.

We can also efficiently find the minimum $\underline{x}_j$ of x_j over $x \in S_\gamma$. By computing the min of $-x_j$, we can also find the maximum $\overline{x}_j$.

2 General Case: What Can Be Computed?

How do we describe imprecise probabilities? The ultimate goal of most estimates is to make decisions. It is known that a rational decision-maker maximizes expected utility $E[u(y)]$.

- For smooth $u(y)$, $y \approx \widetilde{y}$ implies that

$$u(y) = u(\widetilde{y}) + (y - \widetilde{y}) \cdot u'(\widetilde{y}) + \frac{1}{2} \cdot (y - \widetilde{y})^2 \cdot u''(\widetilde{y}).$$

So, to find $E[u(y)]$, we must know moments $E[(y - \widetilde{y})^k]$.
- Often, $u(y)$ abruptly changes: e.g., when pollution level exceeds y_0, the plant has to pay a huge fine; then $E[u(y)]$ is proportional to the cdf:

$$E[u(y)] \sim F(y) \overset{\text{def}}{=} \text{Prob}(y \leq y_0).$$

So, it is enough to know moments and cdf. From the cdf $F(y)$, we can estimate moments, so $F(y)$ is enough.

Imprecise probabilities mean that we don't know $F(y)$ exactly, we only know bounds (p-box) $\underline{F}(y) \leq F(y) \leq \overline{F}(y)$.

What is computable? Computations with p-boxes are practically important. It is thus desirable to come up with efficient algorithms which are as general as possible.

It is known that too general problems are often *not* computable. To avoid wasting time, it is therefore important to find out what *can* be computed.

At first glance, this question sounds straightforward:

- to describe a cdf, we can consider a computable function $F(x)$;
- to describe a p-box, we consider a computable *function interval* $[\underline{F}(x), \overline{F}(x)]$.

Often, we can do that, but we will show that sometimes, we need to go *beyond* computable function intervals. To explain all this, let us recall what computable means in general; see, e.g., [13, 22, 28].

Reminder: what is computable? A real number x corresponds to a value of a physical quantity. We can measure x with higher and higher accuracy. So, we arrive at the following definition:

Definition 2.1. *A real number x is called* computable *if there is an algorithm, that, given k, produces a rational r_k s.t. $|x - r_k| \leq 2^{-k}$.*

A *computable function* computes $f(x)$ from x. We can only use approximations to x. So, an algorithm for computing a function can, given k, request a 2^{-k}-approximation to x. Most usual functions are thus computable.

Not all functions are computable, an exception is a step-function $f(x) = 0$ for $x < 0$ and $f(x) = 1$ for $x \geq 0$. Indeed, no matter how accurately we know $x \approx 0$, from $r_k = 0$, we cannot tell whether $x < 0$ or $x \geq 0$ [13, 22, 28].

Consequences for representing a cdf $F(x)$. We would like to represent a general probability distribution by its cdf $F(x)$. From the purely mathematical viewpoint, this is indeed the most general representation.

At first glance, it makes sense to consider computable functions $F(x)$. For many distributions, e.g., for Gaussian, $F(x)$ is computable.

However, when $x = 0$ with probability 1, the cdf $F(x)$ is exactly the step-function. And we already know that the step-function is not computable. Thus, we need to find an alternative way to represent cdf's – beyond computable functions.

Back to the drawing board. Each value $F(x)$ is the probability that $X \leq x$. We cannot empirically find exact probabilities p. We can only estimate *frequencies* f based on a sample of size N.

For large N, the difference $d \stackrel{\text{def}}{=} p - f$ is asymptotically normal, with $\mu = 0$ and $\sigma = \sqrt{\dfrac{p \cdot (1 - p)}{N}}$. Situations when $|d - \mu| < 6\sigma$ are negligibly rare, so we conclude that $|f - p| \leq 6\sigma$.

For large N, we can get $6\sigma \leq \delta$ for any accuracy $\delta > 0$. We get a sample $X_1, \ldots, X_N$. We don't know the exact values X_i, only measured values $\tilde{X}_i$ such that $|\tilde{X}_i - X_i| \leq \varepsilon$ for some accuracy ε.

So, what we have is a frequency $f = \text{Freq}(\tilde{X}_i \leq x)$.

Resulting definition. Here, $X_i \leq x - \varepsilon$ implies that $\tilde{X}_i \leq x \Rightarrow X_i \leq x + \varepsilon$, so

$$\text{Freq}(X_i \leq x - \varepsilon) \leq f = \text{Freq}(\tilde{X}_i \leq x) \leq \text{Freq}(X_i \leq x + \varepsilon).$$

Frequencies are δ-close to probabilities, so we arrive at the following definition [15]:

Definition 2.2. *A cdf $F(x)$ is called* computable *if there is an algorithm that, given x, $\varepsilon > 0$, and $\delta > 0$, computes a rational number f such that*

$$F(x - \varepsilon) - \delta \leq f \leq F(x + \varepsilon) + \delta.$$

In the computer, to describe a distribution on an interval $[\underline{T}, \overline{T}]$: we select a grid $x_1 = \underline{T}, x_2 = \underline{T} + \varepsilon, \ldots$, and we store the corresponding frequencies f_i with accuracy δ. A class of possible distribution is represented, for each ε and δ, by a finite list of such approximations.

First equivalent definition. It turns out that our definition is equivalent to the following one:

Definition 2.3. *A cdf $F(x)$ is called* computable *if there exists an algorithm that, given x, $varepsilon > 0$, and $\delta > 0$, computes a rational number f which is δ-close to $F(x')$ for some x' such that $|x' - x| \leq \varepsilon$.*

Indeed, here is a proof of equivalence. We know that $F(x + \varepsilon) - F(x + \varepsilon/3) \to 0$ as $\varepsilon \to 0$. So, for $\varepsilon = 2^{-k}$, $k = 1, 2, \ldots$, we take f and f' such that

$$F(x + \varepsilon/3) - \delta/4 \leq f \leq F(x + (2/3) \cdot \varepsilon) + \delta/4$$

$$F(x + (2/3) \cdot \varepsilon) - \delta/4 \leq f' \leq F(x + \varepsilon) + \delta/4.$$

We stop when f and f' are sufficiently close, i.e., when $|f - f'| \leq \delta$. Thus, we get the desired f.

Second equivalent definition. We start with pairs $(x_1, f_1), (x_2, f_2), \ldots$ When $f_{i+1} - f_i > \delta$, we add intermediate pairs

$$(x_i, f_i + \delta), (x_i, f_i + 2\delta), \ldots, (x_i, f_{i+1}).$$

The resulting set of pairs is (ε, δ)-close to the graph

$$\{(x, y) : F(x - 0) \leq y \leq F(x)\}$$

in Hausdorff metric d_H. This metric can be defined as follows.

Definition 2.4. (x, y) and (x', y') are (ε, δ)-close if $|x - x'| \leq \varepsilon$ and $|y - y'| \leq \delta$.

Definition 2.5. The sets S and S' are (ε, δ)-close if for every $s \in S$, there is a (ε, δ)-close point $s' \in S'$; for every $s' \in S'$, there is a (ε, δ)-close point $s \in S$.

Compact sets with metric d_H form a computable compact. So, $F(x)$ is a monotonic computable object in this compact.

What can be computed: a positive result for the 1D case. We are interested in computing the expected value $E_{F(x)}[u(x)]$ for smooth $u(x)$. Our result is as follows:

Proposition 2.1. There is an algorithm that given a computable cdf $F(x)$, a computable function $u(x)$, and accuracy $\delta > 0$, computes $E_{F(x)}[u(x)]$ with accuracy δ.

Comment. For computable classes $\mathscr{F}$ of cdfs, a similar algorithm computes the range of possible values $[\underline{u}, \overline{u}] \stackrel{\text{def}}{=} \{E_{F(x)}[u(x)] : F(x) \in \mathscr{F}\}$.

Proof: main idea. Computable functions are computably continuous: for every $\delta > 0$, we can compute $\varepsilon > 0$ such that $|x - x'| \leq \varepsilon$ implies $|f(x) - f(x')| \leq \delta$. We select ε corresponding to $\delta/4$, and take a grid with step $\varepsilon/4$.

For each x_i, the value f_i is $(\delta/4)$-close to $F(x'_i)$ for some x'_i which is $(\varepsilon/4)$-close to x_i.

The function $u(x)$ is $(\delta/2)$-close to a piece-wise constant function $u'(x) = u(x_i)$ for $x \in [x'_i, x'_{i+1}]$.

Thus, $|E[u(x)] - E[u'(x)]| \leq \delta/2$. Here,

$$E[u'(x)] = \sum_i u(x_i) \cdot (F(x'_{i+1}) - F(x'_i)).$$

Here, $F(x'_i)$ is close to f_i and $F(x'_{i+1})$ is close to f_{i+1}. Thus, $E[u'(x)]$ (and hence, $E[u(x)]$) is computably close to a computable sum

$$\sum_i u(x_i) \cdot (f_{i+1} - f_i).$$

What to do in a multi-D case? For each $g(x)$, y, $\varepsilon > 0$, and $\delta > 0$, we can find a frequency f such that: $|P(g(x) \le y') - f| \le \varepsilon$ for some y' such that $|y - y'| \le \delta$. We select an ε-net $x_1, \ldots, x_n$ for X. Then, $X = \bigcup_i B_\varepsilon(x_i)$, where $B_\varepsilon(x) \overset{\text{def}}{=} \{x' : d(x, x') \le \varepsilon\}$. We select f_1 which is close to $P(B_{\varepsilon'}(x_1))$ for all ε' from some interval $[\underline{\varepsilon}, \overline{\varepsilon}]$ which is close to ε.

We then select f_2 which is close to $P(B_{\varepsilon'}(x_1) \cup B_{\varepsilon'}(x_2))$ for all ε' from some subinterval of $[\underline{\varepsilon}, \overline{\varepsilon}]$, etc.

Then, we get approximations to probabilities of the sets

$$B_\varepsilon(x_i) - (B_\varepsilon(x_1) \cup \ldots \cup B_\varepsilon(x_{i-1})).$$

This lets us compute the desired values $E[u(x)]$.

3 Taking into Account that We Process Physical Data

Computations with real numbers: reminder. From the physical viewpoint, real numbers x describe values of different quantities. We get values of real numbers by measurements. Measurements are never 100% accurate, so after a measurement, we get an approximate value r_k of x. In principle, we can measure x with higher and higher accuracy.

So, from the computational viewpoint, a real number is a sequence of rational numbers r_k for which, e.g., $|x - r_k| \le 2^{-k}$.

By an algorithm processing real numbers, we mean an algorithm using r_k as an "oracle" (subroutine). This is how computations with real numbers are defined in *computable analysis* [13, 22, 28].

Known negative results. The first known negative result that we will use is that no algorithm is possible that, given two numbers x and y, would check whether $x = y$.

Similarly, we can define a computable function $f(x)$ from real numbers to real numbers as a mapping that,

- given an integer n, a rational number x_m and its accuracy 2^{-m},
- produces y_n which is 2^{-n}-close to all values $f(x)$ with $d(x, x_m) \le 2^{-m}$ (or produces nothing)
- so that for every x and for each desired accuracy n, there is an m for which a y_n is produced.

We can similarly define a computable function $f(x)$ on a computable compact set K.

The second negative result that we will use is that no algorithm is possible that, given f, returns x such that $f(x) = \max_{y \in K} f(y)$. (The maximum itself *is* computable.)

From the physicists' viewpoint, these negative results seem rather theoretical. In mathematics, if two numbers coincide up to 13 digits, they may still turn to be different. For example, they may be 1 and $1 + 10^{-100}$. In physics, if two quantities coincide up to a very high accuracy, it is a good indication that they are equal: if an experimentally value is very close to the theoretical prediction, this means that this theory is (triumphantly) true.

This is how General Relativity was confirmed. This is how physicists realized that light is formed of electromagnetic waves: their speeds are very close; see, e.g., [2, 26].

How physicists argue. In math, if two numbers coincide up to 13 digits, they may still turn to be different: e.g., 1 and $1 + 10^{-100}$. In physics, if two quantities coincide up to a very high accuracy, it is a good indication that they are equal. A typical physicist argument is that: while numbers like $1 + 10^{-100}$ (or $c \cdot (1 + 10^{-100})$) are, in principle, possible, they are *abnormal* (not *typical*).

In physics, second order terms like $a \cdot \Delta x^2$ of the Taylor series can be ignored if Δx is small, since:

- while abnormally high values of a (e.g., $a = 10^{40}$) are mathematically possible,
- typical (= not abnormal) values appearing in physical equations are usually of reasonable size.

How to formalize the physicist's intuition of physically meaningful values: main idea. To some physicists, all the values of a coefficient a above 10 are abnormal. To another one, who is more cautious, all the values above 10,000 are abnormal. For every physicist, there is a value n such that all value above n are abnormal.

This argument can be generalized as a following property of the set $\mathcal{T}$ of all physically meaningful elements. Suppose that we have a monotonically decreasing sequence of sets $A_1 \supseteq A_2 \supseteq \ldots$ for which $\bigcap_n A_n = \emptyset$. In the above example, A_n is the set of all numbers $\geq n$. Then, there exists an integer N for which $\mathcal{T} \cap A_N = \emptyset$; see, e.g., [3, 7–12, 14].

How to formalize the physicist's intuition: resulting definition.

Definition 3.1. *We say that $\mathcal{T}$ is a set of physically meaningful elements if:*

- *for every definable decreasing sequence $\{A_n\}$ for which $\bigcap_n A_n = \emptyset$,*
- *there exists an N for which $\mathcal{T} \cap A_N = \emptyset$.*

Comment. Of course, to make this definition precise, we must restrict definability to a *subset* of properties, so that the resulting notion of definability will be defined in formal set theory (ZFC) itself.

Checking equality of real numbers. It is known equality of real numbers is undecidable. For physically meaningful real numbers, however, a deciding algorithm *is* possible.

Proposition 3.1. *For every set $\mathcal{T} \subseteq \mathbb{R}^2$ which consists of physically meaningful pairs (x, y) of real numbers, there exists an algorithm deciding whether $x = y$.*

Proof: We can take $A_n = \{(x, y) : 0 < |x - y| < 2^{-n}\}$. The intersection of all these sets is empty. Hence, $\mathcal{T}$ has no elements from $\bigcap_{n=1}^{N_A} A_n = A_{N_A}$. Thus, for each $(x, y) \in \mathcal{T}$, $x = y$ or $|x - y| \geq 2^{-N_A}$.

Indeed, we can decide which of the two alternatives is true by comparing $2^{-(N_A+3)}$-approximations x' and y' to x and y. Q.E.D.

Finding roots. In general, it is not possible, given a function $f(x)$ attaining negative and positive values, to compute its root. This becomes possible if we restrict ourselves to physically meaningful functions.

Proposition 3.2. *Let K be a computable compact. Let X be the set of all functions $f : K \to \mathbb{R}$ that attain 0 value somewhere on K. Then*

- *for every set $\mathcal{T} \subseteq X$ consisting of physically meaningful functions and for every $\varepsilon > 0$,*
- *there is an algorithm that, given a f-n $f \in \mathcal{T}$, computes an ε-approximation to the set of roots $R \overset{\text{def}}{=} \{x : f(x) = 0\}$.*

In particular, we can compute an ε-approximation to one of the roots.

Optimization. In general, it is not algorithmically possible to find x where $f(x)$ attains maximum. For physically reasonable cases, it is possible:

Proposition 3.3. *Let K be a computable compact. Let X be the set of all functions $f : K \to \mathbb{R}$. Then, for every set $\mathcal{T} \subseteq X$ consisting of physically meaningful functions and for every $\varepsilon > 0$, there is an algorithm that,*

- *given a function $f \in \mathcal{T}$,*
- *computes an ε-approximation to $S = \left\{ x : f(x) = \max_y f(y) \right\}$.*

In particular, we can compute an approximation to an individual $x \in S$.

Proof: by reduction to the roots problem, since $f(x) = \max_y f(y)$ if and only if $g(x) = 0$, where $g(x) \overset{\text{def}}{=} f(x) - \max_y f(y)$.

Computing fixed points. In general, it is not possible to compute all the fixed points of a given computable function $f(x)$. Let K be a computable compact. Let X be the set of all functions $f : K \to K$. Then:

Proposition 3.4. *For every set $\mathcal{T} \subseteq X$ consisting of physically meaningful functions and for every $\varepsilon > 0$, there is an algorithm that,*

- *given a function $f \in \mathcal{T}$,*
- *computes an ε-approximation to the set $\{x : f(x) = x\}$.*

In particular, we can compute an approximation to an individual fixed point.

Proof: reduction to roots, since $f(x) = x$ if and only if $g(x) = 0$, where $g(x) \overset{\text{def}}{=} d(f(x), x)$.

Computing limits. In general, it is not algorithmically possible to find a limit $\lim a_n$ of a convergent computable sequence.

Let K be a computable compact. Let X be the set of all convergent sequences $a = \{a_n\}$, $a_n \in K$. Then:

Proposition 3.5. *For every set $\mathcal{T} \subseteq X$ consisting of physically meaningful functions and for every $\varepsilon > 0$, there exists an algorithm that,*

- *given a sequence $a \in \mathcal{T}$,*
- *computes its limit with accuracy ε.*

Comment. This result enables us to compute limits of iterations and sums of Taylor series (frequent in physics).

Proof (main idea): for every $\varepsilon > 0$ there exists $\delta > 0$ such that when $|a_n - a_{n-1}| \leq \delta$, then $|a_n - \lim a_n| \leq \varepsilon$.

Intuitively: we stop when two consequent iterations are close to each other.

4   How to Take into Account that We Can Use Non-standard Physical Phenomena to Process Data

Solving NP-complete problems is important. In practice, we often need to find a solution that satisfies a given set of constraints. At a minimum, we need to check whether such a solution is possible. Once we have a candidate, we can feasibly check whether this candidate satisfies all the constraints.

In theoretical computer science, "feasibly" is usually interpreted as computable in polynomial time.

The class of all such problems is called NP; see, e.g., [13]. A typical example of such a problem is satisfiability – checking whether a propositional formula like $(v_1 \vee \neg v_2 \vee v_3) \& (v_4 \vee \neg v_2 \vee \neg v_5) \& \ldots$ can be true.

Each problem from the class NP can be algorithmically solved by trying all possible candidates. For example, we can try all 2^n possible combinations of true-or-false values $v_1, \ldots, v_n$.

For medium-size inputs, e.g., for $n \approx 300$, the resulting time 2^n is larger than the lifetime of the Universe. So, these exhaustive search algorithms are not practically feasible.

It is not known whether problems from the class NP can be solved feasibly (i.e., in polynomial time). This is the famous open problem P$\overset{?}{=}$NP.

What we do know is that some problems are *NP-complete*: every problem from NP can be reduced to it. So, it is very important to be able to efficiently solve even one NP-hard problem.

Can non-standard physics speed up the solution of np-complete problems? NP-complete means difficult to solve on computers based on the usual physical techniques. A natural question is: can the use of non-standard physics speed up the solution of these problems?

This question has been analyzed for several specific physical theories, e.g.: for quantum field theory, for cosmological solutions with wormholes and/or casual anomalies.

No physical theory is perfect. If a speed-up is possible within a given theory, is this a satisfactory answer? In the history of physics, always new observations appear which are not fully consistent with the original theory. For example, Newton's physics was replaced by quantum and relativistic theories.

Many physicists believe that every physical theory is approximate. For each theory T, inevitably new observations will surface which require a modification of T. Let us analyze how this idea affects computations.

No physical theory is perfect: how to formalize this idea. We want to formalize a statement that for every theory, eventually there will be observations which violate this theory.

To formalize this statement, we need to formalize what are *observations* and what is a *theory*.

Most sensors already produce *observations* in the computer-readable form, as a sequence of 0s and 1s. Let ω_i be the bit result of an experiment whose description is i. Thus, all past and future observations form a (potentially) infinite sequence $\omega = \omega_1 \omega_2 \ldots$ of 0s and 1s.

A physical *theory* may be very complex. All we care about is which sequences of observations ω are consistent with this theory and which are not.

What is a physical theory? So, a physical theory T can be defined as the set of all sequences ω which are consistent with this theory.

A physical theory must have at least one possible sequence of observations: $T \neq \emptyset$.

A theory must be described by a finite sequence of symbols: the set T must be *definable*.

How can we check that an infinite sequence $\omega = \omega_1 \omega_2 \ldots$ is consistent with the theory? The only way is check that for every n, the sequence $\omega_1 \ldots \omega_n$ is consistent with T; so:

$$\text{if } \forall n \, \exists \omega^{(n)} \in T \, (\omega_1^{(n)} \dots \omega_n^{(n)} = \omega_1 \dots \omega_n) \text{ then } \omega \in T.$$

In mathematical terms, this means that T is *closed* in the Baire metric

$$d(\omega, \omega') \overset{\text{def}}{=} 2^{-N(\omega,\omega')},$$

where

$$N(\omega, \omega') \overset{\text{def}}{=} \max\{k : \omega_1 \dots \omega_k = \omega'_1 \dots \omega'_k\}.$$

A theory must predict something new. So, for every sequence $\omega_1 \dots \omega_n$ consistent with T, there is a continuation which does not belong to T.

In mathematical terms, T is *nowhere dense*. So, we arrive at the following definition.

What is a physical theory: definition.

Definition 4.1. *By a* physical theory, *we mean a non-empty closed nowhere dense definable set T.*

Definition 4.2. *A sequence ω is* consistent *with the no-perfect-theory principle if it does not belong to any physical theory.*

In precise terms, ω does not belong to the union of all definable closed nowhere dense set. There are countably many definable set, so this union is *meager* (= *Baire first category*). Thus, due to Baire Theorem, such sequences ω exist.

How to represent instances of an NP-complete problem. For each NP-complete problem $\mathscr{P}$, its instances are sequences of symbols. In the computer, each such sequence is represented as a sequence of 0s and 1s. We can append 1 in front and interpret this sequence as a binary code of a natural number i.

In principle, not all natural numbers i correspond to instances of a problem $\mathscr{P}$. We will denote the set of all natural numbers which correspond to such instances by $S_{\mathscr{P}}$. For each $i \in S_{\mathscr{P}}$, we denote the correct answer (true or false) to the i-th instance of the problem $\mathscr{P}$ by $s_{\mathscr{P},i}$.

What we mean by using physical observations in computations. In addition to performing computations, our computational device can produce a scheme i for an experiment, and then use the result ω_i of this experiment in future computations.

In other words, given an integer i, we can produce ω_i.

In precise terms, the use of physical observations in computations means corresponds to using ω as an *oracle*.

Main result of this section.

Definition 4.3. *A ph-algorithm $\mathscr{A}$ is an algorithm that uses an oracle ω consistent with the no-perfect-theory principle.*

The result of applying an algorithm $\mathcal{A}$ using ω to an input i will be denoted by $\mathcal{A}(\omega, i)$.

Definition 4.4. *We say that a feasible ph-algorithm $\mathcal{A}$ solves almost all instances of an NP-complete problem $\mathcal{P}$ if:*

$$\forall \varepsilon_{>0} \, \forall n \, \exists N_{\geq n} \left(\frac{\#\{i \leq N : i \in S_{\mathcal{P}} \, \& \, \mathcal{A}(\omega, i) = s_{\mathcal{P},i}\}}{\#\{i \leq N : i \in S_{\mathcal{P}}\}} > 1 - \varepsilon \right).$$

Restriction to sufficiently long inputs $N \geq n$ makes sense: for short inputs, we can do exhaustive search.

Proposition 4.1. *For every NP-complete problem $\mathcal{P}$, there is a feasible ph-algorithm $\mathcal{A}$ solving almost all instances of $\mathcal{P}$.*

This result is the best possible. Our result is the best possible, in the sense that the use of physical observations cannot solve *all* instances:

Proposition 4.2. *If $P \neq NP$, then no feasible ph-algorithm $\mathcal{A}$ can solve all instances of $\mathcal{P}$.*

Can we prove the result for *all* N starting with some N_0? We say that a feasible ph-algorithm $\mathcal{A}$ δ-solves $\mathcal{P}$ if

$$\exists N_0 \, \forall N \geq N_0 \left(\frac{\#\{i \leq N : i \in S_{\mathcal{P}} \, \& \, \mathcal{A}(\omega, i) = s_{\mathcal{P},i}\}}{\#\{i \leq N : i \in S_{\mathcal{P}}\}} > \delta \right).$$

Proposition 4.3. *For every NP-complete problem $\mathcal{P}$ and for every $\delta > 0$,*

- *if there exists a feasible ph-algorithm $\mathcal{A}$ that δ-solves $\mathcal{P}$,*
- *then there is a feasible algorithm $\mathcal{A}'$ that also δ-solves $\mathcal{P}$.*

5 Physical and Computational Consequences

Justification of physical induction. What is physical induction? It means that if a property P is satisfied in the first N experiments, then it is satisfied always.

Comment: N should be sufficiently large.

Proposition 5.1. *For every set $\mathcal{T}$ of physically meaningful sequences $s = s_1 s_2 \ldots$, and for every definable property P, there exists a natural number N such that if $P(s_i)$ holds for all $i \leq N$, then $P(s_i)$ holds for all i.*

Proof: Let us take

$$A_n \overset{\text{def}}{=} \{s : P(s_1) \& \ldots \& P(s_n), \& \exists m \, \neg P(s_m)\}.$$

Then $A_n \supseteq A_{n+1}$ and $\cup A_n = \emptyset$ so $\exists N \, (A_N \cap \mathscr{T} = \emptyset)$.

The meaning of $A_N \cap \mathscr{T} = \emptyset$ is that if $P(s_i)$ holds for all $i \leq N$, then this property holds for all i. Q.E.D.

Ill-posed problem: brief reminder. The main *objectives* of science are to produce:

- *guaranteed* estimates for physical quantities; and
- *guaranteed* predictions for these quantities.

The problem is that estimation and prediction are ill-posed problems, i.e., small changes in the measurement result can lead to drastic changes in the resulting estimates.

Example: measurement devices are inertial, hence they suppress high frequencies ω. So, the signals $\varphi(x)$ and $\varphi(x) + \sin(\omega \cdot t)$ are indistinguishable.

There exist many approaches to solve ill-posed problems: statistical regularization (filtering); Tikhonov regularization (e.g., assuming that $|\dot{x}| \leq \Delta$); expert-based regularization, etc.; see, e.g., [27]. The main problem of all these approaches is that they provide no guaranteed bounds.

On physically meaningful solutions, problems become well-posed. Indeed, let us consider state estimation – an ill-posed problem.

A measurement process is a function f that maps state $s \in S$ into observation $r = f(s) \in R$.

In principle, we can reconstruct s from r as $s = f^{-1}(r)$. The problem is that small changes in r can lead to huge changes in s, i.e., the inverse function f^{-1} *not continuous*.

Proposition 5.2. *Let S be a definably separable metric space. Let $\mathscr{T}$ be a set of physically meaningful elements of S. Let $f : S \to R$ be a continuous 1–1 function. Then, the inverse mapping $f^{-1} : R \to S$ is continuous for every $r \in f(\mathscr{T})$.*

Everything is related: EPR paradox. Due to *Relativity Theory*, two spatially separated simultaneous events cannot influence each other. By their paradox (see, e.g., [2, 26]) Einstein, Podolsky, and Rosen (EPR) intended to show that in quantum physics, such influence is possible.

In formal terms, let x and x' be measured values at these two events. *Independence* means that possible values of x do not depend on x', i.e., $\mathscr{T} = X \times X'$ for some X and X'.

Physical induction implies that the pair (x, x') belongs to a set S of physically meaningful pairs.

Proposition 5.3. *A set $\mathscr{T}$ of physically meaningful pairs cannot be represented as $X \times X'$.*

Thus, everything *is* related – but we probably can't use this relation to pass information (since the set $\mathcal{T}$ isn't computable).

When to stop an iterative algorithm? The following situation is typical in numerical mathematics:

- we know an iterative process whose results x_k are known to converge to the desired solution x, but
- we do not know when to stop to guarantee that $d_X(x_k, x) \leq \varepsilon$.

A usual heuristic approach is to stop when $d_X(x_k, x_{k+1}) \leq \delta$ for some $\delta > 0$.

For example, in physics, if 2nd order terms are small, we use the linear expression as an approximation.

When to stop an iterative algorithm: result.

Definition 5.1. *Let* $\{x_k\} \in \mathcal{T}$, *k be an integer, and* $\varepsilon > 0$ *a real number. We say that* x_k *is* ε-*accurate if* $d_X(x_k, \lim x_p) \leq \varepsilon$.

Definition 5.2. *Let* $d \geq 1$ *be an integer. By a* stopping criterion, *we mean a function* $c : X^d \to R_0^+$ *that satisfies the following two properties:*

- *If* $\{x_k\} \in \mathcal{T}$, *then* $c(x_k, \ldots, x_{k+d-1}) \to 0$.
- *If for some* $\{x_n\} \in \mathcal{T}$ *and k,* $c(x_k, \ldots, x_{k+d-1}) = 0$, *then*

$$x_k = \ldots = x_{k+d-1} = \lim x_p.$$

Proposition 5.4. *Let c be a stopping criterion. Then, for every* $\varepsilon > 0$, *there exists a* $\delta > 0$ *such that if* $c(x_k, \ldots, x_{k+d-1}) \leq \delta$, *and the sequence* $\{x_n\}$ *is physically meaningful, then* x_k *is* ε-*accurate.*

6 Relation with Randomness

Towards relation with randomness. Intuitively, if a sequence s is random, it satisfies all the probability laws such as the law of large numbers. Vice versa, if a sequence satisfies all probability laws, then for all practical purposes we can consider it random. Thus, we can define a sequence to be random if it satisfies all probability laws.

Definition 6.1. *A* probability law *is a statement S which is true with probability 1:* $P(S) = 1$.

So, we arrive at the following definition:

Definition 6.2. *A sequence is* random *if it belongs to all definable sets of measure* 1.

A sequence belongs to a set of measure 1 if and only if it does not belong to its complement $C = -S$ with $P(C) = 0$. So, we arrive at the following equivalent definition:

Definition 6.3. *A sequence is* random *if it does not belong to any definable set of measure* 0.

Randomness and Kolmogorov complexity. Different definabilities lead to different randomness. When definable means computable, the corresponding Kolmogorov-Martin-Löf randomness can be described in terms of Kolmogorov complexity [16], the smallest length of a program that generates a given string:

$$K(x) \overset{\text{def}}{=} \min\{\text{len}(p) : p \text{ generates } x\}.$$

Crudely speaking, an infinite string $s = s_1 s_2 \ldots$ is random if, for some constant $C > 0$, we have $\forall n \, (K(s_1 \ldots s_n) \geq n - C)$.

Indeed, if a sequence $s_1 \ldots s_n$ is truly random, then the only way to generate it is to explicitly print it: $\texttt{print}(s_1 \ldots s_n)$. In contrast, a sequence like $0101\ldots01$ generated by a short program is clearly not random.

From Kolmogorov-Martin-Löf theoretical randomness to a more physical one. The above definition means that (definable) events with probability 0 cannot happen. In practice, physicists also assume that events with a *very small* probability cannot happen.

For example, a kettle on a cold stove will not boil by itself – but the probability is non-zero. If a coin falls head 100 times in a row, any reasonable person will conclude that this coin is not fair.

It is not possible to formalize this idea by simply setting a threshold $p_0 > 0$ below which events are not possible. Indeed, then, for N for which $2^{-N} < p_0$, no sequence of N heads or tails would be possible at all. We cannot have a universal threshold p_0 such that events with probability $\leq p_0$ cannot happen.

However, we know that for each decreasing $(A_n \supseteq A_{n+1})$ sequence of properties A_n with $\lim p(A_n) = 0$, there exists an N above which a truly random sequence cannot belong to A_N. Here is a resulting definition:

Definition 6.4. *We say that $\mathcal{R}$ is a* set of random elements *if for every definable decreasing sequence $\{A_n\}$ for which $\lim P(A_n) = 0$, there exists an N for which $\mathcal{R} \cap A_N = \emptyset$.*

Random sequences and physically meaningful sequences. Let $\mathscr{R}_K$ denote the set of all elements which are random in Kolmorogov-Martin-Löf sense. Then, the following two results hold:

Proposition 6.1. *Every set of random elements consists of physically meaningful elements.*

Proposition 6.2. *For every set $\mathcal{T}$ of physically meaningful elements, the intersection $\mathcal{T} \cap \mathcal{R}_K$ is a set of random elements.*

Proof When A_n is definable, for $D_n \overset{\text{def}}{=} \bigcap_{i=1}^{n} A_i - \bigcap_{i=1}^{\infty} A_i$, we have $D_n \supseteq D_{n+1}$ and $\bigcap_{n=1}^{\infty} D_n = \emptyset$, so $P(D_n) \to 0$. Therefore, there exists an N for which the set of random elements does not contain any elements from D_N. Thus, every set of random elements indeed consists of physically meaningful elements.

7 Proofs of Results Not Proven in the Main Text

A formal definition of definable sets.

Definition 7.1. *Let $\mathcal{L}$ be a theory. Let $P(x)$ be a formula from $\mathcal{L}$ for which the set $\{x \mid P(x)\}$ exists. We will then call the set $\{x \mid P(x)\}$ $\mathcal{L}$-definable.*

Crudely speaking, a set is $\mathcal{L}$-definable if we can explicitly *define* it in $\mathcal{L}$.

All usual sets are definable: the set of natural numbers $\mathbb{N}$, the set of real numbers $\mathbb{R}$, etc.

Not every set is $\mathcal{L}$-definable: indeed,

- every $\mathcal{L}$-definable set is uniquely determined by a text $P(x)$ in the language of set theory;
- there are only countably many texts and therefore,
- there are only countably many $\mathcal{L}$-definable sets; so,
- some sets of natural numbers are not definable.

How to prove results about definable sets. Our objective is to be able to make mathematical statements about $\mathcal{L}$-definable sets. Therefore, in addition to the theory $\mathcal{L}$, we must have a stronger theory $\mathcal{M}$ in which the class of all $\mathcal{L}$-definable sets is a countable set.

For every formula F from the theory $\mathcal{L}$, we denote its Gödel number by $\lfloor F \rfloor$. We say that a theory $\mathcal{M}$ is *stronger* than $\mathcal{L}$ if:

- $\mathcal{M}$ contains all formulas, all axioms, and all deduction rules from $\mathcal{L}$, and
- $\mathcal{M}$ contains a predicate $\text{def}(n, x)$ such that for every formula $P(x)$ from $\mathcal{L}$ with one free variable,

$$\mathcal{M} \vdash \forall y \, (\text{def}(\lfloor P(x) \rfloor, y) \leftrightarrow P(y)).$$

Existence of a stronger theory. As $\mathcal{M}$, we take $\mathcal{L}$ plus all above equivalence formulas.

Is $\mathcal{M}$ consistent? Due to compactness property of first order logic, it is sufficient to prove that for any $P_1(x), \ldots, P_m(x)$, $\mathcal{L}$ is consistent with the equivalences corresponding to $P_i(x)$. Indeed, we can take

$$\mathrm{def}\,(n,\,y) \leftrightarrow (n = \lfloor P_1(x) \rfloor \,\&\, P_1(y)) \vee \ldots \vee (n = \lfloor P_m(x) \rfloor \,\&\, P_m(y)).$$

This formula is definable in $\mathscr{L}$ and satisfies all m equivalence properties. Thus, the existence of a stronger theory is proven.

The notion of an $\mathscr{L}$-definable set can be expressed in $\mathscr{M}$: S is $\mathscr{L}$-definable if and only if

$$\exists n \in \mathbb{N} \,\forall y \,(\mathrm{def}\,(n,\,y) \leftrightarrow y \in S).$$

So, all the statements involving definability become statements from the $\mathscr{M}$ itself, *not* from metalanguage.

Consistency proof.

Proposition 7.1. $\forall \varepsilon > 0$, *there exists a set* $\mathscr{T}$ *of physically meaningful elements for which* $\underline{P}(\mathscr{T}) \geq 1 - \varepsilon$.

Proof. Indeed, there are countably many definable sequences $\{A_n\}$: $\{A_n^{(1)}\}$, $\{A_n^{(2)}\}$, ...For each k, $P\left(A_n^{(k)}\right) \to 0$ as $n \to \infty$. Hence, there exists N_k for which $P\left(A_{N_k}^{(k)}\right) \leq \varepsilon \cdot 2^{-k}$.

We take $\mathscr{T} \overset{\mathrm{def}}{=} - \bigcup_{k=1}^{\infty} A_{N_k}^{(k)}$. Since $P\left(A_{N_k}^{(k)}\right) \leq \varepsilon \cdot 2^{-k}$, we have

$$\overline{P}\left(\bigcup_{k=1}^{\infty} A_{N_k}^{(k)}\right) \leq \sum_{k=1}^{\infty} P\left(A_{N_k}^{(k)}\right) \leq \sum_{k=1}^{\infty} \varepsilon \cdot 2^{-k} = \varepsilon.$$

Hence, $\underline{P}(\mathscr{T}) = 1 - \overline{P}\left(\bigcup_{k=1}^{\infty} A_{N_k}^{(k)}\right) \geq 1 - \varepsilon.$

Proof of Proposition 3.2. To compute the set $R = \{x : f(x) = 0\}$ with accuracy $\varepsilon > 0$, let us take an $(\varepsilon/2)$-net $\{x_1, \ldots, x_n\} \subseteq K$.

For each i, we can compute $\varepsilon' \in (\varepsilon/2, \varepsilon)$ for which $B_i \overset{\mathrm{def}}{=} \{x : d(x, x_i) \leq \varepsilon'\}$ is a computable compact set.

It is possible to algorithmically compute the minimum of a function on a computable compact set. Thus, we can compute $m_i \overset{\mathrm{def}}{=} \min\{|f(x)| : x \in B_i\}$.

Since $f \in T$, similarly to the proof that equality of typical real numbers is decidable, we can prove that

$$\exists N \,\forall f \in T \,\forall i \,(m_i = 0 \vee m_i \geq 2^{-N}).$$

Computing m_i with accuracy $2^{-(N+2)}$, we can check whether $m_i = 0$ or $m_i > 0$.

Let's prove that $d_H(R, \{x_i : m_i = 0\}) \leq \varepsilon$, i.e., that

$$\forall i \,(m_i = 0 \Rightarrow \exists x \,(f(x) = 0 \,\&\, d(x, x_i) \leq \varepsilon))$$

and

$$\forall x \, (f(x) = 0 \Rightarrow \exists i \, (m_i = 0 \,\&\, d(x, x_i) \leq \varepsilon)).$$

Indeed, $m_i = 0$ means that $\min\{|f(x)| : x \in B_i \overset{\text{def}}{=} B_{\varepsilon'}(x_i)\} = 0$.

Since the set K is compact, this value 0 is attained, i.e., there exists a value $x \in B_i$ for which $f(x) = 0$. From $x \in B_i$, we conclude that $d(x, x_i) \leq \varepsilon'$ and, since $\varepsilon' < \varepsilon$, that $d(x, x_i) < \varepsilon$. Thus, x_i is ε-close to the root x.

Vice versa, let x be a root, i.e., let $f(x) = 0$. Since the points x_i form an $(\varepsilon/2)$-net, there exists an index i for which $d(x, x_i) \leq \varepsilon/2$. Since $\varepsilon/2 < \varepsilon'$, this means that $d(x, x_i) \leq \varepsilon'$ and thus, $x \in B_i$. Therefore,

$$m_i = \min\{|f(x)| : x \in B_i\} = 0.$$

So, the root x is ε-close to a point x_i for which $m_i = 0$.

Proof of Proposition 4.1. As $\mathscr{A}$, given an instance i, we simply produce the result ω_i of the i-th experiment.

Let us prove, by contradiction, that for every $\varepsilon > 0$ and for every n, there exists an integer $N \geq n$ for which

$$\#\{i \leq N : i \in S_{\mathscr{P}} \,\&\, \omega_i = s_{\mathscr{P},i}\} > (1 - \varepsilon) \cdot \#\{i \leq N : i \in S_{\mathscr{P}}\}.$$

The assumption that this property is not satisfied means that for some $\varepsilon > 0$ and for some integer n, we have

$$\forall N_{\geq n} \, \#\{i \leq N : i \in S_{\mathscr{P}} \,\&\, \omega_i = s_{\mathscr{P},i}\} \leq (1 - \varepsilon) \cdot \#\{i \leq N : i \in S_{\mathscr{P}}\}.$$

Let

$$T \overset{\text{def}}{=} \{x : \#\{i \leq N : i \in S_{\mathscr{P}} \,\&\, x_i = s_{\mathscr{P},i}\} \leq$$

$$(1 - \varepsilon) \cdot \#\{i \leq N : i \in S_{\mathscr{P}}\} \text{ for all } N \geq n\}.$$

We will prove that this set T is a physical theory (in the sense of the above definition); then $\omega \notin T$.

By definition, a physical theory is a set which is non-empty, closed, nowhere dense, and definable.

- Non-emptiness is easy: the sequence $x_i = \neg s_{\mathscr{P},i}$ for $i \in S_{\mathscr{P}}$ belongs to T.
- One can prove that T is closed, i.e., if $x^{(m)} \in T$ for which $x^{(m)} \to \omega$, then $x \in T$.
- Nowhere dense means that for every finite sequence $x_1 \ldots x_m$, there exists a continuation $x \notin T$. Indeed, for such an extension, we can take $x_i = s_{\mathscr{P},i}$ if $i \in S_{\mathscr{P}}$.
- Finally, we have an explicit definition of T, so T is definable.

Proof of Proposition 4.2. Let us assume that P$\neq$NP; we want to prove that for every feasible ph-algorithm $\mathscr{A}$, it is not possible to have

$$\forall N \ (\#\{i \leq N : i \in S_{\mathscr{P}} \ \& \ \mathscr{A}(\omega, i) = s_{\mathscr{P},i}\} = \#\{i \leq N : i \in S_{\mathscr{P}}\}).$$

Let us consider, for each feasible ph-algorithm $\mathscr{A}$, $T(\mathscr{A}) \overset{\text{def}}{=}$

$$\{x : \#\{i \leq N : i \in S_{\mathscr{P}} \ \& \ \mathscr{A}(x, i) = s_{\mathscr{P},i}\} = \#\{i \leq N : i \in S_{\mathscr{P}}\} \text{ for all } N\}.$$

Similarly to the proof of the main result, we can show that this set $T(\mathscr{A})$ is closed and definable.

To prove that $T(\mathscr{A})$ is nowhere dense, we extend $x_1 \dots x_m$ by 0s; then $x \in T$ would mean P=NP.

If $T(\mathscr{A}) \neq \emptyset$, then $T(\mathscr{A})$ is a theory, so $\omega \notin T(\mathscr{A})$.

If $T(\mathscr{A}) = \emptyset$, this also means that $\mathscr{A}$ does not solve all instances of the problem $\mathscr{P}$ – no matter what ω we use.

Proof of Proposition 4.3. Let us assume that no non-oracle feasible algorithm δ-solves the problem $\mathscr{P}$. Let's consider, for each N_0 and feasible ph-algorithm $\mathscr{A}$,

$$T(\mathscr{A}, N_0) \overset{\text{def}}{=} \{x : \#\{i \leq N : i \in S_{\mathscr{P}} \ \& \ \mathscr{A}(x, i) = s_{\mathscr{P},i}\} >$$

$$\delta \cdot \#\{i \leq N : i \in S_{\mathscr{P}}\} \text{ for all } N \geq N_0\}.$$

We want to prove that $\forall N_0 \ (\omega \notin T(\mathscr{A}, N_0))$.

- Similarly to the proof of the Main Result, we can show that $T(\mathscr{A}, N_0)$ is closed and definable.
- To prove that $T(\mathscr{A}, N_0)$ is nowhere dense, we extend $x_1 \dots x_m$ by 0s.
- If $T(\mathscr{A}, N_0) \neq \emptyset$, then $T(\mathscr{A}, N_0)$ is a theory hence $\omega \notin T(\mathscr{A}, N_0)$.
- If $T(\mathscr{A}, N_0) = \emptyset$, then also $\omega \notin T(\mathscr{A}, N_0)$.

Proof of Proposition 5.2. It is known that if a f is continuous and 1–1 on a compact, then the inverse function f^{-1} is also continuous.

Let us recall that S compact if and only if it is closed and for every ε, it has a finite ε-net, i.e., a finite set such that each element of S is ε-close to one of the elements from the set S.

We assume that the set X is definably separable, i.e., that there exists a definable sequence $s_1, \dots, s_n, \dots$ which is everywhere dense in X.

The solution is to take $A_n \overset{\text{def}}{=} -\bigcup_{i=1}^{n} B_\varepsilon(s_i)$. Since s_i are everywhere dense, we have $\cap A_n = \emptyset$. Hence, there exists N for which $A_N \cap \mathscr{T} = \emptyset$. Since

$$A_N = -\bigcup_{i=1}^{N} B_\varepsilon(s_i),$$

this means $\mathscr{T} \subseteq \bigcup_{i=1}^{N} B_{\varepsilon}(s_i)$. Hence $\{s_1, \ldots, s_N\}$ is an ε-net for $\mathscr{T}$. So, the set $\mathscr{T}$ is pre-compact. Q.E.D.

Proof of Proposition 6.1. Let T consist of physically meaningful elements. Let us prove that $T \cap \mathcal{R}_K$ is a set of random elements.

If $A_n \supseteq A_{n+1}$ and $P\left(\bigcap_{n=1}^{\infty} A_n\right) = 0$, then for $B_m \stackrel{\text{def}}{=} A_m - \bigcap_{n=1}^{\infty} A_n$, we have $B_m \supseteq B_{m+1}$ and $\bigcap_{n=1}^{\infty} B_n = \emptyset$.

Thus, by definition of a set consisting of physically meaningful elements, we conclude that $B_N \cap T = \emptyset$.

Since $P\left(\bigcap_{n=1}^{\infty} A_n\right) = 0$, we also know that $\left(\bigcap_{n=1}^{\infty} A_n\right) \cap \mathcal{R}_K = \emptyset$. Thus, $A_N = B_N \cup \left(\bigcap_{n=1}^{\infty} A_n\right)$ has no common elements with the intersection $T \cap \mathcal{R}_K$. Q.E.D.

Acknowledgements This work was supported in part by the National Science Foundation grants 1623190 (A Model of Change for Preparing a New Generation for Professional Practice in Computer Science), and HRD-1834620 and HRD-2034030 (CAHSI Includes), and by the AT&T Fellowship in Information Technology.

It was also supported by the program of the development of the Scientific-Educational Mathematical Center of Volga Federal District No. 075-02-2020-1478, and by a grant from the Hungarian National Research, Development and Innovation Office (NRDI).

One of the authors (VK) is thankful to all the participants of the 2017 Dagstuhl Seminar "Reliable Computation and Complexity on the Real" for valuable discussions.

References

1. Belohlavek, R., Dauben, J.W., Klir, G.J.: Fuzzy Logic and Mathematics: A Historical Perspective. Oxford University Press, New York (2017)
2. Feynman, R., Leighton, R., Sands, M.: The Feynman Lectures on Physics. Addison Wesley, Boston, Massachusetts (2005)
3. Finkelstein, A.M., Kreinovich, V.: Impossibility of hardly possible events: physical consequences. In: Abstracts of the 8th International Congress on Logic, Methodology, and Philosophy of Science, Moscow, vol. 5, Issue 2, pp. 23–25 (1987)
4. Jaulin, L., Kiefer, M., Didrit, O., Walter, E.: Applied Interval Analysis, with Examples in Parameter and State Estimation, Robust Control, and Robotics. Springer, London (2001)
5. Klir, G., Yuan, B.: Fuzzy Sets and Fuzzy Logic. Prentice Hall, New Jersey (1995)
6. Kosheleva, O., Zakharevich, M., Kreinovich, V.: If many physicists are right and no physical theory is perfect, then by using physical observations, we can feasibly solve almost all instances of each NP-complete problem. Mathematical Structures and Modeling **31**, 4–17 (2014)
7. Kreinovich, V.: Toward formalizing non-monotonic reasoning in physics: the use of Kolmogorov complexity. Revista Iberoamericana de Inteligencia Artificial **41**, 4–20 (2009)
8. Kreinovich, V.: Negative results of computable analysis disappear if we restrict ourselves to random (or, more generally, typical) inputs. Math. Struct. Model. **25**, 100–103 (2012)
9. Kreinovich, V., Finkelstein, A.M.: Towards applying computational complexity to foundations of physics. Notes of Mathematical Seminars of St. Petersburg Department of Steklov Institute of Mathematics 316, 63–110 (2004); reprinted in J. Math. Sci. **134**(5), 2358–2382 (2006)
10. Kreinovich, V., Kunin, I.A.: Kolmogorov complexity and chaotic phenomena. Int. J. Eng. Sci. **41**(3), 483–493 (2003)

11. Kreinovich, V., Kunin, I.A.: Kolmogorov complexity: how a paradigm motivated by foundations of physics can be applied in robust control. In: Fradkov, A.L., Churilov, A.N. (eds) Proceedings of the International Conference Physics and Control PhysCon'2003, Saint-Petersburg, Russia, 20–22 August 2003, pp 88–93 (2003)
12. Kreinovich, V., Kunin, I.A.: Application of Kolmogorov complexity to advanced problems in mechanics. In: Proceedings of the Advanced Problems in Mechanics Conference APM'04, St. Petersburg, Russia, June 24–July 1, pp. 241–245 (2004)
13. Kreinovich, V., Lakeyev, A., Rohn, J., Kahl, P.: Computational Complexity and Feasibility of Data Processing and Interval Computations. Kluwer, Dordrecht (1998)
14. Kreinovich, V., Longpré, L., Koshelev, M.: Kolmogorov complexity, statistical regularization of inverse problems, and Birkhoff's formalization of beauty. In: Mohamad-Djafari, A. (ed.) Bayesian Inference for Inverse Problems, Proceedings of the SPIE/International Society for Optical Engineering, vol. 3459, San Diego, California, pp. 159–170 (1998)
15. Kreinovich, V., Pownuk, A., Kosheleva, O.: Combining interval and probabilistic uncertainty: what is computable? In: Pardalos, P., Zhigljavsky, A., Zilinskas, J. (eds.) Advances in Stochastic and Deterministic Global Optimization, pp. 13–32. Springer Verlag, Cham, Switzerland (2016)
16. Li, M., Vitanyi, P.: An Introduction to Kolmogorov Complexity and Its Applications, Springer (2008)
17. Mayer, G.: Interval Analysis and Automatic Result Verification. de Gruyter, Berlin (2017)
18. Mendel, J.M.: Uncertain Rule-Based Fuzzy Systems: Introduction and New Directions. Springer, Cham, Switzerland (2017)
19. Moore, R.E., Kearfott, R.B., Cloud, M.J.: Introduction to Interval Analysis. SIAM, Philadelphia (2009)
20. Nguyen, H.T., Walker, E.A.: A First Course in Fuzzy Logic. Chapman and Hall/CRC, Boca Raton, Florida (2006)
21. Novák, V., Perfilieva, I., Močkoř, J.: Mathematical Principles of Fuzzy Logic. Kluwer, Boston, Dordrecht (1999)
22. Pour-El, M.B., Richards, J.I.: Computability in Analysis and Physics. Springer, Berlin (1989)
23. Rabinovich, S.G.: Measurement Errors and Uncertainty: Theory and Practice. Springer Verlag, New York (2005)
24. Sheskin, D.J.: Handbook of Parametric and Nonparametric Statistical Procedures. Chapman and Hall/CRC, Boca Raton, Florida (2011)
25. Sun, L., Dbouk, H., Neumann, I., Schoen, S., Kreinovich, V.: Taking into account interval (and fuzzy) uncertainty can lead to more adequate statistical estimates. In: Proceedings of the 2017 Annual Conference of the North American Fuzzy Information Processing Society NAFIPS'2017, Cancun, Mexico, 16–18 October, 2017
26. Thorne, K.S., Blandford, R.D.: Modern Classical Physics: Optics, Fluids, Plasmas, Elasticity, Relativity, and Statistical Physics. Princeptn University Press, Princeton, New Jersey (2017)
27. Tikhonov, A.N., Arsenin, V.Y.: Solutions of Ill-Posed Problems. V.H. Winston & Sons, Washington, DC (1977)
28. Weihrauch, K.: Computable Analysis. Springer-Verlag, Berlin (2000)
29. Zadeh, L.A.: Fuzzy sets. Inf. Control **8**, 338–353 (1965)

Ethical Dilemma of Self-driving Cars: Conservative Solution

Christian Servin, Vladik Kreinovich, and Shahnaz N. Shahbazova[ID]

Abstract When designing software for self-driving cars, we need to make an important decision: When a self-driving car encounters an emergency situation in which either the car's passenger or an innocent pedestrian have a good change of being injured or even die, which option should it choose? This has been a subject of many years of ethical discussions—and these discussions have not yet led to a convincing solution. In this paper, we propose a "conservative" (status quo) solution that does not require making new ethical decisions—namely, we propose to limit both the risks to passengers and risks to pedestrians to their current levels, levels that exist now and are therefore acceptable to the society.

Keywords Ethics · Self-driving car · Conservative solution · Trolley problem

1 Formulation of a Problem

Self-driving cars are expected to be safer than human drivers. Self-driving cars are supposed to provide maximum safety both for the passengers of this car and for all other folks—passengers of other cars, pedestrians, and passers-by. In the nearest

C. Servin
Information Technology Systems Department, El Paso Community College (EPCC), 919 Hunter Dr, El Paso, TX 79915-1908, USA
e-mail: cservin1@epcc.edu

V. Kreinovich (✉)
University of Texas at El Paso, El Paso, TX 79968, USA
e-mail: vladik@utep.edu

Sh. N. Shahbazova
Department of Digital Technologies and Applied Informatics, Azerbaijan State University of Economics, UNEC, Baku AZ1001, Azerbaijan
e-mail: shahbazova@gmail.com; shahnaz_shahbazova@unec.edu.az; shahbazova@cyber.az

Ministry of Science and Education of the Republic of Azerbaijan, Institute of Control Systems, 68 Bakhtiyar Bahabzadeh Street, Baku AZ1141, Azerbaijan

© The Author(s), under exclusive license to Springer Nature Switzerland AG 2023
Sh. N. Shahbazova et al. (eds.), *Recent Developments and the New Directions of Research, Foundations, and Applications*, Studies in Fuzziness and Soft Computing 423,
https://doi.org/10.1007/978-3-031-23476-7_9

future, they are expecting to provide higher level of safety for all these categories than cars operated by human drivers.

Unfortunate situations, while hopefully very rare, cannot be completely avoided. No matter how safe self-driving cars will be, unfortunate situations may still happen, and in such situations, it may not be possible to make everyone safe. For example, if several pedestrians suddenly rush across the road, there may be enough time to stop the car, so the only choices are either hit the pedestrians or swerve this potentially hurting the car's passenger(s) and maybe even passengers of nearby cars. In such situations, what a car will do depends on what algorithm we program into it, and this, in turn, depends on what objective function we use when designing this algorithm; for related discussions, see, e.g., [1–30] and references therein.

Seemingly reasonable idea: social good. At first glance, when designing self-driving cars, we should maximize the overall social good, or, equivalently, minimize the overall social harm. From this viewpoint, if the choice is to harm (or even kill) one passenger or three pedestrians, the proper solution seems to be to harm the smallest number of people—i.e., in this situation, to possibly harm the passenger while trying to avoid harming the pedestrians.

This idea is not as reasonable as it may seem. A detailed analysis, however, shows that such arguments may be oversimplifying and not as reasonable as they may sound at first glance. Following one of the examples proposed by researchers, suppose that a medical doctor in a small town sees a reasonably healthy patient with a healthy heart, healthy liver, and two healthy kidneys, and he/she knows that in this town, there are four patients at risk of dying if they do not get, correspondingly, a new heart, a new liver, and a new kidney. Is is reasonable to kill the first patient and transplant his/her organs to the four dying folks? The argument is the same—shall we save the life of one patient or four patients? However, in this example, the answer to harm the smallest number of people does not seem so reasonable.

To make it even less reasonable, suppose that the first patient is not fully heathy, but had a bad cut and is heavily bleeding—so the patient can die if no medical help is available. Shall the doctor save this patient and let the other four die or shall the doctor save the lives of the four other patients by not attending to the first one?

So what shall we do? This seems like a complex problem for which we need philosophers to argue and to come up with a convincing solution. However, the fact that the philosophers have been discussing this "trolley problem" for many years—probably for many decades—and have not yet come with a convincing solution is, to us, an indication that we should not expect such a solution in the nearest future either. We have to come up with such a solution ourselves.

What we do in this paper. In this paper, we argue that such a convincing solution *is* possible—namely, the solution is to be conservative and to follow the society's accepted norms and practices.

2 How to Solve the Problem: Main Idea

We must be fair to the passenger. At present, a passenger in a car has a certain degree of safety. Some of this safety is provided by technical innovations such as safe and robust car design, airbags, and automatic warnings that inform the driver that another car is too close. Some of the safety is provided by the fact that the driver is in control, and the driver's skills—and the self-preservation instinct—provide safety in complex situations where technical innovations alone cannot help.

It is clearly not fair to the driver if the self-driving cars would provide a smaller degree of safety for the passenger than the degree of safety obtained when this person drives the car. Technological progress is supposed to make all our lives better, not provide advantage to some groups at the expense of others.

We must be fair to others. Similarly, the self-driving cars should provide at least the same level of safety to passengers in other cars, to pedestrians, and to the passers-by, as the current human-driven car.

If the self-driving cars focus only on the safety of their own passengers, this will make it even less probable than now that the car will try to swerve to avoid hitting the pedestrian. In such situation, the increased safety of the passenger will come at the expense of the decreased safety for the passenger.

We must be fair to pedestrians, we must sure that in all situations, their level of safety is at least as high as their current level of safety, in situations when cars are driven by human drivers.

The resulting idea. This fairness is our main idea. Specifically, in situations when the car has the option of either harm its passenger or several pedestrians, it should *not* be concerned only about the passenger—thus increasing the risk to the pedestrian, and it should *not* follow the naively understood social good track idea—this increasing the risk to the passenger. Instead, the car should select proper probabilities of both possible actions—the action that potentially hurts the passenger and the action that potentially hurts the pedestrians—in such a way that for both groups, the level of safety be at least as high as for the current human-driven cars.

3 How to Solve the Problem: Details

What should be the balance between the safety of the passenger and the safety of others. In general, our recommendation is to make sure that the passenger is as safe as when he/she would be driving the car, and others—pedestrians and bystanders—would be at least as safe as when humans drive cars. However, within these two restrictions, there are many possible options. For example:

- If we are pursuing social good idea, we can keep the passenger exactly as safe as when cars are driven by people, and place all the efforts into minimizing the risk for others;
- On the other hand, if we allow customers to select which self-driving cars to buy, customers will naturally want to buy a car that minimizes their risk—while keeping the risk to others at the current level.

Instead of decreasing just one of these risks—risk to the passenger and risk to others—we could try to somewhat decrease both risks. Which strategy should we follow? How should we balance these two risks?

Our idea. Instead of trying to solve a difficult-to-solve (and maybe even impossible-to-solve) ethical problem, why not just follow what people have been doing—and what therefore is socially acceptable? Namely, we can find how the two risks decreased with time and thus, find out what was, in the past, the relation between the two risks—as measured, e.g., by the percentages p_d and p_w of harmful accidents per hour of driving (or being driven) and walking.

In general, these probabilities decrease with time. So, by observing these probabilities $p_{d,i}$ and $p_{w,i}$ at different historic epochs i, we can find the dependence between these two values, i.e., a function $f(p)$ for which $p_{d,i} \approx f(p_{w,i})$ for all i. This function reflects a socially acceptable balance between the two risks. Thus, in the future, when it will be possible to have self-driving cars that decrease both risks, a natural idea is to use the values p_d and p_w for which $p_d = f(p_w)$. This will provide a socially acceptable way to balance the risks.

Caution. Of course, what we propose is what medical doctors call a palliative—a temporary solution that is used in lieu of a better one. At this moment, in the absence of a better more convincing solution, this is what we propose: to follow the current balance between the risks when designing self-driving cars.

This does not mean, of course, that this conservative solution—based on the current and past social understanding—is the only way to go.

- Social moors and opinions have changed many times in the past, they will undoubtedly change again and again, and what is acceptable now will no longer be acceptable—just like the risk level of the original cars is not acceptable nowadays, and if someone wants to drive an ancient car, that car has to be retrofitted with modern safety devices.
- Maybe someone will come up with a convincing solution to the ethical dilemma.

In all these cases, better solutions will be accepted. However, as of now, in the absence of such better solutions, the proposed conservative idea seems to be a reasonable way to proceed.

Acknowledgements This work was supported in part by the National Science Foundation grants 1623190 (A Model of Change for Preparing a New Generation for Professional Practice in Computer Science), and HRD-1834620 and HRD-2034030 (CAHSI Includes), and by the AT&T Fellowship in Information Technology.

It was also supported by the program of the development of the Scientific-Educational Mathematical Center of Volga Federal District No. 075-02-2020-1478, and by a grant from the Hungarian National Research, Development and Innovation Office (NRDI).

References

1. Barcalow, E.: Moral Philosophy: Theories and Issues. Wadsworth, Belmont, CA (2007)
2. Bauman, C.W., McGraw, A.P., Bartels, D.M., Warren, C.: Revisiting external validity: concerns about trolley problems and other sacrificial dilemmas in moral psychology. Soc. Personal. Psychol. Compass **8**(9), 536–554 (2014)
3. Bernhard, R.M., Chaponis, J., Siburian, R., Gallagher, P., Ransohoff, K., Wikler, D., Roy, H., Greene, J.D.: Variation in the oxytocin receptor gene (OXTR) is associated with differences in moral judgment. Soc. Cogn. Affect. Neurosci. **11**(12), 1872–1881 (2016)
4. Bleske-Rechek, A., Nelson, L.A., Baker, J.P., Remiker, M.W., Brandt, S.J.: Evolution and the trolley problem: people save five over one unless the one is young, genetically related, or a romantic partner. J. Soc. Evol. Cultural Psychol. **4**(3), 115–127 (2010)
5. Bonnefon, J.-F., Shariff, A., Rahwan, I.: The social dilemma of autonomous vehicles. Science **352**(6293), 1573–1576 (2016)
6. Ciaramelli, E., Muccioli, M., Làdavas, E., di Pellegrino, G.: Selective deficit in personal moral judgment following damage to ventromedial prefrontal cortex. Soc. Cogn. Affect. Neurosci. **2**(2), 84–92 (2007)
7. Crockett, M.J., Clark, L., Hauser, M.D., Robbins, T.W.: Serotonin selectively influences moral judgment and behavior through effects on harm aversion. Proc. Natl. Acad. Sci. **107**(40), 17433–17438 (2010)
8. Awad, E., Dsouza, S., Kim, R., Schulz, J., Henrich, J., Shariff, A., Bonnefon, J.-F., Rahwan, I.: The Moral Machine experiment. Nature **563**(7729), 59–64 (2018)
9. Francis, H.B., Howard, C., Howard, I.S., Gummerum, M., Ganis, G., Anderson, G., Terbeck, S.: Virtual morality: transitioning from moral judgment to moral action? PLOS One **11**(10), 1–22 (2016)
10. Gogoll, J., Müller, J.F., Julian, F.: Autonomous cars: in favor of a mandatory ethics setting. Sci. Eng. Ethics **23**(3), 681–700 (2017)
11. Greene, J.D., Sommerville, R.B., Nystrom, L.E., Darley, J.M., Cohen, J.D.: An fMRI investigation of emotional engagement in moral judgment. Science **293**(5537), 2105–2108 (2001)
12. Himmelreich, J.: Never mind the trolley: the ethics of autonomous vehicles in mundane situations. Ethical Theor. Moral Pract. **21**(3), 669–684 (2018)
13. JafariNaimi, N.: Our bodies in the trolley's path, or why self-driving cars must *not* be programmed to kill. Sci. Technol. Hum. Values **43**(2), 302–323 (2018)
14. Jarvis Thomson, J.: Killing, letting die, and the trolley problem. The Monist **59**, 204–217 (1976)
15. Jarvis Thomson, J.: The trolley problem. Yale Law J. **94**(6), 1395–1415 (1985)
16. Kahane, G.: Sidetracked by trolleys: Why sacrificial moral dilemmas tell us little (or nothing) about utilitarian judgment. Soc. Neurosci. **10**(5), 551–560 (2015)
17. Kahane, G., Everett, J.A.C., Earp, B.D., Caviola, L., Faber, N.S., Crockett, M.J., Savulescu, J.: Beyond sacrificial harm: a two-dimensional model of utilitarian psychology. Psychol. Rev. **125**(2), 131–164 (2018)
18. Lee, M., Sul, S., Kim, H.: Social observation increases deontological judgments in moral dilemmas. Evol. Hum. Behav. **39**(6), 611–621 (2019)
19. Lim, H.S.M., Taeihagh, A.: Algorithmic decision-making in AVs: understanding ethical and technical concerns for smart cities. Sustainability **11**(20) (2019). Paper 5791
20. Myrna Kamm, F.: Harming some to save others. Philos. Stud. **57**(3), 227–260 (1989)

21. Navarrete, C.D., McDonald, M.M., Mott, M.L., Asher, B.: Virtual morality: emotion and action in a simulated three-dimensional 'trolley problem'. Emotion **12**(2), 364–370 (2021)
22. Patil, I., Cogoni, C., Zangrando, N., Chittaro, L., Silani, G.: Affective basis of judgment-behavior discrepancy in virtual experiences of moral dilemmas. Soc. Neurosci. **9**(1), 94–107 (2014)
23. Rom, S., Paul, C.: The strategic moral self: self-presentation shapes moral dilemma judgments. J. Exp. Soc. Psychol. **74**, 24–37 (2017)
24. Sharp, F.C.: A Study of the influence of custom on the moral judgment. Bull. Univ. Wis. (236), 138 (1908)
25. Sharp, F.C.: Ethics. The Century Co., New York (1928)
26. Skulmowski, A., Bunge, A., Kaspar, K., Pipa, G.: Forced-choice decision-making in modified trolley dilemma situations: a virtual reality and eye tracking study. Front. Behav. Neurosci. **8**(426), 1–16 (2014)
27. Sütfeld, L.R., Gast, R., König, P., Pipa, G.: Using virtual reality to assess ethical decisions in road traffic scenarios: applicability of value-of-life-based models and influences of time pressure. Front. Behav. Neurosci. **11**(122), 1–13 (2017)
28. Unger, P.: Living High and Letting Die. Oxford University Press, Oxford, UK (1996)
29. Valdesolo, P., DeSteno, D.: Manipulations of Emotional Context Shape Moral Judgment. Psychol. Sci. **17**(6), 476–477 (2006)
30. Youssef, F.F., Dookeeram, K., Basdeo, V., Francis, E., Doman, M., Mamed, D., Maloo, S., Degannes, J., Dobo, L.: Stress alters personal moral decision making. Psychoneuroendocrinology **37**(4), 491–498 (2012)

Intelligent Methods and Fuzzy Approach

Transformer-Based Approaches to Sentiment Detection

Olumide Ebenezer Ojo⬤, Hoang Thang Ta⬤, Alexander Gelbukh⬤,
Hiram Calvo⬤, Olaronke Oluwayemisi Adebanji⬤, and Grigori Sidorov⬤

Abstract The use of transfer learning methods is largely responsible for the present breakthrough in Natural Learning Processing (NLP) tasks across multiple domains. In order to solve the problem of sentiment detection, we examined the performance of four different types of well-known state-of-the-art transformer models for text classification. Models such as Bidirectional Encoder Representations from Transformers (BERT), Robustly Optimized BERT Pre-training Approach (RoBERTa), a distilled version of BERT (DistilBERT), and a large bidirectional neural network architecture (XLNet) were proposed. The performance of the four models that were used to detect disaster in the text was compared. All the models performed well enough, indicating that transformer-based models are suitable for the detection of disaster in text. The RoBERTa transformer model performs best on the test dataset with a score of 82.6% and is highly recommended for quality predictions. Furthermore, we discovered that the learning algorithms' performance was influenced by pre-processing techniques, the nature of words in the vocabulary, unbalanced labeling, and the model parameters.

Keywords Transformers · BERT · RoBERTa · DistilBERT · XLNet · Sentiment analysis · Text classification

1 Introduction

Sentiment analysis uses Computational and Natural Language Processing (NLP) techniques to automatically analyze text and classify sentiments regarding a variety of topics. Transformers are deep neural network models that use transfer learning

O. E. Ojo · H. T. Ta · A. Gelbukh · H. Calvo (✉) · O. O. Adebanji · G. Sidorov
Centro de Investigacion en Computación, Instituto Politécnico Nacional, CDMX, Mexico City, Mexico
e-mail: hcalvo@cic.ipn.mx

A. Gelbukh
e-mail: hcalvo@cic.ipn.mx

© The Author(s), under exclusive license to Springer Nature Switzerland AG 2023
Sh. N. Shahbazova et al. (eds.), *Recent Developments and the New Directions of Research, Foundations, and Applications*, Studies in Fuzziness and Soft Computing 423,
https://doi.org/10.1007/978-3-031-23476-7_10

and are based on the attention mechanism. These transformer-based models have been the state-of-theart models for dealing with a variety of NLP tasks. In many sentiment analysis tasks, transformer-based deep learning models have dramatically improved machine learning performance [1, 1–6]. Much study has been conducted on sentiment analysis using social media data [3, 5, 7], however, efforts to investigate disaster detection have attracted less attention. The unpredictability in the language structure of social media texts makes detecting disaster often difficult. Being able to adequately detect disaster information in text has numerous applications as it can assist in making emergency response systems more intelligent.

One of the most effective methodologies for natural language processing has been demonstrated to be deep learning. The ability of algorithms to analyze text has greatly increased as a result of recent developments in deep learning. BERT (Bidirectional Encoder Representations from Transformers) is a breakthrough in the application of Machine Learning that is based on Transformers and serves as a foundation for enhancing the accuracy of machine learning models that handle text. It is a deep learning framework with pre-trained models that have produced cutting-edge results on a number of applications. RoBERTa, DistilBERT, and a large bidirectional neural network architecture known as XLNet are some of the fine-tuned BERT-like versions of frameworks that have been built to solve a variety of tasks. In this sentiment detection task, the approaches utilized and the level of classification can be used to give judgment about the models. Even while transformer-based models have the ability to learn longer-range context dependencies, one of their main drawbacks is that they are constrained to an input of a fixed length of text, usually fewer than a few hundred tokens. As a result, document truncation is unavoidable, resulting in significant data loss. In order to address this problem, different researchers has created extensions to the BERT model. In [8], a training method for extending BERT pretrained models from a general domain to a new pre-trained model for a specific domain with a new additive vocabulary was introduced. The method, called exBERT, was proposed to maximize the use of an intricately pre-trained model for a general domain by enabling the ability of the model to continuously learn and transfer the knowledge of the learnt representation to a new domain with minimal training cost. Other extensions of the transformer model have accelerated natural language processing tasks in different domains.

In this paper, we aim to analyze four transformer-based models for disaster detection in text, as well as to be able to compare their performance. The models were tested on binary labelled data [9], and the their prediction abilities was investigated. The accuracy scores were used to evaluate the models' classification performance. The study is organized as follows: Sect. 2 reviews similar works, Sect. 3 focuses on the models and their implementation. Section 4 describes the model experiments and presents the results. Section 5 focuses on the conclusion and future works.

2 Literature Review

Sentiment analysis has sparked a lot of interest in academia and industry. The use of blogs and social media has led to the phenomenal expansion of shared contents. The problem of automatic classification of disaster-related information in text has been tackled using a variety of existing computational methods [1, 2, 10–12]. The fundamental aim of sentiment detection has been polarity classification. However, by stating the parameters responsible for the models' performance, we can compare their performance. The use of bagof-words embedding, which was employed in some previous methods [7], is no longer relevant which is why we propose using an attention mechanism. The Transformer is an encoder decoder architecture that solely uses the attention mechanism to encode each position by linking two distant words of both the inputs and outputs with respect to itself, and it has proven to be quite useful in aiding the model's comprehension of the language.

The use of Transformer architectures are becoming increasingly popular in the computational linguistic community [3–6, 14]. Vaswani et al. [13] introduced the use of self attention mechanism to overcome difficulties in recurrence issues in sequence-to-sequence problems. The self attention mechanism in the transformer models are used to extract features for each word, and the most well-known modern NLP models are made up of many of such models or variants. BERT (Bidirectional Encoder Representations from Transformers) is a pre-trained encoder-only transformer architecture with deeply bidirectional language representation [15]. It was pre-trained using unlabeled data gathered from BooksCorpus with 800 million words and Wikipedia with 2.5 billion words. BERT also learns to model relationships between words (or subwords) in a text by pre-training on a very simple task that can be built from any text corpus. Classified as bidirectional models, the Transformer encoder reads the full sequence of words in a sentence at once. This property enables the model to deduce the context of a word based on its surroundings. A relation between an aspect and a text can be found using the BERT model. In [6], the authors used BERT model's contextual word representations, combined with a fine-tuning strategy with additional generated text, to find the relationship between an aspect and a text.

The RoBERTa (Robustly Optimized BERT Pre-training Approach) model is a variant of the BERT model proposed by Liu et al. [16]. It extends BERT by modifying crucial hyperparameters, deleting the next-sentence pretraining objective and training with considerably bigger mini-batches and learning rates. As a result, the system is able to predict purposely hidden content within unannotated language instances. RoBERTa's architecture, like BERT's, employs a subword tokenization strategy based on byte pair encoding and a distinct pretraining technique. RoBERTa is part of a larger effort to advance the state-of-the-art in self-supervised systems that don't require time and resources to tag data. The performance of the RoBERTa model was tested in a code completion task on a variety of code completion scenarios, from masked code tokens to entire code blocks [5]. The performance of the model was evaluated and found to be effective.

BERT has cased, uncased, base, large, and multi-lingual models. Due to its enormous size, putting BERT into production is really difficult. Developed by Sanh et al. [17], DistilBERT (a distilled version of BERT) has a distillation technique used to reduce the size of these large models. Knowledge distillation, also referred to as teacher-student learning, is the process of training a small student model to match a larger teacher model as closely as possible. It takes advantage of the model's"dark knowledge", or the uncertainty that arises when models contain non-zero probabilities for wrong classes in classification problems. The authors in [4] combined a machine learning algorithm with DistilBERT to produce a novel hybrid technique for extracting the features of an event in a text classification task. The main advantage is that the specific rules they used were able to improve the results of the machine learning classifier.

Yang et al. [18] introduced XLNet, an auto-regressive language model that uses a transformer architecture with recurrence to enable learning bidirectional contexts. Its training objective is to optimize the predicted likelihood of a word token across all factorization order possibilities. Its goal in training is to determine the probability of a word token based on all permutations of word tokens in a sentence, not just those to the left and right of the target token. The XLNet model was used in [3] to successfully classify personality from text and had the lowest average prediction error. The authors integrated XLNet and the capsule network by extracting emotional aspects from the text information at each point and passing the collected features through the capsule network to extract the personality features further.

3   Methodology

3.1   Data

Appen [9] has donated a lot of research data to many NLP activities to help people focus on skill development and teamwork. We used pre-labeled datasets from Appen's massive database, which has 10,876 comments, to evaluate whether a certain text is about a specific disaster or not. The datasets are well-designed to boost accuracy and overall performance for a variety of purposes. The dataset contains 10,873 comments separated into two classes with 57% categorized as non-disasters and 43% as true disasters. The dataset was split into two parts, one for training and the other for testing. The longest sentence in the entire dataset is 31 words long and matches to the non-disaster label. In order to clean the dataset, we removed all urls, html tags, symbols, punctuation, emoticons, etc., thereby, making our sentences more subjective. We tokenized the input data and reformatted the token sequence before passing it to the model. Figure 1 depicts a sample of the dataset used (Fig. 2).

Fig. 1 Transformer model
architecture [13]

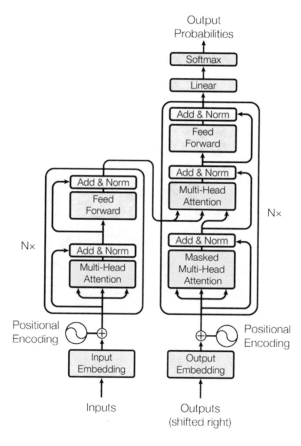

3.2 *Models*

BERT. The model utilized is bert-base-uncased, with a maximum sequence length of 40 to accommodate the longest possible sequence. This model is made up of twelve layered transformer blocks, each with twelve head self-attention layers and 768 hidden layers. The system was built and fine-tuned using the Keras functional API. The maximum length of sequences was taken into account in the input layer, and the Bert model was then supplied with a dropout of 0.1 to reduce overfitting. A dense layer with two neurons was employed to represent the number of classes in our label. This model trained within a short time (because the number of instances in the dataset is very minimal) and gave good performance.

RoBERTa. The RoBERTa-base model, which consists of twelve transformer layers with 768 hidden layers and twelve attention heads was used in this study. The input texts were encoded into tokens and assigned as input ids with the aid of the RoBERTa tokenizer. The features collected from these tokens were then used to classify sentence pairs.

	text	target
0	Our Deeds are the Reason of this #earthquake M...	1
1	Forest fire near La Ronge Sask. Canada	1
2	All residents asked to 'shelter in place' are ...	1
3	13,000 people receive #wildfires evacuation or...	1
4	Just got sent this photo from Ruby #Alaska as ...	1
...	...	...
7608	Two giant cranes holding a bridge collapse int...	1
7609	@aria_ahrary @TheTawniest The out of control w...	1
7610	M1.94 [01:04 UTC]?5km S of Volcano Hawaii. htt...	1
7611	Police investigating after an e-bike collided ...	1
7612	The Latest: More Homes Razed by Northern Calif...	1

Fig. 2 Some samples in the dataset

DistilBERT. The DistilBERT-uncased model with six transformer layers, 768 hidden layers, and twelve attention heads was also part of the models used. The input sentences were tokenized and converted into input ids before being padded and fed into the DistilBERT model for classification.

XLNet. The model used in this study is the one proposed in [18]. Twelve transformer layers with 768 hidden layers and twelve attention head layers were employed in the XLNet-base-cased model. To tokenize the sequences, the XLNet tokenizer was utilized. The XLNet tokenizer addresses open vocabulary problems and assists in selecting the final vocabulary size for training with the aid of the sentence piece library. The tokens were then padded, and classifications were completed.

4   Experiments and Results

The main aim is to classify text as disaster or non-disaster using binary labelled dataset. The dataset contains 10,876 comments, with 7613 for training and 3263 for testing. The extracted features were converted to vector representations and fed into the pre-trained models during the fine-tuning process. The models were trained and their outputs were generated using the input vector transformations. The output was then examined using the test data provided, and the results were analysed. The classification was done by categorizing the sentences into disaster and non-disaster for each of the pre-trained models proposed. We cleaned the unprocessed dataset by

Table 1 Statistics of dataset

Dataset	Number of comments
Training data	7613
Testing data	3263
Total	10,876

Table 2 Performance score of the models

Model	Accuracy on train data (%)	Accuracy on test data (%)
BERT	97.2	81.7
RoBERTa	95.2	**82.6**
DistilBERT	**98.5**	80.5
XLNet	92.4	80.9

Numbers in bold highlight the best result

removing html elements, symbols, punctuation, emojis, and other things that could mislead the algorithm. The input layer takes into account the maximum length of sequences, which was subsequently fed to all of the models.

The models were optimized with the Adam optimizer, with Categorical Cross-Entropy (CCE) as the loss parameter. The models had a learning rate of 6e-5 and a weight decay of 0.01. The BERT model has the following characteristics: a validation size of 25%, a batch size of 64, train and test accuracy of 97.2% and 81.7%, respectively, and a training time of 10 epochs. Batch size 64, train and test accuracy of 95.2% and 82.6%, validation size of 25%, and training time of 10 epochs were all features of the RoBERTa model. The DistilBERT model has a batch size of 64, train and test accuracy of 98.5% and 80.5%, validation size of 25%, and was trained for up to 10 epochs. The XLNet model has train and test accuracy of 92.4% and 80.9%, respectively, as well as a 25% validation size, and a training time of 10 epochs. Tables 1, 2 and 3 show the dataset's statistics, performance metrics and parameters of the models.

5 Conclusion

Several researches have been conducted using the transformer architecture to identify sentiment in text. In this research, we evaluated the performance of four transformer based models in the task of disaster detection in text. Furthermore, we fine-tuned the parameters of each model used in order to better understand the aspects that contributed to each score we acquired. Tuning the parameters of the models and performance metrics for each of the models summarizes our findings. The accuracy score on the test data is used to evaluate the transformer models' performance. As

Table 3 Parameter of the models

Parameters	BERT	RoBERTa	DistilBERT	XLNet
Learning rate	6e-5	6e-5	6e-5	6e-5
Weight decay	0.01	0.01	0.01	0.01
Validation size (%)	25	25	25	25
Batch size	64	64	64	64
Epochs	10	10	10	10
Optimizer	Adam	Adam	Adam	Adam
Clipnorm	1.0	1.0	1.0	1.0
Loss	CCE	CCE	CCE	CCE

shown in Table 3 and Fig. 3, all of the models performed well on the dataset. On the test dataset, the RoBERTa model outperforms other models, whereas the DistilBERT model outperforms other models on the train dataset. This suggests that RoBERTa is capable of appropriately transferring its knowledge and is well-suited to the task of disaster identification in text. It is our expectation that this exploratory research has established an approach for disaster detection in text. We plan to investigate multi-label instances of the problem in the future and to test on more data, where the additional information will be even more beneficial.

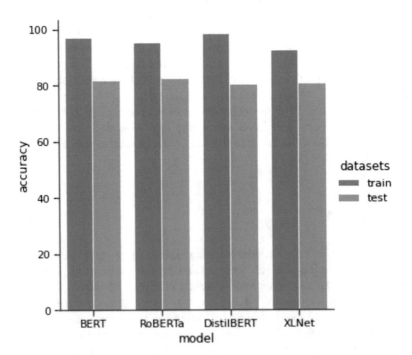

Fig. 3 Accuracy of the models on the training and testing datasets

Acknowledgements The work was done with partial support from the Mexican Government through the grant A1-S-47854 of the CONACYT, Mexico, and by the Secretaría de Investigacion y Posgradó of the Instituto Politecnico Nacional, Mexico, under Grants 20211884, 20220859, and 20220553, 20230140, EDI; and COFAA-IPN.

References

1. Congcong, W., Nulty, P., Lillis, D.: Transformer based Multi-task Learning for Disaster Tweet Categorisation CoRR, abs/2110.08010 (2021). https://arxiv.org/abs/2110.08010
2. Zhang, T., Schoene, A.M., Ananiadou, S.: Automatic identification of suicide notes with a transformer-based deep learning model Elsevier, vol. 25, pp. 100422, ISSN: 2214–7829 (2022). https://doi.org/10.1016/j.invent.2021.100422
3. Wang, Y., Zheng, J., Li, Q., Wang, C., Zhang, H., Gong, J.: XLNet-caps: personality classification from textual posts electronics. ISSN. **10**(11), 2079–9292 (2021). https://doi.org/10.3390/electronics10111360
4. Dogra, v., Verma, S., Singh, A., Kavita, Talib, M.N., Humayun, M.: Banking news-events representation and classification with a novel hybrid model using DistilBERT and rule-based features Turkish J. Comput. Math. Educ. (TURCOMAT) **12**(10), 3039–3054 (2021). https://doi.org/10.17762/turcomat.v12i10.4954
5. Ciniselli, M., Cooper, N., Pascarella, L., Poshyvanyk, D., Di Penta, M., Bavota, G.: An Empirical Study on the Usage of BERT Models for Code Completion CoRR, vol. abs/2103.07115 (2021). https://arxiv.org/abs/2103.07115
6. Hoang, M., Bihorac, O.A. and Rouces, J.: Aspect based sentiment analysis using BERT Proceedings of the 22nd Nordic Conference on Computational Linguistics, Linkoping University Electronic Press, pp. 187–196 (2019). https://aclanthology.org/W19-6120
7. Ojo, O.E., Gelbukh, A., Calvo, H., Adebanji, O.O.: Performance study of n-grams in the analysis of sentiments. J. Nigerian Soc. Phys. Sci. 477–483 (2021). https://doi.org/10.46481/jnsps.2021.201
8. Tai, W., Kung, H.T., Dong, X., Comiter, M., Kuo, C-F.: exBERT: Extending pre-trained models with domain-specific vocabulary under constrained training resources findings of the association for computational linguistics: EMNLP 2020. Association for Computational Linguistics, pp. 1433–1439 (2020). https://doi.org/10.18653/v1/2020.findings-emnlp.129
9. [Appen Limited. https://appen.com/open-source-datasets/Level6/9Help.St.Chatswood NSW 2067, Australia, 2022.
10. Ojo, O.E., Ta, T.H., Adebanji, O.O., Gelbukh, A., Calvo, H., Sidorov, G.: Automatic Hate Speech Detection Using Deep Neural Networks and Word Embedding Computacíon y Sistemas **26**, 1 (2022)
11. Mustafa R.U., Ashraf N., Ahmed F.S., Ferzund J., Shahzad B., Gelbukh A.A.: Multiclass depression detection in social media based on sentiment analysis In: Latifi, S. (eds) 17th International Conference on Information Technology–New Generations (ITNG 2020). Advances in Intelligent Systems and Computing, vol. 1134. Springer, Cham.https://doi.org/10.1007/978-3-030-43020-7-89
12. Ashraf, N., Zubiaga, A., Gelbukh, A.: Abusive language detection in Youtube comments leveraging replies as conversational context PeerJ Computer. Science **7**, e742 (2021). https://doi.org/10.7717/peerj-cs.742
13. Vaswani, A., Shazeer, N., Parmar, N., Uszkoreit, J., Jones, L., Gomez, A.N., Kaiser, Ł., Polosukhin, I.: Attention is all you need Advances in Neural Information Processing Systems, pp. 5998–6008 (2017). http://arxiv.org/abs/1706.03762
14. Butt, S., Ashraf, N., Siddiqui, M.H.F., Sidorov, G. and Gelbukh, A.: Transformer-Based Extractive Social Media Question Answering on TweetQA. Comput. Sistemas **25** (1) (2021). https://doi.org/10.13053/cys-25-1-3897

15. Devlin, J., Chang, M.W., Lee, K., Toutanova, K.: BERT: Pre-training of Deep Bidirectional Transformers for Language Understanding CoRR, vol. abs/1810.04805 (2018). http://arxiv.org/abs/1810.04805

16. Liu, Y., Ott, M., Goyal, N., Du, J., Joshi, M., Chen, D., Levy, O., Lewis, M., Zettlemoyer, L. and Stoyanov, V.: RoBERTa: a robustly optimized BERT pretraining approach CoRR, vol. abs/1907.11692, pp. 487–493 (2019). http://arxiv.org/abs/1907.11692

17. Sanh, V., Debut, L., Chaumond, J., Wolf, T.: DistilBERT, a distilled version of BERT: smaller, faster, cheaper and lighter CoRR, vol. abs/1910.01108 (2019). http://arxiv.org/abs/1910.01108

18. Yang, Z., Dai, Z., Yang, Y., Carbonell, J., Salakhutdinov, R.R. and Le, Q.V.: XLNet: Generalized Autoregressive Pretraining for Language Understanding CoRR, vol. abs/1906.08237 (2019). http://arxiv.org/abs/1906.08237

19. Ojo, O.E., Gelbukh, A., Calvo, H., Adebanji, O.O.: Sentiment detection in economics texts mexican international conference on artificial intelligence, pp. 271–281. Springer 2021. https://doi.org/10.1007/978-3-030-60887-3-24

Automatic Humor Classification: Analysis Between Embeddings and Models

Victor Manuel Palma Preciado, Grigori Sidorov, and Alexander Gelbukh

Abstract The present work aims to present different methods for detecting humor in One-liners, in general humor is everything that causes us gracefulness or that in an ingenious way presents us with a humorous punchline. In this way, it is common to find humor that its construction tends to be related to the topic being discussed, its premises / punchlines. To classify humor in texts there are several aspects: A powerful embedding that correctly represents the meaning we want to obtain, a model robust enough that it is easy to learn with our data and finally a combination of both an embedding / robust model capable of successfully carrying out the expectations of the given task. As a primary approach, pre-trained embeddings were used in a basic CNN in contrast to the paradigm of Tranformers. Obtaining good results in both areas for both embedding and pre-trained transformer models, with a qualification above 99 of the F1-Score.

Keywords Humourism · Humor identification · Transformers · Pretrained models · Embeddings

1 Introduction

1.1 Humor Us

An on-liner is a joke that is transmitted in a single line, within the same short jokes there are different constructions:

Riddles: In this type of short joke, in which the actor tends to ask the interlocutor a question, where the answer is usually comic or witty.

What do you call a turtle without its shell? Dead.

V. M. P. Preciado (✉) · G. Sidorov · A. Gelbukh
Centro de Investigación en Computación, Instituto Politécnico Nacional, CDMX, México, México
e-mail: victorpapre@gmail.com

© The Author(s), under exclusive license to Springer Nature Switzerland AG 2023   111
Sh. N. Shahbazova et al. (eds.), *Recent Developments and the New Directions of Research, Foundations, and Applications*, Studies in Fuzziness and Soft Computing 423,
https://doi.org/10.1007/978-3-031-23476-7_11

Differences: It can be understood as the comparison between X and Y where normally the answer tends to be a clue given by the question, which points to something humorous or shocking [1].

How are music and candy similar? we throw away the rappers.

Unary expression: it is a short joke that does not require an answer and that the delivery of the humorous sentence is immediate, given in a few words, without much explanation.

Ugh, i just spilled red wine all over the inside of my tummy.

In this aspect, most of the short jokes usually use a resource called anaphora, which consists of repeating concepts or words [2]. In this sense, the anaphora functions as a repetitive or self-reflective referent, since in the One-liner the object of humor is referred to in the sentence or question, as well in the punch line.

Which allows us to understand certain patterns within short jokes and how they are formed. Given that humor has different variables that can make its detection difficult, it is important to focus on a written humorous set that allows us to know part of humor but that its construction is not so complicated as to make it difficult to detect, precisely the One-liners enter into this category since the premises are usually accompanied by a punch line that accordingly denotes its intention, therefore it was decided to take the One-liners as a data set.

Based on the premises of the experiment, it was decided to take the approach of "Tranformers vers Strong Embeddings".

In the case of BERT, it has become an extremely powerful model as it is part of the state of the art in different tasks in natural language and DestilBERT, its lightest version, a sure reference for BERT-like models, against the power of embeddings such as ELMO [3] and USE (Universal Sentence Encoder) [4] along with a simple convolutional network, for which we use the data set of [1] which has 100,000 One-liners and 100,000 No-liners.

2   Dataset

2.1   Characteristics and Curiosities

The data set with which it was experimented [5] was given a general treatment with which to obtain a better result in the embeddings phase, a small analysis was also done, to know the characteristics of the data. Some curiosities of the One-liners were, words that are widely used to make comparisons, since they have a wide occurrence Fig. 1.

We could say that [1] are right when describing the structure of the One-liners, the 4 g of the data set just help to strengthen this idea.

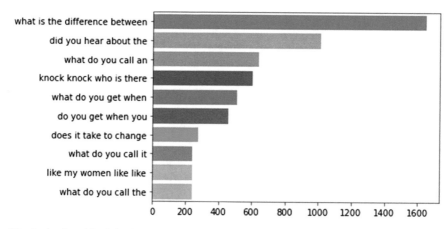

Fig. 1 4-g from bit of the dataset

The data that served as inputs for the embedding models and for those of Transformers, were lemmatized and contractions were removed, thus allowing the comparison to be correct in the sense of giving the same characteristics for one model as for the other (Tables 1 and 2).

A list of contractions was used to substitute them in our data set, for example, going from an "isn´t" to "is not". In order to have better embedding, stopwords were removed in USE and ELMO, but not before lemmatizing, which consists of passing from the inflected form of the word (plural, conjugated or gender) to its representation that can be found in the dictionary.

Table 1 Tranformers versus strong embeddings

Tranformer paradigm	Strong embeddings
BERT	CNN + ELMO
DestilBERT	CNN + USE

Table 2 Parameter for the transformer models

Models	Learning Rate	Batch size	Characteristics	Epochs
BERT	0.00002	16	250	2
DistilBERT	0.00008	6	500	2

3   Models and Parameters

3.1   Transformer Models

For the BERT model, an average time per epoch of 3 h 15 min was had, for a total execution time of 6 h 40 min. On the other hand, with the DestilBERT model the average time decreased to 2 h 41 min per epoch, for a total time of 5 h 26 min it is logical to think that the time will decrease since it is considered that the DestilBERT execution is much lighter than the BERT model is about 18% faster for the classification task, but not in the same way in the performances that we will talk about next.

For the BERT model, an average time per epoch of 3 h 15 min was had, for a total execution time of 6 h 40 min. On the other hand, with the DistilBERT model the average time decreased to 2 h 41 min per epoch, for a total time of 5 h 26 min it is logical to think that the time will decrease since it is considered that the DistilBERT execution is much lighter than the BERT model is about 18% faster for the classification task, but not in the same way in the performances that we will talk about next.

3.2   CNN Model with Embeddings

On the other hand, a CNN-type network was used for classification using the embeddings, in this case the time it took for the embeddings process was added to have an approximate time that the entire system actually takes. Since the embedding were processed separately in case it was necessary to test with other architectures that allow other results, in terms of performance (Tables 3 and 4).

Table 3 Parameters for CNN + Embeddings

Models	Lerning Rate	Batch size	Characteristics	Epochs
CNN + USE	0.06	250	512	20
CNN + ELMO	0.06	250	1024	20

Table 4 Confusion matrix

Models	True positive (%)	False positive (%)	False negative (%)	True negative (%)
CNN + USE	49.06	1.14	1.14	48.65
DistilBERT	49.52	0.41	0.67	49.38
CNN + ELMO	46.22	1.03	3.98	48.76
BERT	49.56	0.39	0.63	49.39

As can be seen in general, what changed in the CNN model is the use of different embedding and the number of characteristics inherent to them. The same results were quite short time for the execution of the networks, but combining the time of the embeddings competes with the amount of time required by the models of the Transformers paradigm.

The characteristics obtained varied within the two different embeddings on the ELMO side, 1024 characteristics were obtained in 80 batches of 2500 sentences with an average time of 15 h and in the case of USE, the 200,000 sentences were processed obtaining 512 characteristics per sentence in a estimated time of 30 min. So, if we add these times, they give us 15 h and 12 min of total execution time to complete the classification task with ELMO embedding and around 42 min of execution for USE embedding, being able to observe that USE is much faster to mount and apply than ELMO. Almost rivaling the average time of a BOW model run for our data set.

In this case we can see that the use of a network coupled with USE embeddings, is faster than most of the methods applied in the next section we can observe the behavior, if it is fast and really good enough to contend against those models that are slower, both in efficiency and effectiveness.

4   Results

4.1   Tabular Results

Under the different methods applied, a very interesting comparison was obtained compared to a conventional model added to strong embeddings in contrast to the Tranformers, as the results of the classification can be seen in the following table.

4.2   Classification Examples

As can be seen below, each of the models classifies the different data with some confidence, so it is logical to think that each of them do better in different situations and depending on the humor to be classified. If this were the case, it would verify that the networks are able to discern the humor in One-liners and under the theory of its structure.

The predictions made by the models yielded some reliability in the One-liners classification:

"Michael cera probably apologizes and gives back cars in grand theft auto." Which really is a One-liner.

The BERT model effectively classified it as a One-liner with 99.9% confidence that it really was.

On the other hand, we can see that DistilBERT when trying to classify the phrase: "Does this headline look blue to you? then it might also feel like a triangle." It has problems since it is not a One-liner but even so the model gives it a 98.9% rating of being one, but under human thought, due attention is paid to it, the phrase could well be a One-liner given its construction and use of the absurd. So perhaps the model did interpret the characteristics of the phrase as laughable and therefore classify it as such. For the CNN classifiers plus ELMO and USE the phrases were used:
"I invented a new word! Plagiarism.!".

When your only tool is a hammer, all problems start looking like nails.

They were effectively classified as One-liners with 93% confidence for ELMO and 95% for USE. As we can see generally, the models behave in a good way in the face of humor in the One-liners and can discern it to a good extent.

On the other hand, as we could observe previously, the Transformer paradigm is better, therefore, we are interested in knowing in those cases in which it does not detect correctly, both the cases that were really One-liner and those that were not.

4.3   ELI5 Prospection

Using the Eli5 library to find out where the attention of those who were not classified as it should go, the next No-liner despite being labeled as such was classified as a Oneliner by BERT.

How do we keep alias generation off facebook? permanent mittens

As we can see the previous No-liner when using the library, we see that the parts that contributed to identify were those marked with tones, but even so we could argue that it was not really wrong, since at first glance it seems like a joke since the word "mittens" can be used in multiple contexts, making the answer seem like a punchline. This coupled with the fact that it gives you a 99.5% confidence. We will take the opposite case in which it is really a One-liner and it was not classified as such.

The CIA finally succeeded in killing Fidel Castro using the innovative, old age technique

Clearly the One-liner can pass as the headline of a newspaper or a phrase that has a quite dark context such as death, but when we look at it we realize that it is a joke, BERT takes as the words CIA, innovative and technique like attention in the same way Fidel, using the and old age that seem to indicate a certain logic on their own.

On the other hand, DestilBERT appears to be performing well like its more robust BERT counterpart, but still falters when it comes to detecting ambiguous One-liner and No-liners.

World's smallest porpoise inches closer to extinction. there are now just 30 of them left.

The previous example is not a One-liner and could really be the headline of a biology magazine or a newspaper, although also understand that you can make jokes with a metric sense even with the sentence as it is. On the other hand, in the following Oneliner, what we could see as something humorous, the model does not take it as such, the model focuses mainly on "man on" and "with flint and steel" which does not seem to give us much idea until we notice that he also pays attention to the words "spark" and "outrage" where it makes sense for him to look at flint and steel. But it fails to actually decree and it seems logical that it would not detect it to some extent given its ambiguity.

Very offensive man on the loose with flint and steel sparks outrage.

The difference between the previous examples is a bit tenuous since they can and really do, confuse whether they are really a joke or not, since we commonly understand a joke, when we are really sure that the proposed sentence is incoherent in a certain way.

5 State of Art

The task of detecting humor is a difficult task, due to the existence of multiple types of humor, different connotations and their rhythm. Therefore, it is not uncommon to find works concerning the detection of humor and even more the detection of humor in small texts, headlines, news and One-liners. The use of CNN for the detection of a humor in text proposed by [6, 7] is a clear example that with a sufficiently robust embedding, quite interesting results can be found, this in addition to a previous work by [8] in which also One-Liners was used as its focal point. Also, the use of BERT-like models can be found with quite promising results in humor tasks and also in their creation, such as [9] and [5] precisely the latter being the state of the art of the data set with which tests were done (Tables 5 and 6).

Obtaining better results when testing with pre-trained models [10], although the use of One-liners in other models can be missed.

Please note that the first paragraph of a section or subsection is not indented. The first paragraphs that follows a table, figure, equation etc. does not have an indent, either.

Subsequent paragraphs, however, are indented.

Table 5 F1-score, precision and recall

Models	Recall	Precision	F1-score
CNN + USE	0.9372	0.9319	0.9346
CNN + ELMO	0.9753	0.9833	0.9793
DistilBERT	0.9853	0.9772	0.9813
BERT	0.9894	0.9916	0.9905

Table 6 Comparison between humorous pre-trained BERT and DistilBERT versus state of art

Models	Precision	Recall	F1-Score
Modelo [5]	0.990	0.974	0.982
XLNet [5]	0.872	0.973	0.920
BERT	0.9916	0.9894	0.9905
DistilBERT	0.9772	0.9853	0.9813

As we can see, Bert works much better than the model proposed by [5] that only uses BERT as embedding. DistilBERT [11] on the other hand almost achieves results very similar to [5] but almost in half the time of BERT. Therefore, it should be noted that the used models are pre-trained and the characteristics were explained previously.

6   Conclusions

As it was previously presented, the models based on the Transformer paradigm are quite good at classifying humor in One-liners, but even so we cannot stop creating strong embedding to correct certain areas, since it cannot be forgotten that despite classifying in a less exact way these models tend to be fast, the idea of using the Transformer model can be seen as a good solution to the humor issue since with [5] it can be observed that BERT embeddings are the good enough to grant a high percentage of confidence, nor should we ignore the classical methods, since as suggested by [6], [8, 12] the implementation of convulsion networks also solves this task satisfactorily. Also, the fact that obtaining better characteristics will allow obtaining improvements in the way that the models classify [9, 13] should not be ignored.

References

1. Miller, K.E.L.: The unuttered punch line: Pragmatic Incongruity and the Parsing of 'What's the Difference' Jokes"REAL ACADEMIA ESPAÑOLA: Diccionario de la lengua española, 23.ª ed., [versión4 en línea]. <https://dle.rae.es> [25/03/2021].
2. Annamoradnejad, I.: ColBERT: Using BERT sentence embedding for humor detection. ArXiv, abs/2004.12765 (2020)
3. Peters, M.E., Neumann, M., Iyyer, M., Gardner, M., Clark, C., Lee, K., & Zettlemoyer, L.: Deep contextualized word representations. NAACL-HLT.
4. Cer, D.M., Yang, Y., Kong, S., Hua, N., Limtiaco, N., John, R.S., Constant, N., Guajardo-Cespedes, M., Yuan, S., Tar, C., Sung, Y., Strope, B., Kurzweil, R.: Universal Sentence Encoder. ArXiv, abs/1803.11175 (2018)
5. Devlin, J., Chang, M., Lee, K., Toutanova, K.: BERT: Pre-training of deep bidirectional transformers for language understanding. NAACL-HLT (2019)

6. Kim Y.: Convolutional neural networks for sentence classification. In: Proceedings of the 2014 Conference on Empirical Methods in Natural Language Processing, EMNLP 2014, October 25–29, 2014, Doha, Qatar, A meeting of SIGDAT, a Special Interest Group of the ACL, pp. 1746–1751 (2004)
7. Chen P.-Y., y Soo V.-W.: Humor recognition using deep learning. in: Proceedings of the 2018 Conference of the North American Chapter of the Association for Computational Linguistics: Human Language Technologies, Vol. 2 (Short Papers).
8. Mahurkar S., Patil R., LRG at SemEval-2020 Task 7: Assessing the Ability of BERT and Derivative Models to Perform Short-Edits based Humor Grading (2020)
9. Ortega-Bueno, R., Muñiz-Cuza Carlos E., Medina Pagola José E., Rosso P.: UO-UPV: Deep Linguistic Humor Detection in Spanish Social Media (2018)
10. Maiya, A.S.: Ktrain: A low-code library for augmented machine learning. ArXiv, abs/2004.10703 (2020)
11. Sanh, V., Debut, L., Chaumond, J., & Wolf, T. DistilBERT, a distilled version of BERT: smaller, faster, cheaper and lighter. ArXiv, abs/1910.01108 (2019)
12. Mihalcea, R., Strapparava, c.: Making computers laugh: investigations in automatic humor recognition. In: Conference on Human Language Technology and Empirical Methods in Natural Language Processing, pp. 531–538 (2005)
13. Yang, D., Lavie, A., Dyer, C., Hovy E.: Humor recognition and humor anchor extraction. In: Conference on Empirical Methods in Natural Language Processing, pp. 2367–2376 (2015)

Skin Cancer Diagnosis Enhancement Through NLP and DNN-Based Binary Classification

Joshua R. G. Guerrero-Rangel, Christian E. Maldonado-Sifuentes, M. Cristina Ortega-García, Grigori Sidorov, and Liliana Chanona-Hernandez

Abstract In present work, we enhance the diagnosis of skin-cancer by applying Natural Language processing and Machine Learning techniques to a specific classification problem. Our experimentation with different architectures yielded noteworthy results over a specific corpora of medical notes. Using a BoW/TF-IDF representation of the corpus and a 1-dimensional Convolutional Neural Network(CNN1D) architecture, we achieved an F1-measure of 0.95, which is better than state of the art.

1 Introduction

1.1 About Cancer

What is cancer The term Cancer is used to refer to a group of diseases in which cells divide without control and can spread into another tissues. The human body is made up of trillions of cells and almost any of them can start the carcinogenic process. Normally, when these cells age or suffer significant damage, they die and are replaced by new cells, however, sometimes this process fails, allowing cells abnormal or damaged cells to reproduce forming tumors which can be benign or malignant (cancerous) [1].

J. R. G. Guerrero-Rangel · C. E. Maldonado-Sifuentes · G. Sidorov (✉) · L. Chanona-Hernandez
Center for Computing Research (CIC), Instituto Politécnico Nacional (IPN), Mexico City, Mexico
e-mail: sidorov@cic.ipn.mx

J. R. G. Guerrero-Rangel
e-mail: joshuaguerrero@live.com.mx

C. E. Maldonado-Sifuentes
e-mail: cmaldonados2018@cic.ipn.mx

M. C. Ortega-García
Augmented Innovation Research, Transdisciplinary Research for Augmented Innovation - Laboratory (TRAI-L), Mexico City, Mexico
e-mail: cristina.ortega@trai-l.com

Sh. N. Shahbazova et al. (eds.), *Recent Developments and the New Directions of Research, Foundations, and Applications*, Studies in Fuzziness and Soft Computing 423, https://doi.org/10.1007/978-3-031-23476-7_12

Table 1 Types of cancer with the highest incidence of new cases in 2020 (Worldwide) [1]

Type of cancer	Incidence (Million cases)
Breast	2.26
Lung	2.21
Colorectal	1.93
Prostate	1.41
Non-melanoma (skin)	1.20
Gastric	1.09

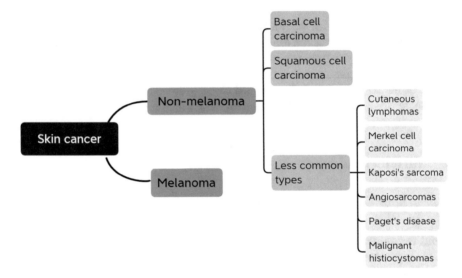

Fig. 1 Classification of types of skin cancer. Created by the authors with information from [4]

About cancer prevalence According to the World Health Organization (WHO) "Cancer is the leading cause of death worldwide" almost 10 million deaths in 2020 [2]. They report 6 types of cancer with the highest incidence of new cases in 2020, amongst which we find non-melanoma skin cancer [2] (Table 1).

"Malignant skin tumors are the most frequent [they] appear on [...]the face, around the eyes and mouth, ears, among other [sites]" [3]. According to the WHO, a significant part of malignant and non-malignant skin cancers will develop serious health problems ranging from disfiguring treatments to death.

Types of skin cancer According to Suárez et al., skin cancer is mainly classified into two types: non-melanoma skin cancer and melanoma [3]. Non-melanoma skin cancer mainly includes keratinocytic neoplasms (basal cell and squamous cell carcinoma) and less frequent tumors [4] as shown in Fig. 1.

Non-melanoma skin cancer prevalence It is estimated that the number of new cases of non-melanoma cancer around the world is 1,198,073 new cases. [5]. On the other

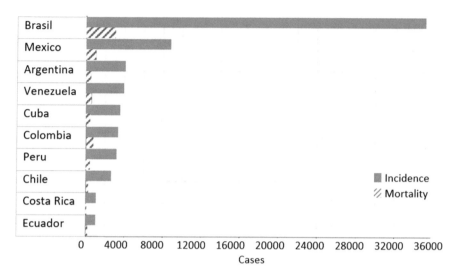

Fig. 2 Latin America estimated number of incident cases and deaths non-melanoma skin cancer, both sexes, all ages [5]

hand, with regard to Latin America, the countries with the highest incidence and deaths from non-melanoma skin cancer are Brazil, Mexico, Argentina, Venezuela, Cuba, Colombia, Peru and Chile (Fig. 2). As can be seen, according to information from GLOBOCAN, Mexico ranks second in incidence and mortality [5].

Importance of early detection According to the ACS: Most cases can be cured, if treated early [6]. Furthermore, as noted in the WHO fact sheet on cancer "Early diagnosis improves cancer outcomes by providing care at the earliest possible stage [...] When cancer care is delayed or inaccessible there is a lower chance of survival [7]."

1.2 Machine Learning Approaches for Cancer Detection

In recent years multiple advances have been done in the improvement of early detection and diagnosis through the application of Machine Learning (ML) algorithms and techniques to the task of image classification for skin cancer detection.

Description of our work For the present work we collected a corpus of real-life medical notes provided by a governmental dermatological clinic in Mexico City, Mexico, through a collaboration agreement between the main author and the institution. A transdisciplinary team was assembled consisting of programmers, computational scientists, linguists, and health professionals. The experimentation consisted on a matrix of three different language representations in the Bag-of-Words (BoW) format namely: *TF, TF-IDF and word n-grams* [8, 9] and three different ML algorithms: SVM, RF, and CNN.

Contributions of this study are as follows. The collection, preprocessing and publication of corpus of 775,000 real-world dermatological medical notes in Spanish. To our understanding, the fist of its kind and the development of a successful experiment on the combination of NLP and ML techniques for the binary classification of non-melanoma skin cancer and subsequently a method for the same task.

2 State of the Art

In this section we present both the State of the Art (SotA) on the tasks related to cancer detection.

2.1 Text-Based Methods for Machine Learning Detection of Skin Cancer

We proceed to move towards research using Natural Language Processing techniques rather than imaging for the application of Machine learning to the classification of this type of disease. In particular there are several works that use NLP and ML to classify for different types of cancer including breast, lung and pancreas cancer, which we lest because of they text-based approach; we finish by listing the works that closely resemble our task: classifying for skin cancer using NLP techniques.

Breast, pancreas and lung cancer classification We begin this subsection with a special mention to an seminal article, not on the SotA, from the 1997 "Identification of findings suspicious for breast cancer based on natural language processing of mammogram reports" from Jain and Friedman, in which they created the system MedLEE using the assistance of an expert system and encoding the results using simple NLP techniques. They improved the cancer diagnosis by 16% above simple electronic logbooks [10].

In 2015, Roch et al. used NLP techniques to process 566,233 medical record of patients of a clinical institution, the team achieve a 98% Sensitivity and specificity using an expert system and an curated list of synonyms to classify patients at risk of developing pancreatic cancer [11].

Krunkaran et al. present a work describing Geisinger's Close-the-Loop clinical program which uses NLP applied on radiology notes from pulmonary scans that were looking for other affections. This work is applied in a real-life setting in which care managers using the system contact patients at risk. They analyze millions of records with an accuracy of 90%"[12].

In the work Banerjee et al. (2019), the authors use the OncoSHARE database, with 8956 women diagnosed with breast cancer with the participation of experts they curated a dictionary. They compared an expert system vs a NLP/ML approach. The NN model achieved a sensitivity of 0.83 and a specificity of 0.73, greatly outperforming the expert system [13].

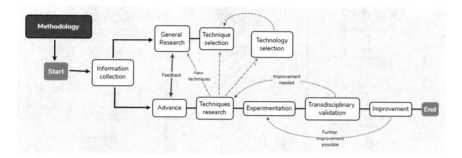

Fig. 3 Diagram of our methodology

Zeng and Banerjee use the Stanford Cancer Institute Research in 2021 consisting of 483,782 clinical notes, in this case, with multiple cancer types.0 Using NLP-ML techniques they obtain scores above 83% in all cases [14].

Skin-cancer classification In the work of Gong et al., the authors used a NLP-CNN architecture. Their work is close to ours in approach since they use similar architectures and with a similar purpose, obtaining, an accuracy of 0.8393 [15]. Although similar their work differs from ours in the scope, for them is all skin cancers, for us non-melanoma, this makes them not fully comparable, but the task and approach are similar enough to compare with our work.

3 Proposed Method

In this section we describe our methodology and the specific method followed to achieve our best result.

3.1 Proposed Methodology

We start with information collection, in this step we tackle the laborious process of gathering real world information. We then proceed to divide our research efforts in general research about the topic and advanced or state-of-the-art techniques.

General Research leads to the general understanding of the problem and its context, in this respect we conformed a transdisciplinary study group. The health professionals secured the task was useful, ethically responsible and compliant with current regulations as well as providing the the public health statistics and the health implications.

From the findings of general research the main technology line is selected which then leads to connect with the Techniques research, if new techniques appear during this process a bidirectional until a final set of suitable technique candidates is selected.

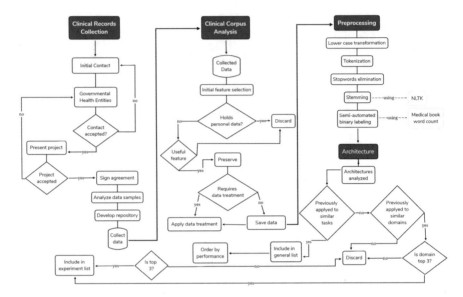

Fig. 4 Diagram of our method

We then proceed to apply said techniques to the data and start experimentation with a loop to the Techniques research step to reduce the search space. Upon success the health and linguistics experts validate the results, and we iterate our experiments until the best results are obtained. Our methodology is shown in Fig. 3.

3.2 Description of the Method

Our classification method is highly correlated with the practices of the NLP community at the time of the development of this work. We start with our data collection, in this case clinical medical records. We move on to analyze the structure and contents of the dataset or corpus.

After determining, through analysis and research which are the best features we move on to pre-process our corpus applying a set of NLP techniques concordant with the literature. We proceed to select the algorithms and to define and program the algorithms to begin with the search of the optimal algorithm/data combination. A graphical simplification of our method is shown in Fig. 4.

Clinical records collection Data collection is known to be, both, one of the most important elements of research for applied Artificial Intelligence and, probably, the most difficult and laborious. Inter-institutional relationships and human factors tend to difficult the collection of real-world data.

We propose a simple iterative model where multiple institutions with the possibility of providing the contextually correct data. The institutions are approached formally with support from our center of ascription and later contacted personally for a presentation of our project by the researchers until a suitable institution agrees to interact with us.

When the agreement with the health institution is accepted we formalize the agreement while analyzing samples of their data, developing the adequate connections and repositories to receive their data. The final step is to collect the raw data for further analysis and preprocessing.

Clinical corpus analysis Here we analyze and curate the features of our data. As the first step we work with the astringent goal of protecting the identities and avoid any possible sensitive data of the patients. In the first step all this data is expunged. After which a careful analysis process is conducted as we move on to decide on a limited set of vectorial representations and determine if the selected features require preprocessing techniques, saving the data for prerpocessing as shown in Fig. 4.

Preprocessing We transform al entries to lowercase, then we apply tokenization, stop word elimination and stemming using NLTK. We finally proceed to label each note according to a dictionary processing specific non-melanoma skin cancer sections doing word counts and removing stop words. The final list of the most used terms is presented to the Health experts which realize a final curation of the most representative words. Using TF we proceed to pre-label each record and have the experts scan a 2% randomized sample of 750 positive diagnoses and 750 negative diagnoses.

NN architecture design The purpose of this step of the method is to prevent the brute-force approach to experimentation by reducing the search space applying both research and heuristics. We create a list of known architectures checking if it was applied to a similar task, since similar tasks are scarce we take all the proposed approaches as candidate architectures, order them by performance and select the top three. We then move on to architectures applied to a similar domains.

4   Results

In this section we present our corpus and describe the results of our experimentation.

Corpus Our corpus is composed by 775,000 medical records, deprived of all personal data including, but not limited to: name, location, gender, age and other features. Notes are strongly structured.The final representation of the corpus is set of 62,000 dimensional vectors, in the form of a BoW and applying TF/IDF metrics. We have made available a significant portion of our corpus as well as the weights and biases of the Convolutional Neural Network[1].

---

[1] Public Corpus + Weights and biases of CNN https://drive.google.com/drive/folders/1kt9g0rfCxPS-uWAUHgDHqxEfvvgkNhOk?usp=sharing.

Table 2 BoW/TF results

Algorithm	Precision		Recall		F1
	0	1	0	1	
SVM+TF	68.1%	77.2%	73.2%	73.4%	73.2%
RF+TF	70.2%	77.1%	72.9%	75.3%	74.1%
CNN+TF	86.6%	90.2%	90.4%	86.1%	88.2%
SVM+TF/IDF	0%	100%	0%	58%	58%
RF+TF/IDF	63.8%	81.3%	75.2%	72.1%	73.4%
CNN+TF/IDF	90.2%	93.1%	92.9%	89.3%	91.1%

Table 3 Comparison with Cheng et al.

Algorithm	Recall		Avg
	0	1	
Chen et al.	82.3%	85.6%	83.9%
Guerrero et al.	92.9%	89.3%	91.1%

Performance In the Table 2, we present the results of the BoW with TF features with the selected algorithms. In this case our best result was achieved with the CNN and TF/IDF at 0.911 on the F1 measure.

5 Comparison with the State of the Art

Finally We compare our work with the work of Cheng et al. which uses CNN and NLP representations over medical record to predict a single binary category of skin cancer while our work classifies for non-melanoma skin cancer. While both works are not fully comparable they are similar indeed. Given this we can contend we hold the state of the art for the specific task. This is shown in the Table 3.

6 Conclusions

The task of the usage of ML and NLP to classify for cancer is neither new, nor unique still little work has been done with these techniques with skin cancer. It is important to continue developing and applying these techniques to create more opportunities to take advantage of these techniques that have proved useful in many types of cancer, even in real world settings. Artificial Intelligence can be taken as another technology, the role of these techniques is not to replace the judgment of a physician bit to aid with additional data.

In the future, we intend to classify different skin diseases and lesions from the same medical notes but also we intend to improve the scope and specialization of our work by combining ML techniques that have been successful on image databases.

Acknowledgements We thank the Instituto Politecnico Nacional of Mexico, through the BEIFI Project 20211168 and the CONACyT scholarship 001624. We acknowledge the collaboration of experts: Violeta Jaurez-Estrada, Mercedes Aguilar-Chagoyan, Josue Morado-Enriquez and Rafael Alvarado-Corona from TRAI-L.com. Special thanks to Osiris Juárez for her support on technical issues.

References

1. National Cancer Institute: What is cancer? **1** (2021)
2. World Health Organziation: Cáncer: Nota descriptiva (2022)
3. Suárez Jiménez, A., Fernández Allegues, Y., Anasagasti Angulo, L.: Estudio de los costos de cuatro modalidades de tratamiento de los carcinomas de células basales y espinosas de la piel en el Instituto Nacional de Oncología y Radiobiología (INOR). Cofin Habana **11**(2), 354–366 (2017)
4. Lobos, B.P., Lobos, S.A.: Cáncer de piel no-melanoma. Rev. Med. Clin. Las Condes **22**(6), 737–748 (2011)
5. Sung, H., Ferlay, J., Siegel, R.L., Laversanne, M., Soerjomataram, I., Jemal, A., Bray, F.: Global cancer statistics 2020: GLOBOCAN estimates of incidence and mortality worldwide for 36 cancers in 185 countries. CA Cancer J. Clin. **71**(3), 209–249 (2021)
6. Siegel, R., Miller, K., Jemal, A.: Cancer Facts & Figures 2016, American Cancer Society, pp. 1–72. Atlanta, Georgia (2016)
7. World Health Organziation: Promoting cancer early diagnosis (2022)
8. Arroyo-Fernández, I., Méndez-Cruz, C.F., Sierra, G., Torres-Moreno, J.M., Sidorov, G.: Unsupervised sentence representations as word information series: revisiting TF-IDF. Comput. Speech Lang. **56**, 107–129 (2019)
9. Sidorov, G.: Syntactic N-grams in Computational Linguistics. Springer, Berlin (2019)
10. Jain, N.L., Friedman, C.: Identification of findings suspicious for breast cancer based on natural language processing of mammogram reports. In: Proceedings of the AMIA Annual Fall Symposium. p. 829. American Medical Informatics Association (1997)
11. Roch, A.M., Mehrabi, S., Krishnan, A., Schmidt, H.E., Kesterson, J., Beesley, C., Dexter, P.R., Palakal, M., Schmidt, C.M.: Automated pancreatic cyst screening using natural language processing: a new tool in the early detection of pancreatic cancer. Hpb **17**(5), 447–453 (2015)
12. Karunakaran, B., Misra, D., Marshall, K., Mathrawala, D., Kethireddy, S.: Closing the loop-finding lung cancer patients using NLP. In: 2017 IEEE international conference on big data (big data). pp. 2452–2461. IEEE (2017)
13. Banerjee, I., Bozkurt, S., Caswell-Jin, J.L., Kurian, A.W., Rubin, D.L.: Natural language processing approaches to detect the timeline of metastatic recurrence of breast cancer. JCO Clin. Cancer Inform. **3**, 1–12 (2019)
14. Zeng, J., Banerjee, I., Henry, A.S., Wood, D.J., Shachter, R.D., Gensheimer, M.F., Rubin, D.L.: Natural language processing to identify cancer treatments with electronic medical records. JCO Clin. Cancer Inform. **5**, 379–393 (2021)
15. Gong, X., Xiao, Y.: A skin cancer detection interactive application based on CNN and NLP. In: Journal of Physics: Conference Series. vol. 2078, p. 012036. IOP Publishing (2021)

Analysis of Fake News Detection Methods

Maaz Amjad, Oxana Vitman, Grigori Sidorov, Alisa Zhila,
and Alexander Gelbukh

Abstract The fact that it might be difficult to distinguish between real news and fake news is of the utmost significance. It is difficult, time-consuming, and expensive to manually review enormous volumes of digital data in order to detect false news, which is information that is not real in line with the facts. There is a new piece of fake news published every second; hence, the creation of automated systems is absolutely necessary in order to detect instances of fake news. This study provides a comprehensive analysis of various machine and deep learning approaches, as well as how the systems perform in identifying fake Urdu news, which are submitted in the shared tasks named UrduFake@FIRE2020 and UrduFake@FIRE2021. It was hoped that the shared tasks would attract new researchers to come up with algorithms that could detect Urdu fake news articles available in the digital media. More than 50 teams from ten different nations participated to find a solution to this binary classification issue. The objective of the shared tasks was to classify a given Urdu news instance as either real or fake, and the teams proposed numerous algorithms to achieve this goal. Among the several methods of text representation that have been investigated in this study are count-based BoW features and word vector embeddings. Traditional neural and non-neural network approaches, including BERT, XLNet and RoBERTa are also examined, and their analysis is presented. Moreover, precision, recall, $F1_{Real}$, $F1_{Fake}$, and $F1_{Macro}$ are used for evaluation. The winning system in 2021 employed the linear classifier function and achieved 0.679 $F1_{Macro}$, while the highest performing system in 2020 was based on BERT and obtained 0.907 $F1_{Macro}$. Thus, the results of this

M. Amjad · O. Vitman · G. Sidorov · A. Gelbukh (✉)
Center for Computing Research (CIC), Instituto Politécnico Nacional, Mexico City, Mexico
e-mail: gelbukh@gelbukh.com

M. Amjad
e-mail: maazamjad@phystech.edu

G. Sidorov
e-mail: sidorov@cic.ipn.mx

A. Zhila
Ronin Institute for Independent Scholarship, Montclair, USA
e-mail: alisa.zhila@roninstitute.org

© The Author(s), under exclusive license to Springer Nature Switzerland AG 2023
Sh. N. Shahbazova et al. (eds.), *Recent Developments and the New Directions of Research, Foundations, and Applications*, Studies in Fuzziness and Soft Computing 423, https://doi.org/10.1007/978-3-031-23476-7_13

research show that machine learning classifiers underperformed in detecting Urdu fake news than deep learning techniques.

Keywords Text classification · Fake news detection · Analysis of methods

1 Introduction

The process of producing and disseminating digital content has been transformed in recent years, and because of developments in both the internet and technology, the method of creating and distributing digital material has undergone a sea change in two decades. As social media platforms have grown in popularity, users now have more options for exchanging messages, exchanging information, and conversing with one another. According to an article published on Forbes,[1] the emergence of social media and the effect it has had on our daily lives have both had a big impact on how we consume news, and this trend is only likely to increase in the coming years. A Forbes study found that 64.5

Various researches have presented multiple definitions of fake news. For example, a news is called fake news if it contains factually incorrect information and the news is spread intentionally [6]. Conspiracy theories and fake news induce various social and societal issues that directly influence our society. For example, BBC[2] that some people were sucked into anti-vaccine conspiracy theories, which put them in severe health conditions. According to another study [29], individuals are more willing to share and repost fake news than true news, particularly political news. There is also a connection between fake news and the volatility of the stock market as well as massive transactions, which makes economies susceptible to the propagation of fake news. As a consequence of this, the use of automated technologies is crucial for identification of fake news instances and reduce their speed of propagation in digital media.

Numerous linguistic studies discussed fake news in well-known languages, such as the English [26], Spanish [22]. Urdu and other low-resource languages lack studies on how to recognize fake news instances. This study is to provide the findings of numerous classification approaches that were presented in shared tasks "Urdu-Fake@FIRE2020" and "UrduFake@FIRE2021" for classifying Urdu fake news. It is vital to emphasize that these shared tasks were organized in order to entice other academics to encourage and attract them to the development of automated systems for detection and classification of Urdu fake news instances in digital media.

The paper is summarized and divided into five sections. In the second section, we analyze the significance of spotting fake news in Urdu, and provide a review of recent literature. In Sect. 3, we discuss the process of collecting and assembling both

---

[1] https://www.forbes.com/sites/nicolemartin1/2018/11/30/how-social-media-has-changed-how-we-consume-news/?sh=19bb0343c3ca.

[2] https://www.bbc.com/news/uk-wales-58103604.

real and fake news instances, as well as the annotation process is demonstrated. The analysis of neural networks and non-neural networks based algorithms is presented, in addition to this, their results, and an evaluation of the classifiers are also described in Sect. 4. Finally, the study's conclusion is presented in Sect. 5, which is the final section of the paper.

2 Literature Review and Fake News Detection in Urdu

In the field of natural language processing, Urdu is regarded as a language with a poor availability of resources and has attracted only a small amount of research attention (NLP). Urdu is one of the top 10 most extensively spoken languages in the world, with a native speaker population of over 230 million people[3] and Urdu is one of the official languages of Pakistan. On the other side, Pakistan has had a hard time identifying fake news articles disseminated on the internet, which is proving to be a significant problem. Therefore, Pakistan's government is now developing ways to reduce and counteract fake news. For example, the Sindh government (a Pakistani province) recently warned the public that those found guilty of spreading false information and propaganda about coronavirus vaccines on social media would face severe consequences.[4] Furthermore, when propaganda news against the polio vaccination was published in 100 local Urdu newspapers, a spike in the number of polio infections was discovered in the propaganda-affected areas [18]. Therefore, it is critical to use automated methods for reducing the harm caused by fake news in indigenous communities.

Scientific works in Urdu also did not attract other researchers for false news identification. Furthermore, Urdu is a low-resource language, meaning it does not have a lot of NLP tools, structured and labeled corpus. Recently, the first dataset for Urdu fake news detection was proposed [6]. Therefore, the shared tasks aimed to encourage NLP researchers to build tools and techniques for verifying the validity of Urdu news articles' validity and social media postings.

Different classification methods [3, 5] have been proposed for various text classification tasks, such automatically detecting fake news [21, 22]. The majority of work focused on English [21, 24], with some works in other languages, such as Spanish [22, 24], German[28], Arabic [2, 25], Persian [31], Indonesian [23]. Moreover, some shared tasks were also organized in different public forums, such as SemEval 2017 task 8 RumourEval for English [9], SemEval-2019 task 7 RumourEval for English [13], and profiling of Fake News Spreaders on Twitter at PAN 2020 [24] in English.

Emotional data has been investigated as one of the aspects that helps in detecting fake news [11, 12]. One study reported that Linguistic features [22] are important for fake news detection, while another study [21] revealed that the differences of writing

---

[3] https://www.statista.com/statistics/266808/the-most-spoken-languages-worldwide/.

[4] https://www.geo.tv/latest/353706-sindh-warns-against-spreading-false-information-and-propaganda-about-coronavirus-vaccines-on-social-media.

styles between fake and real news had been recognized as beneficial in identifying fake news [22].

The shared tasks to identify Urdu false news at FIRE 2020 and Fire 2021, to our knowledge, were the first shared task. It is essential to highlight that PAN@CLEF 2020 [24], MediaEval3, and RumourEval [13] are three past forums that have influenced the shared task at FIRE 2020 and at FIRE 2021 in Urdu. English [13, 32], Spanish [21] and Arabic [25] languages were also included in the shared tasks. The studies usually concentrate on resource-rich languages like English [21], Spanish [22], German [28], and Chinese [32]. Finally, fake news are still proliferating on the internet and social media; thus, low-resource languages like Urdu require automatic methods for fake news detection.

3   Datasets Collection and Annotation

This section presents an overview of two datasets prepared for the shared tasks. The datasets contained five kinds of news: (v) Technology, (ii) Showbiz (entertainment), (iii) Business, (iv) Health, and (v) Sports. These news domains are comparable to the dataset [21] used to identify fake news in English. Nonetheless, only one news sector, namely, education-related news, is missing from our dataset since it was challenging to obtain education-related real news in Urdu.

A smaller version of the dataset named "Bend-The-Truth," was presented in a recent research [6], along with extensive details regarding the data collection and annotation description. A first dataset (ShardTask 1) was created for Urdu-Fake@FIRE2020 by presenting 900 news articles, 500 of which were real news and 400 of which were fake news. These articles were taken from the previously disclosed "Bend-The-Truth" dataset for the development of the algorithms, and a new test set for UrduFake@FIRE2020 was acquired that contained 400 news articles (250 real news, 150 fake news), all of which had never been seen or published before. This test set was used to check the robustness of the proposed models and determine how well the submitted systems performed on the previously unexplored dataset. Therefore, this dataset for UrduFake@FIRE2020 shared task contained 1300 news articles in total(750 real and 550 fake).

For the purpose of training and developing the systems that were submitted into the UrduFake@FIRE2021 competition, a second dataset (ShardTask 2) was created for the UrduFake@FIRE2021 competition by presenting 1300 news articles (750 real news and 550 fake news) from the previously disclosed "UrduFake@FIRE2020" dataset 1 for the purpose of the competition. In the same manner as the "Urdu-Fake@FIRE2020" test set, a new test set was produced, which consisted of 250 news instances: 150 real news articles and 100 fake news articles, both of which had never been seen or published before. Moreover, this test set was used for checking the effectiveness of the methods that were submitted in "UrduFake@FIRE2021" shared task. Finally, this dataset consisted of 1550 news stories in their entirety (900 real news articles, 650 fake news articles) for UrduFake@FIRE2021 shared task.

3.1 Real News Collection

"Newspaper",[5] a Python library was used to scrape news articles from the websites of several national and international mainstream media outlets. Moreover, rigorous annotation guidelines [6] for labeling news articles were followed, and all the news were manually labeled. For example, the news article should be published in a reputable newspaper or by a well-known news organization, and the details, such as the location, image, and date of the incident, should be verified by other reputable and trustworthy newspaper organizations.

We further checked to see whether there was a relationship between the title of the news article and its contents in order to guarantee that the title adequately summarized the content of the article. If any of these conditions were not met by a news piece, the news article was immediately removed from the annotation process. It is worth mentioning that each news article is different in length. This was due to the fact that each news organization has its own way of writing news articles. A news article on BBC Urdu, for example, had an average of more than 1500 words. As a result, we carefully annotated all the real news articles.

3.2 Professional Crowdsourcing of Fake News

It is challenging to collect and annotate fake news articles that correlate to real news articles for several reasons. First, finding and validating the inaccuracy of fake news in the same domain as the accessible actual news articles is a difficult task, which requires a significant amount of time and human resources. As a result, manually analyzing hundreds of thousands of news stories for verification is crucial. Second, unlike the English language, most of the news verification in Urdu is done manually due to the lack of digital services that provide news validation. As a result, in order to gather false news that corresponded to the actual news, we sought the assistance of professional journalists working for a variety of Pakistani news organizations. Finally, Fake news pieces were developed by the journalists for use in training and testing scenarios for both shared tasks UrduFake@FIRE2020 and UrduFake@FIRE2021.

This strategy is similar to creating a fake news dataset for the English language [22], in which the authors gathered fake news with the use of professional crowdsourcing. Moreover, it is worth noting that the style and language of news articles differ based on the news domain. For example, news in Sports, business, education, technology, and the showbiz domain require relevant topics. Therefore, we assigned news articles based on the journalists' knowledge of the relevant topic.

---

[5] https://newspaper.readthedocs.io/en/latest.

3.3   Training and Testing Datasets

Training, Validation, and Testing for ShardTask 1: The training set for identifying fake news in Urdu for shared task UrduFake@FIRE2020 was provided to the participants so that they may train their algorithms. The training set consisted of 900 news articles: 500 actual news stories, gathered from January 2018 through December 2018, and 400 fake news stories that corresponded to the actual news stories. In addition, the training set for shared task UrduFake@FIRE2020 was openly available for the participants to use for training, including for validation, development, and parameter tuning. Finally, a new test test set was released without the ground truth labeling for shared task UrduFake@FIRE2020. This new test set included 250 pieces of real news gathered between January 2019 and June 2020, and 150 pieces of fake news that corresponded to the real news.

Training, Validation, and Testing for ShardTask 2: UrduFake@FIRE21 used the training set that was released in UrduFake@FIRE20 (sharedTask 1) as a part of its training and testing set. In other words, we combined both training and testing datsets of UrduFake@FIRE20 and released as training dataset for UrduFake@FIRE21. Like UrduFake@FIRE20 shared task, a new test set was released without the ground truth labeling for UrduFake@FIRE21. The new test set consisted of 150 real news articles gathered from January 2021 to August 2021 and 100 fake news articles that corresponded to the real news. The purpose of this test set was to evaluate the robustness of the algorithms proposed in the shared task UrduFake@FIRE21. Finally, the corpus distribution of news articles by domains for both shared tasks is shown in Table 1.

Table 1   Data set distribution by topic

Fake news distribution by topic

Category	Training set				Testing set			
	Real		Fake		Real		Fake	
	Task 1	Task 2	Task 1	Task 2	Task 1	Task 2	Task 1	Task 2
Business	100	150	50	80	50	50	30	20
Health	100	150	100	130	50	50	30	20
Showbiz	100	150	100	130	50	50	30	20
Sports	100	150	50	80	50	50	30	20
Technology	100	150	100	130	50	50	30	20
Total	500	750	400	550	250	250	150	100

4 Methods for Fake News Detection

This section describes the methods used for detection of fake news in Urdu language.

4.1 Features Creation

N-grams: Text representation approaches based on n-grams are considerably used in natural language processing (NLP). In a vector space model, n-grams are usually represented as features. The recurrence of the n-grams are used to calculate the values of these traits, some weights might be applied to it.

Traditional n-grams are element sequences that emerge in the literature. These components may be words, characters, POS tags, phonemes, syllables, letters, or other text elements. The number of items in a sequence is usually represented by the letter "n" in n-grams [27]. In our research, we implemented character 1–6 g, as well as a combination of character and word n-grams.

Words Embeddings: Word embeddings are a sort of word representation that encodes the meaning of a word using a real-valued vector. Words with comparable meanings may be represented in the same way using this method. This study used transformer embeddings, Word2Vec, GloVe, and fastText. Word2Vec technique obtained by two Neural Network methods: Skip Gram and Common Bag Of Words (CBOW). Skip Gram works effectively with limited amounts of data and accurately represents unusual words, according to Mikolov [20]. CBOW, on the other hand, is faster and provides more accurate representations of common words. Global Vectors (Glove) is an acronym that stands for "global vectors". It's a technique for generating word embeddings from a corpus by aggregating global word co-occurrence matrices. The main concept is to use statistics to build a dependence model. FastText is a word embedding strategy that is based on a modified version of the Word2Vec model. Instead of learning vectors for words, FastText encodes each word as an n-gram of letters. This allows embeddings to detect suffixes and prefixes as well as the meaning of shorter words. Table 2 shows different text representations and feature weighting schemes.

4.2 Machine Learning Algorithms

Machine Learning Algorithms, a subset of AI, use a number of precise, probabilistic, and improved strategies to help computers learn from data and discover difficult-to-perceive patterns in vast, noisy, or complicated datasets.[6]

---

[6] https://www.sciencedirect.com/topics/engineering/machine-learning-algorithm.

Table 2 Text representations and feature weighting scheme

Text representation
n-grams
Char n-grams
Word2Vec
Combination of character and word n-grams
Embeddings
Word2Vec
GloVE
FastText
BERT pretrained embeddings
Feature weighting scheme
TF-IDF

Logistic Regression: It implies statistical methods for predicting the outcome of a given situation. Regression models are based on the probability notion and are utilized as a predictive analytic technique. The logistic regression (LR) model employs features to build a linear model using multinomial logistic regression with ridge estimator.

Random Forest: This is a classification, regression, and ensemble learning strategy that operates by generating a multitude of decision trees. Each decision tree is created by picking attributes at random from a dataset. The output of the random forest algorithm is based on the decision trees' predictions. It creates predictions by averaging the output of a large number of trees. As the number of trees grows, the accuracy of the result improves. The nodes of the tree are selected to improve data gathering, and the most often used criteria are GINI and Entropy.

Support Vector Machines: SVM is a machine learning classifier utilized to tackle problems that are linear, nonlinear, or regression. This approach creates a hyperplane or set of them in a multi-dimensional space. Nonlinear classification problems are performed using a kernel approach that implicitly transforms the input into a more dimensional feature space. SVM has been widely used for fake news detection tasks [14, 15, 17]. SVM outperforms a number of supervised machine learning algorithms for fake news identification when content-based attributes (e.g. linguistic and visual data) are included.

AdaBoost: The AdaBoost classifier is an iterative ensemble technique using the adaptive boosting algorithm. Each instance's weights are re-allocated, with larger weights applied to incorrectly classified instances. AdaBoost is ideal for improving decision tree performance on binary classification issues. Decision trees with one level are the most often used AdaBoost algorithm, and it produces better results when dealing with less noisy datasets; otherwise, it overfits [6].

4.3 Deep Learning Algorithms

Techniques that are based on neural networks are often used in the process of detecting fake news [1, 8, 16]. In this paper, we employed three different models that were based on neural networks: (i) Multilayer Perceptron (MLP), (ii) Convolutional Neural Network (CNN), and (iii) Long Short-Term Memory Networks (LSTM).

Multilayer Perceptron: A feedforward artificial neural network known as a multilayer perceptron (often abbreviated as MLP) is deployed for the purpose of resolving regression and classification tasks. Back-propagation is used to train the model via the use of this supervised learning technique. Furthermore, an MLP has three distinct layers: an input layer, a hidden layer, and an output layer. All data instances are received by the input layer, while the output layer is in charge of functions like classification and prediction. The dot product of the input instances and weights is computed by the neural network's hidden lawyers.

Convolutional Neural Network: A Convolutional Neural Network, sometimes also known as a CNN, is a kind of deep neural network that has found use in the field of computer vision, namely for the purposes of picture identification and classification. Moreover, CNN comprises many layers, including a convolutional layer, a non-linearity layer, a pooling layer, and a fully-connected layer. The convolutional layer is the inner product of the convolutional kernel and the layer's input matrix. The model is trained using a back-propagation method. In comparison to feedforward networks, CNN is computationally efficient since it comprises neurons with learnable weights. CNN has grown its popularity due to its ability to retrieve vital information quickly and accurately.

Long Short-Term Memory Networks: The Long Short-Term Memory (LSTM) network is yet another kind of deep neural network that is capable of handling sequence data, and these tasks may include order dependency. The LSTN neural network has four linear layers (MLP layers), and in each cell, these layers are activated at each and every time step in the sequence. Moreover, LSTM neural networks have feedback connections, in contrast to the more common feedforward neural networks. Not only individual data points (like photos), but also entire data sequences, may be handled. Examples of this include: (such as speech or video). Long short-term memories, or LSTMs, are useful for a wide variety of applications, including undifferentiated linked handwriting detection, voice recognition, network traffic analysis, and anomaly detection in intrusion detection systems (IDSs).

4.4 Transfer Learning

Unlike machine and deep learning classifiers, which are aimed to solve certain tasks, transfer learning takes the knowledge needed to train a model for one task and applies it to another. Transfer learning resolves the isolated learning paradigm by assigning the weights of the old model to initialize the weights for the new model, which means the new model does not have to learn everything from scratch. For example, predicting the next task given the word in a phrase is a typical goal of transfer learning on numerous complex NLP tasks, such as Emotion Classification, dangerous language [4], and threatening language detection [7], transformer-based techniques exceeded the state-of-the-art.

BERT: BERT is a bidirectional Transformer language model. BERT was pre-trained on language modeling (15% of tokens are masked and BERT was taught to predict tokens from context) as well as prediction of the following sentence (based on the first sentence) [10]. BERT produces contextual embeddings as a result of its training. It also can be fine-tuned after pre-training with fewer resources to improve the performance of certain tasks.

XLNet: XLNet is a large bidirectional transformer. It uses revised training techniques, more data, and computational resources to overcome the limitations of BERT. XLNet employs permutation language modeling, predicting all tokens in a arbitrary order. On the contrary, BERT's masked language model predicts only masked (15%) tokens. It also distinguishes XLNet from traditional language models that expect sequential tokens rather than random ones. This enables the model to learn bidirectional links, improving the handling of word dependencies in the model. In addition, the Transformer XL design was used as the basic architecture, resulting in excellent performance without sequence-based training. XLNet was trained for 2.5 d using 130 GB of textual data and 512 TPU chips, which is much greater than BERT [30].

RoBERTa: The vigorously optimized BERT method, RoBERTa, is an upgrade version that has refined the training process, increased data by 1000%, and increased computing power by 1000%. RoBERTa uses BERT's language masking strategy. This trains the system to look for data that is deliberately hidden inside otherwise unannotated language occurrences. RoBERTa trains with a much larger mini-batch and learning rate while removing BERT's next set pre-training goal. As a result, RoBERTa outperforms BERT with masked language modeling goals and improves the performance of downstream tasks. For example, RoBERTa is pre-loaded with 160 GB of material, including BERT's 16 GB Books Corpus and English Wikipedia. Other data sets included the CommonCrawl News dataset (63 million articles, 76 GB), Web text corpus (38 GB), and Common Crawl Stories (31 GB) [19].

4.5 Evaluation Metrics

We used the standard evaluation metrics, such as *Precision* (P), *Recall* (R), *Accuracy*, and two *F1-scores*, namely $F1_{Real}$ for the prediction of label "real" out of all news, and $F1_{Fake}$ for the prediction of label "fake" out of all news, to evaluate the classification performance. This is due to the fact that the dataset contains two equally important classes in a binary classification problem. We also employed the macro-averaged

Table 3 Algorithms for fake news detection

Classic machine learning techniques

Algorithm	Fake class		Real class	
	Precision	Recall	Precision	Recall
RF, Adaboost, MLP, SVM	0.709	0.733	0.837	0.82
linear SVM with SGD	0.754	0.4	0.757	0.935
XGBoost+LightGBM+AdaBoost	0.634	0.33	0.729	0.905
linSVM+LR+MLP+XGB+RF	0.821	0.23	0.716	0.975
MLP+AdaBoost+GraidentBoost+RF	0.85	0.17	0.703	0.985
SVM	0.72	0.18	0.701	0.965

Deep learning techniques

Algorithm	Fake class		Real class	
	Precision	Recall	Precision	Recall
Dense Neural Network	0.89	0.593	0.797	0.956
Bi-directional GRU model	0.881	0.593	0.796	0.952
ULMFiT model	0.783	0.627	0.8	0.896
textCNN	0.592	0.48	0.762	0.835
CNN	0.48	0.49	0.742	0.735
MuRIL	0.96	0.24	0.723	0.995
DNN	0.793	0.23	0.356	0.715

Transfer learning

Algorithm	Fake class		Real class	
	Precision	Recall	Precision	Recall
CharCNN-Roberta	0.89	0.86	0.918	0.936
XLNet pre-trained model	0.836	0.713	0.842	0.916
BERT-base	0.454	0.1	0.676	0.94
RoBERTa-urdu-small	0.266	0.12	0.654	0.835

$F1_{Macro}$. It is an average of $F1_{Real}$ and $F1_{Fake}$ and it represents an account for the bias related to the real class, which prevails since its number of samples is greater than the fake news class. The final ranking is determined by the $F1_{Macro}$ score. The results of the proposed systems for detecting fake news in Urdu are summarized in Tables 3 and 4.

5 Conclusions

The challenge of automatically identifying fake news is crucial, especially in languages with a limited amount of resources. In this study, several different approaches to machine learning are investigated. These approaches range from various text rep-

Table 4 Prediction scores of the machine learning techniques.

Classic machine learning techniques

Algorithm	F1_Fake	F1_Real	F1_Macro	Accuracy
RF, Adaboost, MLP, SVM	0.721	0.828	0.774	0.787
linear SVM with SGD	0.522	0.836	0.679	0.756
XGBoost+LightGBM+AdaBoost	0.434	0.808	0.621	0.713
linSVM+LR+MLP+XGB+RF	0.359	0.826	0.592	0.726
MLP+AdaBoost+GraidentBoost+RF	0.283	0.82	0.552	0.713
SVM	0.288	0.812	0.55	0.703

Deep learning techniques

Algorithm	F1_Fake	F1_Real	F1_Macro	Accuracy
Dense Neural Network	0.712	0.869	0.791	0.82
Bi-directional GRU model	0.709	0.867	0.788	0.818
ULMFiT model	0.696	0.845	0.77	0.795
textCNN	0.53	0.797	0.663	0.716
CNN	0.485	0.738	0.611	0.653
MuRIL	0.384	0.837	0.61	0.743
DNN	0.356	0.823	0.59	0.59

Transfer learning

Algorithm	F1_Fake	F1_Real	F1_Macro	Accuracy
CharCNN-Roberta	0.874	0.926	0.9	0.908
XLNet pre-trained model	0.769	0.877	0.823	0.84
BERT-base	0.163	0.786	0.475	0.66
RoBERTa-urdu-small	0.165	0.734	0.449	0.596

resentation techniques, such as pre-trained embeddings, contextual representation, and end-to-end neural network-based methods. Other methods include crafting traditional features and applying standard machine learning and deep learning techniques including CNN, and non-Urdu specialized Transformers (BERT, XLNet, RoBERTa) as well as Urdu-specialized Transformers (MuRIL, RoBERTa-urdu-small) were employed for the aim of identifying Urdu fake news.

The effectiveness of the machine and deep learning algorithms was evaluated using a variety of evaluation measures, including accuracy, recall, $F1_{Real}$, $F1_{Fake}$, $F1_{Macro}$. For the shared task one UrduFake@FIRE20, the system that performed the best was based on BERT; it received $0.907\ F1_{Macro}$. For second shared task UrduFake@FIRE21, the system that performed the best employed the linear classifier function; it obtained $0.679\ F1_{Macro}$. Therefore, the conclusion that can be drawn from this research is that contextual representation and methodologies using huge neural networks perform better than standard feature-based models.

References

1. Abusaa, M., Diederich, J., Al-Ajmi, A., et al.: Machine Learning, Text Classification and Mental Health. HIC 2004: Proceedings, p. 102 (2004)
2. Alkhair, M., Meftouh, K., Smaïli, K., Othman, N.: An arabic corpus of fake news: collection, analysis and classification. In: International Conference on Arabic Language Processing, pp. 292–302. Springer (2019)
3. Ameer, I., Ashraf, N., Sidorov, G., Gómez Adorno, H.: Multi-label emotion classification using content-based features in twitter. Computación y Sistemas **24**(3), 1159–1164 (2020)
4. Ameer, I., Sidorov, G., Gómez-Adorno, H., Nawab, R.M.A.: Multi-label emotion classification on code-mixed text: data and methods. IEEE Access **10**, 8779–8789 (2022). https://doi.org/10.1109/ACCESS.2022.3143819
5. Ameer, I., Sidorov, G., Nawab, R.M.A.: Author profiling for age and gender using combinations of features of various types. J. Intel. Fuzzy Syst. **36**(5), 4833–4843 (2019)
6. Amjad, M., Sidorov, G., Zhila, A., Gómez-Adorno, H., Voronkov, I., Gelbukh, A.: Bend the truth: benchmark dataset for fake news detection in urdu language and its evaluation. J. Intel. Fuzzy Syst. **39**(2), 2457–2469 (2020)
7. Amjad, M., Zhila, A., Sidorov, G., Labunets, A., Butt, S., Amjad, H.I., Vitman, O., Gelbukh, A.: Urduthreat@ fire2021: shared track on abusive threat identification in urdu. FIRE 2021, pp. 9–11. Association for Computing Machinery, New York, USA (2021). https://doi.org/10.1145/3503162.3505241, https://doi.org/10.1145/3503162.3505241
8. Coppersmith, G., Dredze, M., Harman, C., Hollingshead, K., Mitchell, M.: Clpsych 2015 shared task: depression and ptsd on twitter. In: Proceedings of the 2nd Workshop on Computational Linguistics and Clinical Psychology: From Linguistic Signal to Clinical Reality, pp. 31–39 (2015)
9. Derczynski, L., Bontcheva, K., Liakata, M., Procter, R., Hoi, G.W.S., Zubiaga, A.: Semeval-2017 task 8: Rumoureval: determining rumour veracity and support for rumours. arXiv preprint arXiv:1704.05972 (2017)
10. Devlin, J., Chang, M.W., Lee, K., Toutanova, K.: Bert: Pre-training of deep bidirectional transformers for language understanding. arXiv preprint arXiv:1810.04805 (2018)
11. Ghanem, B., Rosso, P., Rangel, F.: An emotional analysis of false information in social media and news articles. ACM Trans. Internet Technol. (TOIT) **20**(2), 1–18 (2020)
12. Giachanou, A., Rosso, P., Crestani, F.: Leveraging emotional signals for credibility detection. In: Proceedings of the 42nd International ACM SIGIR Conference on Research and Development in Information Retrieval, pp. 877–880 (2019)
13. Gorrell, G., Kochkina, E., Liakata, M., Aker, A., Zubiaga, A., Bontcheva, K., Derczynski, L.: Semeval-2019 task 7: Rumoureval, determining rumour veracity and support for rumours. In: Proceedings of the 13th International Workshop on Semantic Evaluation, pp. 845–854 (2019)
14. Gravanis, G., Vakali, A., Diamantaras, K., Karadais, P.: Behind the cues: a benchmarking study for fake news detection. Exp. Syst. Appl. **128**, 201–213 (2019)
15. Hiramath, C.K., Deshpande, G.: Fake news detection using deep learning techniques. In: 2019 1st International Conference on Advances in Information Technology (ICAIT), pp. 411–415. IEEE (2019)
16. Hussain, M.G., Hasan, M.R., Rahman, M., Protim, J., Hasan, S.A.: Detection of Bangla Fake News Using MNB and SVM Classifier. arXiv preprint arXiv:2005.14627 (2020)
17. Jain, A., Shakya, A., Khatter, H., Gupta, A.K.: A smart system for fake news detection using machine learning. In: 2019 International Conference on Issues and Challenges in Intelligent Computing Techniques (ICICT), vol. 1, pp. 1–4. IEEE (2019)
18. Kausar, S., Tahir, B., Mehmood, M.A.: Prosoul: a framework to identify propaganda from online urdu content. IEEE Access **8**, 186039–186054 (2020)
19. Liu, Y., Ott, M., Goyal, N., Du, J., Joshi, M., Chen, D., Levy, O., Lewis, M., Zettlemoyer, L., Stoyanov, V.: Roberta: A Robustly Optimized Bert Pretraining Approach. arXiv preprint arXiv:1907.11692 (2019)

20. Mikolov, T., Chen, K., Corrado, G., Dean, J.: Efficient Estimation of Word Representations in Vector Space. arXiv preprint arXiv:1301.3781 (2013)
21. Pérez-Rosas, V., Kleinberg, B., Lefevre, A., Mihalcea, R.: Automatic Detection of Fake News, pp. 3391–3401 (2018)
22. Posadas-Durán, J.P., Gómez-Adorno, H., Sidorov, G., Escobar, J.J.M.: Detection of fake news in a new corpus for the spanish language. J. Intel. Fuzzy Syst. **36**(5), 4869–4876 (2019)
23. Pratiwi, I.Y.R., Asmara, R.A., Rahutomo, F.: Study of hoax news detection using naïve bayes classifier in indonesian language. In: 2017 11th International Conference on Information & Communication Technology and System (ICTS), pp. 73–78. IEEE (2017)
24. Rangel, F., Giachanou, A., Ghanem, B.H.H., Rosso, P.: Overview of the 8th author profiling task at pan 2020: profiling fake news spreaders on twitter. In: CEUR Workshop Proceedings. vol. 2696, pp. 1–18. Sun SITE Central Europe (2020)
25. Rangel, F., Rosso, P., Charfi, A., Zaghouani, W., Ghanem, B., Sánchez-Junquera, J.: On the author profiling and deception detection in arabic shared task at fire. In: Proceedings of the 11th Forum for Information Retrieval Evaluation, pp. 7–9 (2019)
26. Reis, J.C., Correia, A., Murai, F., Veloso, A., Benevenuto, F.: Supervised learning for fake news detection. IEEE Intell. Syst. **34**(2), 76–81 (2019)
27. Sidorov, G., Velasquez, F., Stamatatos, E., Gelbukh, A., Chanona-Hernández, L.: Syntactic n-grams as machine learning features for natural language processing. Exp. Syst. Appl. **41**(3), 853–860 (2014)
28. Vogel, I., Jiang, P.: Fake news detection with the new german dataset "germanfakenc". In: International Conference on Theory and Practice of Digital Libraries, pp. 288–295. Springer (2019)
29. Vosoughi, S., Roy, D., Aral, S.: The spread of true and false news online. Science **359**, 1146–1151 (2018). https://doi.org/10.1126/science.aap9559
30. Yang, Z., Dai, Z., Yang, Y., Carbonell, J., Salakhutdinov, R.R., Le, Q.V.: Xlnet: Generalized autoregressive pretraining for language understanding. Adv. Neural Inform. Proces. Syst. **32** (2019)
31. Zarharan, M., Ahangar, S., Rezvaninejad, F.S., Bidhendi, M.L., Pilevar, M.T., Minaei, B., Eetemadi, S.: Persian stance classification data set. In: TTO (2019)
32. Zhang, X., Cao, J., Li, X., Sheng, Q., Zhong, L., Shu, K.: Mining dual emotion for fake news detection. In: Proceedings of the Web Conference 2021, pp. 3465–3476 (2021)

Decision Making in a Distributed Intelligent Personnel Health Management System on Offshore Oil Platform

Masuma Mammadova[ID] and **Zarifa Jabrayilova**[ID]

Abstract This paper explores the decision-making problems in a geographically distributed intelligent health management system for oil workers working in offshore industry. The decision-making methodology is based on the concept of a person-centered approach to managing the health and safety of personnel, which implies the inclusion of employees as the main component in the control loop. This paper develops a functional model of the health management system for workers employed on offshore oil platforms and implements it through three phased operations, that is monitoring and assessing the health indicators and environmental parameters of each employee, and making decisions. These interacting operations, as the links of a single decision-making process, combine the levels of a distributed intelligent system for managing the health of workers and ensure its functioning as a whole. The paper offers the general principles of functioning of a distributed intelligent system for managing the health of workers in the context of structural components and computing platforms. It presents appropriate approaches to the implementation of decision support processes and describes one of the possible methods for evaluating the generated data and making decisions using fuzzy pattern recognition.

Keywords Offshore oil platforms · Distributed intelligent health management system · Decision making

1 Introduction

Today, a significant number of skilled workers and specialists (oilmen) are involved in various operations and production processes implemented at geographically distributed offshore facilities and structures. Their functional duties and work activities are associated with potential hazards and health risks. Therefore, the tasks of

---

M. Mammadova (✉) · Z. Jabrayilova
Institute of Information Technology, Azerbaijan National Academy of Science, B.Vahabzade Street, Baku 9A, AZ1141, Azerbaijan
e-mail: mmg51@mail.ru

© The Author(s), under exclusive license to Springer Nature Switzerland AG 2023
Sh. N. Shahbazova et al. (eds.), *Recent Developments and the New Directions of Research, Foundations, and Applications*, Studies in Fuzziness and Soft Computing 423, https://doi.org/10.1007/978-3-031-23476-7_14

improving the efficiency of employees' health management and ensuring a safe working and living environment are the most important part of the oil and gas industry at all stages of the life cycle of oil and gas business. The problem of employees' health management is of particular relevance in offshore oil and gas industry, which is classified as a high-risk segment [1, 2]. Note that the concept and tools of "Industry 4.0" provide an opportunity to develop IoT-based cyber-physical systems [3], allowing to further partially or completely depersonalize the operational processes implemented in offshore oil and gas segment (OOGS) of the industry [4, 5]. However, development and application of IoT solutions to eliminate possible representation of the human factor and to support the health and safety of workers in oil and gas industry and, particularly, the offshore industry has been poorly studied yet [6, 7].

2 Methodological Approaches to the Development of a Health Managing System for Human Resources Employed on Offshore Oil Platforms

2.1 Concept of a Person-Centered Approach

An analysis of the professional activities of workers involved in the offshore development and operation of oil and gas fields, through the prism of the impact of working conditions, everyday life and external factors on their health, shows that offshore development and operation of oil and gas fields take place in difficult and often extreme working and living conditions. An analysis of the causes of accidents shows that many of them are associated with an unforeseen health deterioration of workers. Available rules and standards of labor safety fixed in regulatory documents mainly include the requirements for the safety of workplaces, the environment, and equipment. However, despite the constant improvement of regulatory documents considering technological innovations, the number of incidents caused by the human factor remains quite high (more than 70% of accidents and incidents in the oil and gas industry). This circumstance, recently recognized by large oil and gas companies, prompts the latter to focus on the human's role in ensuring the workers' health at all stages of life cycle of industry. Human factor on OOP refers to the possibility of person committing erroneous actions under certain conditions or making wrong decisions caused an incident. In such situations, the subjectivity of nature and the psychophysiological characteristics of a person are manifested. Therefore, the human factor in hazardous production begins to pose danger rather than the production itself. Based on this, we assume that the likelihood of making erroneous decisions by any employees directly depends on the state of health affects his/her behavior, as well as on the nature of his/her actions and activity during the shift on platform. This actualizes the need for systematic remote monitoring of health and safety of workers in their working and living environment. Basing on IoT and e-health solutions, the works [8, 9] develop the concept and architecture of a distributed intelligent system

(DIS) for managing the health and safety of workers employed in OOPs. The main idea of the concept is to improve the safety of oil workers through the introduction of a person-centered approach to managing their health. "Placing" a person at the center of the personnel health and safety management system enables linking the vital health indicators of each employee with the context of the environment and reasonably assessing the criticality of current situation.

2.2 Architecture DIS for Managing the Health of Shift Workers

Architecture of intelligent health management system for shift workers in OOS has a hierarchical structure, in which each of the three geographically distributed layers is a target intelligent information system with particular purpose and functions (Fig. 1) [8]. All three layers are integrated into a single decision support process and ensure the functioning of system as a whole.

Application of IoT-based platform is capable of simultaneously transmitting sensed data to various situational control centers (servers) located both horizontally (at the same layer) and vertically along the control hierarchy. In this case, the OOP personnel acts as a biological object equipped with body-worn and/or wearable devices generating different information in accordance with the purpose. These devices provide user interaction with the environment and are capable of recording, accumulating, processing and transmitting data. The use of wearable devices with built-in sensors and address identification of the monitored parameters ensures to obtain context-sensitive information about each employee's health status with reference to a specific location, date and time.

The IoT technological ecosystem, functioning in conjunction with physical devices, computing platforms and analytical tools, integrates entire work processes in the proposed architecture of the OOP personnel health management system into hierarchically distributed computing levels: Dew computing, Fog computing and Cloud computing.

The goal of this work is to develop the principles of functioning of the DIS for personnel's health status monitoring and to realize decision-making processes based on the concept of a person-centered approach.

2.3 Principles of Functioning of Distributed Intelligent System Determining Approaches to the Implementation of Decision-Making Processes

The functional model of the health management system of OOP personnel is implemented by tracking vital indicators of the physiological state and parameters of the

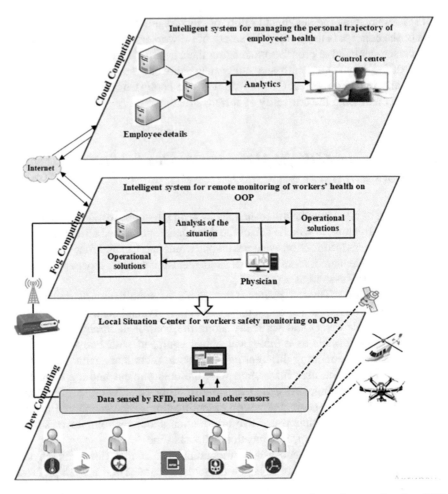

Fig. 1 The architecture of an intelligent health management system for workers employed on OOP

environment of each employee, their assessment and analysis. These interacting operations distributed across the DIS levels, are the links of decision-making process. The principles of functioning of DIS in the context of structural layers, proposed in the work, are as follows:

1. All three layers of DIS along with many specific applications are equipped with a unique IoT application (software) for each of them. This application is an intelligent information system based on a functional model of health management of personnel employed on OOP.
2. Modules of IIS database include digitized ranges of changes in normative, edge and critical values of each health indicator (temperature, pulse, pressure, heart rate, etc.), information on standards (reference images) of activity and behavior within the framework of technological requirements and restrictions, authorized

and prohibited formats and coordinates of access to hazardous geo-zones (in accordance with the map of drilling rig, working and residential sites, explosive zones on OOP, etc.), permissible limits and level of excess environmental toxicity.

3. IIS knowledge base contains cognitive information linking the expert assessments and decisions with granules of possible values of various indicators and parameters, including critical ones, provoking the emergency situations on OOP.

4. The process of continuous health and safety monitoring of workers employed on OOP generates a huge amount of data, which is problematic to analyze through traditional methods. Therefore, it is assumed that the analytical block of DIS computing platforms based on IoT solutions includes high-performance algorithms and intelligent analytical tools (Decision support tools, Softcomputing, Big Data, Machine Learning).

5. IoT monitors in parallel the streams of sensed data of all workers on OOP, compares them with the normative (reference) health status templates, behavioral patterns, geolocation and environmental parameters pre-recorded in IIS databases and knowledge bases, and identifies the deviation rate of a particular indicator and parameter in real time. Instant comparison of normative (reference) and current (real) data by the IoT system identifies the degree of their similarity or deviation by all parameters.

6. IoT, instantly analyzing the current situation, reveals the deviation of certain indicators and parameters from the norm and analyzes the current situation. Depending on the criticality of the situation, the degree of its compliance with already known typical models, or the identification of new patterns, decision can be made according to two scenarios: 1) automatic formation of a control action by the system; 2) real time data redirecting to emergency response services to make an operational decision.

3 Decision Support in a Distributed Intelligent Health Management System for OOP Personnel

We propose one of the possible implementation options for a distributed intelligent health management system for OOP personnel in the context of structural components and modern computing platforms.

First layer of DIS focuses on health and safety monitoring of workers on OOP. This layer takes urgent measures to organize the rescue of an employee at scene of accident and provide first aid. Data collection, processing and analysis is implemented through Dew computing, which provides real-time decision making ensuring low latency in data processing. Targeted data of workers recorded by sensors and RFID through wearable device and smartphone used as a gateway is transmitted via wireless or wired communication to the Local Situation Center for Emergency Response (LSCER) on OOP. LSCER is a computerized workplace of persons responsible for health and safety of workers on OOP. Physically, this is a local computer (Dew data center) designed to receive and analyze incoming data streams on health and safety of workers during the shift. IoT continuously compares the normative (refer-ence), initial (pre-shift) and current (real) values of monitored health indicators and parameters of the contextual environment of workers. As long as all data of workers and their environments is within acceptable limits, nothing is transferred to local computer (Dew data center). As soon as the values of any health indicators and/or coordinates and parameters recorded by sensors go beyond the typical range, these data are sent to local IoT application for processing, analysis and decision-making. IoT application, equipped with special analytical tools and intelligent algorithms, identifies changes in the health of each employee and deviations of environmental parameters from standards and offers solutions for their elimination.

Second layer in the network architecture of DIS is designed for remote health and safety monitoring of workers employed on OOP from the nearest coastal situa-tion center. Reception of data generated at sensor layer, their analytical processing, decision making and temporary storage are implemented in real time through *Fog computing*.

The intelligent IoT platform DIS implements the following services in *Fog* envi-ronment: 1) accepting and processing actual data coming from Dew layer in the absence of direct communication between OOP and Cloud; 2) based on Fog analytics results, making decision on each received situation and transfer the appropriate control action for execution to Dew layer; 3) sending data to Cloud DPC that are critically deviated from the standards.

Third layer of the network architecture of DIS, that is *Cloud Computing* is designed to manage the personal health trajectory of shift workers. Solution of this problem is based on the regular data collection and accumulation from various sources on the dynamics of health and safety of workers and the formation of repre-sentative data bases on the chronology of changes in the vital physiological indi-cators of each worker. These data bases are stored in *Private Cloud* and serve for decision-making at management level of offshore oil and gas company.

4 Discussion

Geographically distributed IoT-based intelligent health management system auto-matically (without human intervention) analyzes the data and synthesizes diagnosis decisions. Depending on the situation, i.e., the degree of deviation of certain indi-cators and parameters from the norm, different methods can be used, and in the case of typical situations, the task of decision-making can be reduced to the prob-lems of recognition of fuzzy patterns. In this case, decision is made regarding the current situation and the criticality of their health status by identifying the degree of similarity between the reference and real (current) fuzzy images of each employee [10]. In this regard, all indicators and parameters are brought to a fuzzy universal scale. The advantage of using fuzzy universal scales in assessing the health status of workers employed in OOPs is the possibility of sharing both qualitative and quantitative parameters expressed in a single term-set of linguistic variables [11]. A fuzzy universal scale of some linguistic variable for a given parameter (for example, temperature, pulse, etc.) of health status of workers employed in OOP can be built using the following procedure:

1. Based on the normative (reference) data, the minimum and maximum values of the subject scale X, i.e., upper $(x_{u.l.})$ and lower $(x_{l.l.})$ limits of parameter values are determined;
2. Based on the formula $x_{ref} = (x_{u.l.} + x_{l.l.})/2$, the reference values of the parameters are determined;
3. Taking into account the accepted inclusion of limits and the equality of the two situations, the lower and upper limits of parameter change range are assigned a certain value from the interval [0; 1] (for example, 0,5).
4. Segments $[x_{l.l.}; x_{ref}]$ and $[x_{ref}; x_{u.l.}]$. . are divided into several parts (according to the gradations of linguistic variables) and, in accordance with the parameter values, membership functions are constructed from the interval [0, 5; 1].

In this case, as a proximity measure of any two events, the degree of fuzzy equality (equivalence) or fuzzy inclusion, determined by formulas (1), (2), respectively, can be used [10]:

$$\mu(A, B) = \underset{x \in X}{\&} (\mu_A(x) \rightarrow, \mu_B(x))$$

$$= \underset{x \in X}{\min} = [\min (1 - \mu_A(x), \mu_B(x)), \max (\mu_A(x), 1 - \mu_B(x)))] \quad (1)$$

$$\mu(A, B) = \underset{x \in X}{\&} (\mu_A(x), \mu_B(x)) = \underset{x \in X}{\&} (\max (1 - \mu_A(x), \mu_B(x)))$$

$$= \underset{x \in X}{\min} = \max (1 - \mu_A(x), \mu_B(x))) \quad (2)$$

For example, [12] uses the degree of fuzzy inclusion to assess the psycho-physiological state of sea workers, 16 indicators of the Cattell test are used as linguistic variables.

To recognize the real state of employee's health, first of all, the similarity degree of current and normative state is determined. The change range in the degree of fuzzy equality can be divided into several fuzzy intervals depending on its severity degree, i.e., semantic interpretation of verbal gradations (terms), the expert assessment method. These intervals reflect the change area in the membership functions of fuzzy sets of verbal gradations of the linguistic variable "similarity of the current and normative state" δ_i, defined on the set of real numbers R_δ in the display $\mu_{\delta_i} : R_\delta \rightarrow [0, 1]$ (Table 1).

According to Table 1, the following rules are introduced into the knowledge base:

1. IF $(\mu_{\delta_i} : R_\delta \rightarrow [0, 8; 1])$ THEN "employee's health status is good";
2. IF $(\mu_\delta : R_\delta \rightarrow [0, 65; 0, 8))$ THEN "employee's health status is satisfactory";
3. IF $(\mu_{\delta_i} : R_\delta \rightarrow [0, 5; 0, 65))$ THEN "employee's health status requires consideration";

Table 1 Range of change of membership functions of fuzzy sets of linguistic variables "similarity of the current and normative state" of health of workers

Linguistic variables	Terms of linguistic variable "similarity of the current and normative state"	Range of change of terms on the non-similarity scale
Similarity of the current and normative state	Optimal (normative) similarity	[0.8;1]
	Minimum deviation	[0.65;0.8)
	Permissible deviation	[0.5; 0.65)
	Critical deviation	[0; 0.5]

4. IF $(\mu_{\delta_i} : R_\delta \rightarrow [0; 0, 5])$ THEN "employee's health status is critical and requires urgent intervention" etc.

Periodic observation of the latter rules requires an analysis of other environmental parameters (geolocation indicators, equipment requirements, toxicity parameters of the environment, activity, position in space, employee's behavior – compliance labor safety standards) and the solution for investigating the causes of adverse health effects (these issues are among the perspective research areas of the authors).

In non-standard situations, all relevant information and IoT solutions automatically proposed by intelligent decision system in real time are provided to interested services and their authorized persons (supervisors, doctors, occupational safety specialists, heads of relevant departments, experts). This enables the latter to find out the reasons for deviations of indicators from the standard values and make informed decisions to eliminate hazards to health and possible incidents, thereby minimizing the impact of the human factor. In this case, decision-making can be solved by taking into account different types of functional and distributed knowledge related to the situation (or rather, employee's health) and by making decision-making methods based on a distributed knowledge base that allows geographically distributed decision-makers to make collective decisions.

5   Conclusion

The continuous monitoring of the health and safety of personnel employed on OOP based on IoT and e-health ensures systematic data collection and accumulation of personalized information, the formation of a sufficiently representative and regularly updated database on their health dynamics for a certain period through targeted interaction with each worker. Embedding this base in the architecture of an intelligent personnel health management system as a dynamic database module and joint analytical processing of current and retrospective data will allow to objectively assess the changes' tendency in the health status of each employee, make informed and objective decisions to eliminate problems negatively affecting the personnel's health in the short, medium and long term.

References

1. Niven, K., McLeod, R.: Offshore industry: management of health hazards in the upstream petroleum industry. Occup. Med. **59**(5), 304–309 (2009)
2. Cumberland, S.: The human factor of IoT in safety, www.plantengineering.com/articles/the-human-factor-of-iot-in-safety/, last accessed 2019/02/13.
3. TKh., Fataliyev, Mehdiyev, Sh.A.: Analysis and New Approaches to the Solution of Problems of Operation of Oil and Gas Complex as Cyber-Physical System. Information Technology and Computer Science **10**(11), 67–76 (2018)
4. Khan, W.Z., Aalsalem, M.Y., et al.: Reliable IoT based Architecture for Oil and Gas Industry. In: 19th International Conference on Advanced Communication Technology Proceedings, pp. 705–710, IEEE, South Korea (2017).
5. Wanasinghe, T.R., Gosine, R.G., et al.: The Internet of Things in the Oil and Gas Industry: A systematic Review. IEEE Internet Things J. **7**(9), 8654–8673 (2020)
6. Rahmani, A.M., Gia, T.N., et al.: Exploiting smart e-Health gateways at the edge of healthcare Internet-of-Things: A fog computing approach. Futur. Gener. Comput. Syst. **78**(1), 641–658 (2018)
7. Majumder, S., Mondal, T., Deen, M.J.: Wearable Sensors for Remote Health Monitoring. Sensors **17**(1), 1–5 (2017)
8. Mammadova, M.H., Jabrayilova, Z.G.: Conceptual approaches to intelligent human factor management on offshore oil and gas platforms. ARCTIC Journal **74**(2), 19–40 (2021)
9. Mammadova, M. H., Jabrayilova, Z. G.: Conceptual approaches to IoT-based personnel health management in offshore oil and gas industry. Proceedings of the7th International Conference on Control and Optimization with Industrial Applications (COIA 2020), v.1, pp. 257–259. Baku, Azerbaijan (2021).
10. Melikhov, A. N., Bernshtein L. S., Korovin, S. Ya.: In book: Situational advising systems with fuzzy logic. Location: Moscow: Nauka, 1990.
11. Zadeh, L.A.: The concept of a linguistic variable and its application to approximate reasoning. Inf. Sci. **8**(3), 199–249 (1975)
12. Mammadova, M. H., Jabrayilova, Z. G.: The intelligent monitoring and evaluation of the psychophysiological state of the ship crew in maritime transport. In: Proceedings of the International Conference on problems of Logistics, Management and Operation in the East-West Transport Corridor (PLMO 2021), pp. 257–259. IEEE, Baku, Azerbaijan (2021).

Methods and Applications in Medicine

Balance-Based Fuzzy Logic Approach for the Classification of Liver Diseases Due to Hepatitis C Virus

Dursun Ekmekci⊙ **and Shahnaz N. Shahbazova**⊙

Abstract Fuzzy logic can be successfully applied in the many fields of artificial intelligence, such as control, classification, clustering, and prediction. However, assigning the optimal values for the control parameters and designing the optimal fuzzy rule table is vital to the method's performance. Besides, increasing the number of variables increases the solution space combinatorically. This study proposes a novel method that uses the fuzzy logic approach. The method positions the inputs and outputs on the unit circle and aims to balance them at the circle origin. The efficiency and performance of the method were investigated in the problem of classification of Liver Diseases Due to Hepatitis C Virus. The results prove that the method can design successful systems for classification problems.

Keywords Evolutionary fuzzy classifier · Genetic algorithm · Traction power system

D. Ekmekci (✉)
Karabuk University, Karabuk 78050, Turkey
e-mail: dekmekci@karabuk.edu.tr

Sh. N. Shahbazova
Ministry of Science and Education of the Republic of Azerbaijan, Institute of Control Systems, 68 Bakhtiyar Bahabzadeh Str., Baku AZ1141, Azerbaijan
e-mail: shahbazova@gmail.com; shahnaz_shahbazova@unec.edu.az; shahbazova@cyber.az

Department of Digital Technologies and Applied Informatics, Azerbaijan State University of Economics, UNEC, Baku AZ1001, Azerbaijan

© The Author(s), under exclusive license to Springer Nature Switzerland AG 2023   157
Sh. N. Shahbazova et al. (eds.), *Recent Developments and the New Directions of Research, Foundations, and Applications*, Studies in Fuzziness and Soft Computing 423,
https://doi.org/10.1007/978-3-031-23476-7_15

1 Introduction

A fuzzy system or multi-valued logic is an approach that models the relationship between inputs and outputs by applying "if–then" fuzzy rules. Based on the fuzzy set theory proposed by Zadeh [1], information is modeled in linguistic form, and systematic calculations are made using Membership Functions (MFs) [2]. One of the fields in which the method is actively used in the literature is classification problems. Fuzzy logic was hybridized to design effective fuzzy classifiers with many intelligent methods such as Evolutionary Algorithms (EAs) [3], neural networks [4], and clustering techniques [5]. Contrary to the other machine learning methods, EAs provide a means to encode and evolve rule antecedent aggregation operators, different rule semantics, rule base aggregation operators, and defuzzification methods. EAs remain among the fewest knowledge acquisition schemes available to optimize the fuzzy system, allowing decision-makers to decide which components are fixed and which evolve according to the performance measures [6]. Due to these advantages, EAs have been widely preferred to optimize fuzzy classifiers in recent years.

The hybrid model, defined as the Evolutionary Fuzzy Classifier (EFC), is fundamentally a fuzzy system powered by a learning process that includes evolutionary computation strategies. Designing a fuzzy system with EAs is equivalent to coding it as a parameter structure and finding the parameter values that give the optimum fitness function. In this context, the first step in designing an efficient EFC is to decide which parts of the fuzzy model are subjected to evolutionary computation. Since designing of optimal fuzzy rule table depends on the relationship between inputs and outputs, these operations can be defined as "*learning*" processes. Determining the optimal parameter values for the input and output variables can be interpreted as the "*tuning*" process.

The main focus of many EFC designs is the problem of obtaining a compact and precise Fuzzy Rule-Based System (FRBS) from the examination data [7]. The inductive learning of classical FRBSs suffers from the exponential growth of the fuzzy rule search space. In this context, FRBS design can be considered a searching or optimization problem for proper solutions that only require an ample search space and a performance measure. Figure 1 illustrates the effectiveness of the Genetic Algorithm (GA) in choosing the optimum fuzzy rules required for efficient learning in FRBS design.

The tuning process, on the other hand, is a process that considers the whole information obtained. Operations do not focus on fuzzy rules or linguistic expressions but the appropriate MFs for the determined fuzzy sets and the parameter optimization of these MFs [8].

Many intelligent solutions have been proposed to problems in different engineering disciplines using fuzzy logic. EA-based computational intelligence techniques are often preferred in the optimal design of the systems. These techniques generally focus on two leading solutions, such as parameter optimization or optimization of fuzzy rule tables. However, considering both together may be more advantageous in understanding the selected problem's general characteristics. [1]

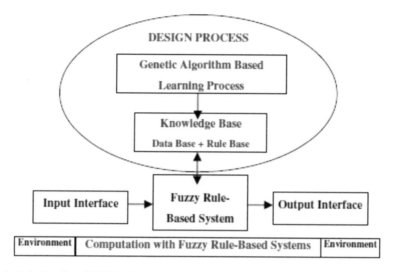

Fig. 1 Hybridization of FRBS with Genetic Algorithm [6]

presents a solution technique prepared in this direction and capable of producing successful results. This study presents a fuzzy logic-based solution that considers input and output variables together and aims to balance them all on a unit circle on the same plane. In the proposed method, firstly, weights are determined for each input. These weights are then positioned north of the unit circle, and the resultant weight is calculated relative to the circle's origin. The weight in the south, which balances this resultant weight at the circle's origin, is calculated as the output. The iterative operations search for the parameters of the membership functions that determine the inputs' weights, and the optimal positions of these weights on the circle. The method has been used to classify liver diseases due to hepatitis C. The results from the experiments have proven that the method produces successful results at negligible error levels for classification.

The remainder of the paper is designed as follows: In Sect. 2, EFCs designed with different computational intelligence-based techniques are presented, with emphasis on EA-based suggestions. Section 3 briefly explains the effect of the hepatitis virus on liver diseases and categorizes liver diseases due to this virus. In Sect. 4, the proposed method is explained in detail, and in Sect. 5, it is applied to the classification problem of hepatitis virus-related liver diseases. In Sect. 6, the study is concluded.

2 Related Studies: A Brief Review

One of the widely used heuristics in EFC design is GA. This hybrid approach has brought intelligent solutions to classification problems in many fields such as engineering, software, healthcare, commerce, government, and education [3]. Some of these studies, while fixing MFs, [9, 10] focused on rule optimization for data mining, control problems [11], modeling [12], and classification [13]. On the other hand, some studies [14, 15] sought a solution to MF tuning by fixing the rules. In [16], adaptive inference systems have been developed to achieve higher cooperation among the fuzzy rules and, in this context, more accurate fuzzy models without losing linguistic rule interpretability. On the other hand, the most used technique in practice due to its good performance and efficiency is the defuzzification method, [6] where the defuzzification function is applied to each derived fuzzy rule set, and their weighted average operator is calculated [17]. However, fuzzy methods are systems that must be considered together with MFs and rules. Therefore, some studies [18] have optimized these two problems together.

Shi et al. [19] proposed a scalable solution for complex problems that determine the optimal number of rules, selects the MFs type and tunes their parameters. In some studies [20, 21], the fuzzy model is handled hierarchically, and the modules of the system's architecture, the input variables in each module, the interactions between the modules, and the rules of each module are defined [22].

3 Liver Diseases Due to Hepatitis C Virus

Hepatitis C virus (HCV) is an RNA virus that was first identified by cloning in 1989 [23]. Microbiologists emphasize that the Hepatitis C Virus (HCV) is the most crucial cause of hepatitis, cirrhosis, and liver cancer worldwide. Less than 5% of those with HCV infection under the age of 40 develop cirrhosis within 20 years, while the probability of developing cirrhosis in those over 40 is 20% [24]. As a result, 25% of patients with chronic HCV hepatitis develop cirrhosis, and a significant proportion develops Hepatocellular Carcinoma (HCC). Worldwide, 27% of cirrhosis and 25% of HCC are HCV-associated [25]. An accurate definition of the liver fibrosis stage is important in planning the treatment and in the follow-up of chronic liver disease [26]. The biopsy is the standard method used to determine the stage of fibrosis. However, the biopsy is an invasive procedure with a risk of complications [27]. In addition, it may give false results in some patients [28]. These reasons limit the routine use of biopsy in the follow-up of the liver fibrosis stage.

4 Proposed Method

This section introduces a proposed new EFC method. The basic approach of the method, its mathematics, the selected MF type, in terms of evolutionary computation, the possible solution vector, and its components are explained.

4.1 Balance-Based Fuzzy Logic Approach

The main objective of the proposed method is to ensure that the inputs and output are balanced. In the method, firstly, the weight of each input is determined by MFs of gauss2mf type. These weights are then placed north of the unit circle, and the Horizontal Resultant (HR) and Vertical Resultant (VR) weights are calculated. Finally, the opposite value to the resultant horizontal and vertical weights, which balances all these inputs at the circle's origin, is defined as the output. The proposed method is illustrated in Fig. 2.

The "*Gaussian combination membership function (gauss2mf)*" [29] type MF is used to compute the input weights. The gauss2mf is obtained as a combination of two Gaussian functions whose formula is given in Eq. (1). Each Gaussian function defines the shape of one side of the MF. gauss2mf is a more successful MF compared to others [30, 31].

$$f(x : \sigma, c) = e^{\frac{-(x-c)^2}{2\sigma^2}} \tag{1}$$

In (1), σ represents the standard deviation, and c represents the Gaussian function's average. The membership value is calculated for x. Usage of gauss2mf in MATLAB is as shown in (2).

$$y = gauss2mf(x, [\sigma_1 c_1 \sigma_2 c_2]) \tag{2}$$

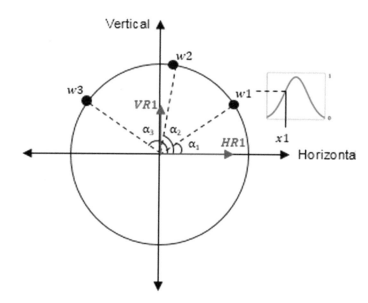

Fig. 2 Illustration of the proposed method

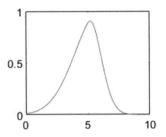

Fig. 3 gauss2mf with the parameters $\sigma_1 = 2, c_1 = 6, \sigma_2 = 1, c_2 = 5$

σ_1 and c_1 in (2) are the parameters for the left curve, while σ_2 and c_2 are used for the right curve. Figure 3 shows the gauss2mf plotted with parameters $\sigma_1 = 2, c_1 = 6, \sigma_2 = 1, c_2 = 5$.

The calculated weights for all inputs are randomly positioned on the north of the unit circle. Then, the HR and VR weights are calculated. The effects of w1 shown in Fig. 2 on the HR and the VR are given in (3) and (4), respectively.

$$HR1 = \cos \alpha_1 \tag{3}$$

$$VR1 = \sin \alpha_1 \tag{4}$$

In a system with n inputs, the HR and VR weights are calculated with (5) and (6), respectively.

$$HR = \sum_{i=1}^{n} \cos \alpha_i \tag{5}$$

$$VR = \sum_{i=1}^{n} \sin \alpha_i \tag{6}$$

In Fig. 4, the resultant weights and output are schematized. In Fig. 4, the resultant of HR and VR is shown as the hypotenuse. The hypotenuse is calculated by (7).

$$hypotenuse = \sqrt{HR^2 + VR^2} \tag{7}$$

After calculating the hypotenuse, the Beta between the hypotenuse and VR is calculated by (8).

$$\beta = \arcsin(HR/hypotenuse) \tag{8}$$

In the method, while the inputs are located in the north, the output that will balance the common resultant of all inputs is located in the south. In this context, the output lies in the hypotenuse symmetrical concerning the origin.

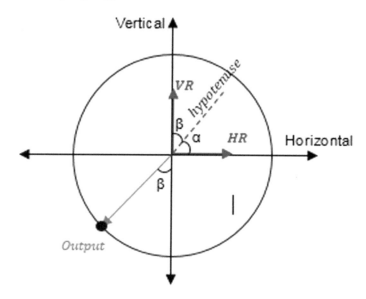

Fig. 4 Example resultant weights and the calculated output according to them

While determining the output value, the VR value is considered. The projection of the output on the vertical axis should balance the VR. Therefore, the output is calculated by (9).

$$output = VR/\cos\beta \qquad (9)$$

The outputs obtained by the model are compared with the targets, and the total error for m samples is calculated by (10).

$$E = \sum_{i=1}^{m} abs(output(i) - t\arg et(i)) \qquad (10)$$

4.2 Artificial Chromosome Structure for GA Procedures

In the proposed balance-based fuzzy logic approach, the gauss2mf parameters and the positions of the input weights on the circle are assigned by GA.

Accordingly, the artificial chromosome (V) structure designed for the balance-based fuzzy logic model in solving a problem with n inputs is presented in Fig. 5.

As seen in Fig. 5, each V consists of $5*n$ items. The first n items of V determine the positions of the input weights on the circle. In this context, the input weights are positioned with the angle values calculated by (9).

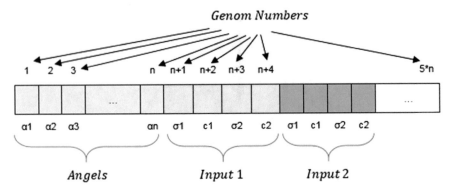

Fig. 5 An artificial chromosome example designed for n inputs

$$\alpha_i = 2 * \pi * V_i \quad i = 1, 2, 3, \ldots, n \tag{11}$$

The objective function of the method is to minimize the total error calculated by (8).

5 Experimental Study

In this section, the implementation of the proposed method for classifying liver diseases due to the hepatitis C virus is presented in detail. First, the selected dataset and the preliminaries applied to the dataset are explained. Then, the design of the proposed method for solving the problem, the parameter settings for GA, and the running environment are explained. Finally, the results from the experiments are presented in tables and discussed in detail.

5.1 Prepared Dataset

The data set used in the study was taken from the link https://archive.ics.uci.edu/ml/datasets/HCV+data. Data were recruited as controls from four hundred healthy volunteer blood donors at the Hannover Medical School. Volunteers were thoroughly examined to rule out health problems. The data set contains laboratory values of blood donors and Hepatitis C patients and demographic values like age. To detail the data set, [32] and [33] can be investigated. In the study, 40 samples selected from the data set were used.

The following eleven parameters in the dataset are designated as inputs: age, albumin (ALB), alkaline phosphatase (ALP), alanine aminotransferase (ALT), aspartate aminotransferase (AST), bilirubin (BIL), choline esterase (CHE), cholesterol

Table 1 The lower bounds, upper bounds, and center points defined for target	Values	Hepatitis	Fibrosis	Cirrhosis
	Lover bound	0	0.3333	0.6667
	Upper bound	0.3333	0.6667	1
	Center	0.1667	0.5	0.8333

(CHOL), CREA, γ-glutamyl-transferase (GGT), total protein (PROT). The target attribute for classification is Category ('just' Hepatitis C, Fibrosis, Cirrhosis).

Each input parameter and target are normalized with (10) and reduced to the range of [0–1].

$$x_{norm} = \frac{x - \min(X)}{\max(X) - \min(X)} \tag{12}$$

where X is a numerical array, and x is any item of the X.

5.2 Implementation of the Proposed Method to the Problem

The proposed technique was coded in the MATLAB R2017b and run on a machine having the Intel(R) Core (TM) i7-4710MQ 2.50 GHz processor with 8 GB RAM and Windows 8 operating system.

The target must correspond to either Hepatitis, Fibrosis, or Cirrhosis. For these linguistic expressions, the lower bound, upper bound, and center points determined in the range [0–1] are given in Table 1. In the proposed method, the aim is that the output produced for classification reaches the center point determined for the target.

As explained in the previous section, eleven inputs are in the prepared data set.

A solution vector consisting of 55 items is required for the gauss2mf parameter values that determine the weights of each input and the angles expressing the position of these weights on the circle.

The artificial chromosome structure for GA is shown in Fig. 6. All elements of the artificial chromosome are in the range of [0–1]. The coding prepared in the MATLAB platform and used as a fitness function is shown in Fig. 7.

5.3 Parameter Settings

In the study, MATLAB's "*optimtool*" toolbox was used for GA. The values of the algorithm parameters are given in Table 2.

The application was run 30 times independently with the parameter values given in Table 2, and the obtained results were saved. The maximum cycle number (MCN) in each experiment assign to 1000.

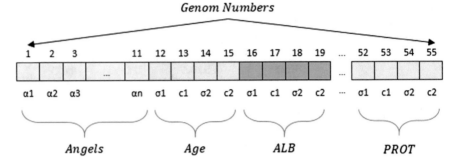

Fig. 6 An artificial chromosome used in the proposed method

```
function y = Balance(solution)
    % size of solution = [1x55]
    % size of DataSet=[40x12]
    DataSet=[...];

    angles(1:11)=solution(1:11);
    error=0;
    for i=1:40
        Vv=0;
        Hv=0;
        for j=1:11
            ParamsOfMF=[solution((j-1)*4+12) solution((j-1)*4+13)...
                solution((j-1)*4+14) solution((j-1)*4+15)];
            Vv=Vv+gauss2mf(DataSet(i,j),ParamsOfMF)*sin(pi*angles(j));
            Hv=Hv+gauss2mf(DataSet(i,j),ParamsOfMF)*cos(pi*angles(j));
        end;

        hypotenuse=sqrt(Vv^2+Hv^2);
        beta=acos(Vv/hypotenuse);
        output(i)=Vv/cos(beta);

        error=error+abs(output(i)-DataSet(i,12));
    end;
    y=error;
end
```

Fig. 7 The Fitness function prepared in MATLAB

Table 2 Parameter settings of GA

Procedure	Type	Value
Population	Double vector	200
Fitness scaling	Rank	
Elite count		10
Crossover	Constraint dependent	0.8
Mutation	Constraint dependent	0.1

5.4 *Results*

In this subsection, the best result of the proposed model in 30 trials is shared and discussed in detail, and the method's performance is interpreted.

The best EFC design obtained from the experiments achieved a total error value of 1.5361 for 40 samples. In the design, the positions of the weights of the 11 inputs in the north of the circle are shown in Fig. 8.

The *gauss2mf* parameters calculating the input weights given in Fig. 8 are given in Table 3.

The results obtained by the proposed method for the samples used in training are shown in Fig. 9. In Fig. 9, blue colored circles represent targets, red colored stars output, and notice that green lines represent the borders of the targets.

As seen in Fig. 9, the proposed method has achieved successful classification results. Only 3 out of 40 classification samples are incorrect. The 10th sample, which should be hepatitis, was classified as fibrosis, and the 17th and 21st samples, which should be fibrosis, were classified as hepatitis.

In the classification made with the proposed method, the total error value calculated by Eq. (10) for 40 samples is 1.5361. The mean error is 0.0384, and the standard

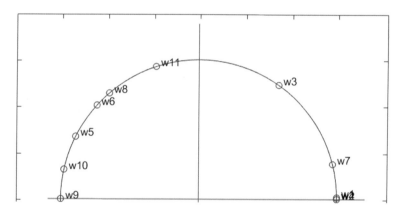

Fig. 8 The positions of the inputs on the unit circle

Table 3 The gauss2mf parameters calculating the input weights in the EFC design

Input No	$\sigma 1$	$c1$	$\sigma 2$	$c2$	Input No	$\sigma 1$	$c1$	$\sigma 2$	$c2$
1	0.6409	0.8603	0.0043	0.0039	7	0.6194	0.0595	0.0011	0.2844
2	0.6699	0.0564	0.9917	0.5883	8	0.0159	0.3753	0.0539	0.2823
3	0.2110	0.8780	0.2527	0.0657	9	0.9933	0.0920	0.0869	0.7446
4	0.0759	0.8304	0.1196	0.8409	10	0.3923	1.0000	0.0143	0.2327
5	0.4205	0.9992	0.5236	0.0118	11	0.0432	1.0000	0.0531	0.8729
6	0.0065	0.5625	0.6969	0.0458					

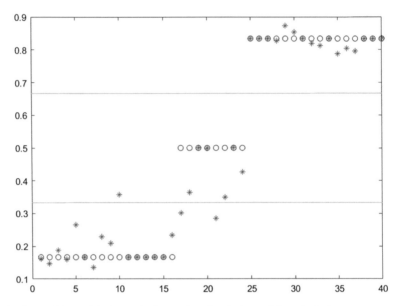

Fig. 9 The positions of the inputs in the northern hemisphere of the unit circle

deviation is 0.0595. Determined the maximum error was 0.2142, and an error less than 0.0001 was obtained in 14 samples.

6 Conclusion and Future Studies

In the study, a new fuzzy classification method is introduced. The method considers all input and output variables together. For the interaction between the inputs, the positions at the northern arc of the circle and the balance at the circle's origin are considered for the interaction between the input–output. The method eliminates the difficulty of designing the fuzzy rule table.

Experiments to classify liver diseases due to the hepatitis C virus show that the method can make successful classifications. Achieved negligible errors in the experiments.

In future studies, the method's performance in an artificial intelligence problem with multiple outputs is being investigated, where the classical fuzzy logic approach can be applied. In this context, the outputs will be positioned in the southern arc of the circle. In addition, the problem will be considered a multi-objective optimization problem, and a solution will be sought so that each output represents a different objective.

References

1. Zadeh, L.A.: Fuzzy sets. Inf Control **8**, 338–353 (1965)
2. Jang, J.S.R., Sun, C.T., Mizutani, E.: Neuro-fuzzy and soft computing-a computational approach to learning and machine intelligence [book review]. IEEE Trans Automat Contr **42**, 1482–1484 (1997). https://doi.org/10.1109/TAC.1997.633847
3. Dennis, B., Muthukrishnan, S.: AGFS: adaptive genetic fuzzy system for medical data classification. Appl Soft Comput **25**, 242–252 (2014). https://doi.org/10.1016/j.asoc.2014.09.032
4. Chakraborty, D., Pal, N.R.: A neuro-fuzzy scheme for simultaneous feature selection and fuzzy rule-based classification. IEEE Trans. Neural Netw. **15**, 110–123 (2004). https://doi.org/10.1109/TNN.2003.820557
5. Lee, C.-Y., Lin, C.-J., Chen, H.-J.: A self-constructing fuzzy CMAC model and its applications. Inf. Sci. (NY) **177**, 264–280 (2007). https://doi.org/10.1016/j.ins.2006.03.010
6. Herrera, F.: Genetic fuzzy systems: taxonomy, current research trends and prospects. Evol. Intell. **1**, 27–46 (2008). https://doi.org/10.1007/s12065-007-0001-5
7. Beloufa, F., Chikh, M.A.: Design of fuzzy classifier for diabetes disease using modified artificial bee colony algorithm. Comput. Methods Programs Biomed. **112**, 92–103 (2013). https://doi.org/10.1016/j.cmpb.2013.07.009
8. Casillas, J., Cordon, O., del Jesus, M.J., Herrera, F.: Genetic tuning of fuzzy rule deep structures preserving interpretability and its interaction with fuzzy rule set reduction. IEEE Trans. Fuzzy Syst. **13**, 13–29 (2005). https://doi.org/10.1109/TFUZZ.2004.839670
9. Thrift, P. R.: Fuzzy logic synthesis with genetic algorithms. In ICGA, pp. 509-513 (1991)
10. Wu, C.J., Liu, G.Y.: Design of fuzzy logic controllers using genetic algorithms. In: Proceedings of the IEEE International Conference on Systems, Man and Cybernetics, pp. 1383–1388 (1999)
11. Britain, G., Press, P., Akron, A.: Model-based Fuzzy Control. In: Advanced Control of Industrial Processes, pp. 33–106. Springer, London (2007)
12. Pedrycz W: Fuzzy Modelling, 1996th edi. Springer US, Boston, MA (1996)
13. Au, W.-H., Chan, K.C.C., Wong, A.K.C.: A fuzzy approach to partitioning continuous attributes for classification. IEEE Trans. Knowl. Data Eng. **18**, 715–719 (2006). https://doi.org/10.1109/TKDE.2006.70
14. Karr, C.L., Gentry, E.J.: Fuzzy control of pH using genetic algorithms. IEEE Trans. Fuzzy Syst. **1**, 46 (1993). https://doi.org/10.1109/TFUZZ.1993.390283
15. Botta, A., Lazzerini, B., Marcelloni, F., Stefanescu, D.C.: Context adaptation of fuzzy systems through a multi-objective evolutionary approach based on a novel interpretability index. Soft Comput. **13**, 437–449 (2009). https://doi.org/10.1007/s00500-008-0360-6
16. Alcalá-Fdez, J., Herrera, F., Márquez, F., Peregrín, A.: Increasing fuzzy rules cooperation based on evolutionary adaptive inference systems. Int. J. Intell. Syst. **22**, 1035–1064 (2007). https://doi.org/10.1002/int.20237
17. Kim, D., Choi, Y.-S., Lee, S.-Y.: An accurate COG defuzzifier design using Lamarckian co-adaptation of learning and evolution. Fuzzy Sets Syst. **130**, 207–225 (2002). https://doi.org/10.1016/S0165-0114(01)00167-1
18. Homaifar, A., McCormick, E.: Simultaneous design of membership functions and rule sets for fuzzy controllers using genetic algorithms. IEEE Trans. Fuzzy Syst. **3**, 129–139 (1995). https://doi.org/10.1109/91.388168
19. Shi, Y., Eberhart, R., Chen, Y.: Implementation of evolutionary fuzzy systems. IEEE Trans Fuzzy Syst. **7**, 109–119 (1999). https://doi.org/10.1109/91.755393
20. Cordón, O., Herrera, F., Zwir, I.: A hierarchical knowledge-based environment for linguistic modeling: models and iterative methodology. Fuzzy Sets Syst. **138**, 307–341 (2003). https://doi.org/10.1016/S0165-0114(02)00388-3
21. Joo, M.G., Lee, J.S.: A class of hierarchical fuzzy systems with constraints on the fuzzy rules. IEEE Trans. Fuzzy Syst. **13**, 194–203 (2005). https://doi.org/10.1109/TFUZZ.2004.840096
22. Shukla, P.K., Tripathi, S.P.: A survey on interpretability-accuracy (I-A) trade-off in evolutionary fuzzy systems. In: 2011 Fifth International Conference on Genetic and Evolutionary Computing. IEEE, pp. 97–101 (2011)

23. Barut, H.Ş, Günal, Ö.: Dünyada ve Ülkemizde Hepatit C Epidemiyolojisi. Klimik Derg **22**, 38–43 (2009)
24. Lavanchy, D.: The global burden of hepatitis C. Liver Int. **29**, 74–81 (2009). https://doi.org/10. 1111/j.1478-3231.2008.01934.x
25. Alter, M.J.: Epidemiology of hepatitis C virus infection. World J. Gastroenterol **13**, 2436 (2007). https://doi.org/10.3748/wjg.v13.i17.2436
26. Davoudi, Y., Layegh, P., Sima, H., et al.: Diagnostic value of conventional and doppler ultrasound findings in liver fibrosis in patients with chronic viral hepatitis. J. Med. Ultrasound **23**, 123–128 (2015). https://doi.org/10.1016/j.jmu.2014.10.002
27. Söker, G., Gülek, B., Öztürk, A.B., et al.: The value of fibrosis index in discrimination of chronic hepatitis and cirrhosis. Haseki Tıp Bülteni **55**, 212–215 (2017). https://doi.org/10.4274/haseki. 57966
28. Bernatik, T., Strobel, D., Hahn, E.G., Becker, D.: Doppler measurements: a surrogate marker of liver fibrosis? Eur. J. Gastroenterol Hepatol. **14**, 383–387 (2002). https://doi.org/10.1097/ 00042737-200204000-00008
29. Matlab: Fuzzy Logic Toolbox ^{TM} User's Guide R 2021a, 1995th–2021st ed. The MathWorks, Inc. (2014)
30. Zheng, H., Jiang, B., Lu, H.: An adaptive neural-fuzzy inference system (ANFIS) for detection of bruises on Chinese bayberry (Myrica rubra) based on fractal dimension and RGB intensity color. J. Food Eng. **104**, 663–667 (2011). https://doi.org/10.1016/j.jfoodeng.2011.01.031
31. Marjani, A., Babanezhad, M., Shirazian, S.: Application of adaptive network-based fuzzy inference system (ANFIS) in the numerical investigation of Cu/water nanofluid convective flow. Case Stud. Therm. Eng. **22**, 100793 (2020). https://doi.org/10.1016/j.csite.2020.100793
32. Lichtinghagen, R., Pietsch, D., Bantel, H., et al.: The enhanced liver fibrosis (ELF) score: normal values, influence factors and proposed cut-off values. J. Hepatol. **59**, 236–242 (2013). https://doi.org/10.1016/j.jhep.2013.03.016
33. Hoffmann, G., Bietenbeck, A., Lichtinghagen, R., Klawonn, F.: Using machine learning techniques to generate laboratory diagnostic pathways—a case study. J. Lab. Precis. Med. **3**, 58–58 (2018). https://doi.org/10.21037/jlpm.2018.06.01

Scale-Invariance-Based Pre-processing Drastically Improves Neural Network Learning: Case Study of Diagnosing Lung Dysfunction in Children

Nancy Avila, Julio Urenda, Nelly Gordillo, Vladik Kreinovich, and Shahnaz N. Shahbazova ⓘD

Abstract To adequately treat different types of lung dysfunctions in children, it is important to properly diagnose the corresponding dysfunction, and this is not an easy task. Neural networks have been trained to perform this diagnosis, but they are not perfect in diagnostics: their success rate is 60%. In this paper, we show that by selecting an appropriate invariance-based pre-processing, we can drastically improve the diagnostic success, to 100% for diagnosing the presence of a lung dysfunction.

Keywords Lung dysfunction in children · Scale invariance · Neural networks

1 Formulation of the Problem

Lung dysfunctions. One of the major lung dysfunctions is asthma, a long-term inflammatory disease of the airways of the lungs. It is characterized by recurring airflow obstruction, bronchospasms, wheezing, coughing, chest tightness, and short-

N. Avila · J. Urenda · V. Kreinovich (✉)
University of Texas, 500 W. University, El Paso, Texas 79968, USA
e-mail: vladik@utep.edu

N. Avila
e-mail: nsavila@utep.edu

J. Urenda
e-mail: jcurenda@utep.edu

N. Gordillo
Department of Electrical and Computer Engineering, Universidad Autónoma de Ciudad Juárez, Ave. del Charro No. 450 Norte, Ciudad Juárez, Chihuahua 32310, Mexico
e-mail: nelly.gordillo@uacj.mx

Sh. N. Shahbazova
Department of Digital Technologies and Applied Informatics, Azerbaijan State University of Economics, UNEC, Baku AZ1001, Azerbaijan
e-mail: shahbazova@gmail.com; shahnaz_shahbazova@unec.edu.az; shahbazova@cyber.az

Ministry of Science and Education of the Republic of Azerbaijan, Institute of Control Systems, 68 Bakhtiyar Bahabzadeh Street, Baku AZ1141, Azerbaijan

171

Sh. N. Shahbazova et al. (eds.), *Recent Developments and the New Directions of Research, Foundations, and Applications*, Studies in Fuzziness and Soft Computing 423,
https://doi.org/10.1007/978-3-031-23476-7_16

ness of breath. These episodes may occur a few times a day or a few times per week [22].

Asthma may be preceded by Small Airway Impairment (SAI), a chronic obstructive bronchitis. If inflammation persists during SAI, it could cause asthma.

SAI, in its turn, may be preceded by a less severe condition that medical doctors classify as Possible Small Airways Impairment (PSAI).

Diagnostics of different lung dysfunctions is difficult but important. All lung dysfunctions lead to similar symptoms like wheezing, coughing, etc. As a result, it is difficult to distinguish between these dysfunctions—and it is also difficult to distinguish these chronic dysfunctions from a common short-term respiratory disease.

However, the diagnosing of these diseases is very important, because in general, for different diseases, different treatments are efficient.

How different dysfunctions are diagnosed now. Since it is difficult to diagnose different dysfunctions solely based on the symptoms, the corresponding diagnostics involves measuring airflow in different situations. The most effective diagnostic comes from active measurements—spirometry. A patient is asked to deeply inhale, to hold their breath, and to exhale as fully as possible—and the corresponding instrument is measuring the airflow following all these instructions. Based on these measurements, symptoms, and clinical history, medical doctors come up with a diagnosis of different dysfunctions.

Children diagnostics: a serious problem. Unfortunately, the spirometry technique described above does not work with little children, especially children of pre-school age, since it not easy to make them follow the corresponding instructions; see, e.g., [16]. The same problem occurs with elderly patients and patients with certain limitations.

An additional problem is that even when children follow instructions during the spirometry testing, spirometry results are not sensitive enough to detect obstruction of small airways (2 mm or less in diameter); see, e.g., [6, 11, 12, 15, 19].

How can we diagnose children: main idea of the corresponding measurements. Since we cannot use active measuring techniques, techniques that require children's active participation, we have to rely on passive techniques, i.e., techniques that do not require such participation. What we can do is change the incoming airflow and measure how that affects the outcoming airflow.

Passive measurements: details. The easiest way of changing the airflow is to switch a certain extra amount of airflow on or off. This is the main idea behind the Impulse Oscillometry System (IOS); see, e.g., [6]. In a usual IOS, the additional airflow is switched on and off with a period 5 Hz, meaning that we have a 0.1 sec period with extra flow, 0.1 sec period without, then again a 0.1 sec period with extra flow, etc. The system then measures the resulting outflow $y(t)$.

In real-life clinical environment, the measurement result are affected by noise. Because of this noise, the measured values $\widetilde{y}(t)$ somewhat deviate from the actual (unknown) flow results $y(t)$. The deviations $\widetilde{y}(t) - y(t)$ measured at different moments of time t are usually caused by different factors and are, thus, statistically independent. As a result of this independence, the noise is heavily oscillating—i.e., changes with high frequency. To decrease the effect of this noise, it is therefore reasonable to take a Fourier transform and ignore high-frequency components of this transform—since these components are heavily corrupted by noise.

Since the input signal is periodic, with the period 5 Hz, we expect the output signal to also be periodic, with the same frequency. In general, when we perform Fourier transform on a signal which is periodic with frequency f, we only get components corresponding to multiples of f, i.e., to f, $2f$, $3f$, etc. In our case, this means that we will have components corresponding 5, 10 Hz, etc. In practice, it was discovered that components above 25 Hz are too noisy to be useful—actually, the most informative values are one corresponding to 5–15 Hz range. The values corresponding to 20 and 25 Hz are also useful, but they are somewhat less informative that the 5–15 Hz values. So, the system returns the components corresponding to 5, 10, 15, 20, 25 Hz. To be on the safe side, the system also returns the component corresponding 35 Hz, which sometimes adds some additional information.

Another problem is related to the fact that while it is relatively easy to implement the on-off switching of the input airflow, the actual values of the on- and off-case airflows may change with time: the pressure in the system may decrease, etc. In principle, it is possible to maintain the exact airflow values, but this will make the system too complicated and thus, too expensive. Instead, the existing systems rely on the fact that while it is not easy to maintain the input airflow $a(t)$ at some pre-defined level a_0, it is possible to accurately measure this airflow.

The added airflow $a(t) - a_0$ is relatively small, so when estimating the reaction $y(t)$ of a human breathing system to this airflow, we can safely ignore terms which are quadratic and of higher order in terms of $a(t) - a_0$ and conclude that the dependence is linear: $y(t) = \int c(t, s) \cdot a(s) \, ds$ for some coefficients $c(t, s)$.

The system does not change much during the time when measurements are performed. So if we start the experiment t_0 seconds earlier, i.e., if we take $\tilde{a}(t) = a(t + t_0)$ instead of $a(t)$, then the output should change accordingly, to $\tilde{y}(t) = y(t + t_0)$. So, on the one hand, we have

$$\tilde{y}(t) = \int c(t, s) \cdot \tilde{a}(s) \, ds = \int c(t, s) \cdot a(s + t_0) \, ds,$$

which, if we introduce a new variable $\tilde{s} = s + t_0$ for which $d\tilde{s} = ds$, leads to

$$\tilde{y}(t) = \int c(t, \tilde{s} - t_0) \cdot a(\tilde{s}) \, d\tilde{s}.$$

On the other hand,

$$\tilde{y}(t) = y(t + t_0) = \int c(t + t_0, s) \cdot a(s) \, ds.$$

So, for all inputs $a(t)$, we should have

$$\int c(t + t_0, s) \cdot a(s) \, ds = \int c(t, s - t_0) \cdot a(s) \, ds.$$

Two linear functions coincide if the coefficients at all the unknown (in this case, $a(s)$) coincide, so we must have $c(t + t_0, s) = c(t, s - t_0)$. In particular, for every two values v_1 and v_2, we can take $s = t_0 = v_2$ and $t = v_1 - v_2$ and conclude that $c(v_1, v_2) = c(v_1 - v_2, 0)$, i.e., that $c(v_1, v_2) = z(v_1 - v_2)$, where we denoted $z(a) \stackrel{\text{def}}{=} c(a, 0)$. Substituting this expression for $c(v_1, v_2)$ into the formula that describes the relation between $a(t)$ and $y(t)$, we conclude that $y(t) = \int z(t - s) \cdot a(s) \, ds$. This is called a *convolution* of functions $z(t)$ and $a(t)$.

It is known that the Fourier transform of the convolution is equal to the product of Fourier transforms. Thus, in this case, for the corresponding Fourier transforms $Z(f)$, $Y(f)$, and $A(f)$, we get $Y(f) = Z(f) \cdot A(f)$. We are interested in the values $Z(f)$ that do not depend on the inputs. In our case, we have computed the values $Y(f)$, so we can also compute the Fourier coefficients $A(f)$, and return the ratios $Z(f) = Y(f)/A(f)$. This is exactly what the IOS system returns: the complex numbers $Z(f) = R(f) + i \cdot X(f)$ that correspond to six frequencies $f = 5, 10, \ldots, 35$. In analogy with electric circuits, the complex value $Z(f)$ is called the *impedance*, its real part $R(f)$ is called *resistance*, and its imaginary part $X(f)$ is called *reactance*.

These six complex numbers—or, equivalently, the six real parts and the six imaginary parts—are what we can use to properly diagnose the lung dysfunction.

It is not easy to make a diagnosis based on IOS data. If we plot the IOS data corresponding to patients with different diagnoses, we see that the corresponding ranges of values $R(f)$ and $X(f)$ have a huge intersection; see, e.g., Figs. 1 and 2. This shows that it is not easy to diagnose a patient based on IOS data.

How IOS-based diagnosis is performed now. There are no exact formulas that describe the diagnosis based on the six complex values $Z(f)$; the research about the clinical applications of the IOS parameters is still ongoing. However, we do have several patient records for which, on the one hand, we know the corresponding values $Z(f)$, and, on the other hand, we have a diagnosis provided by a skilled medical doctor. It is therefore reasonable to use machine learning and train the system to be able to diagnose a patient.

This has indeed been done: researchers have used either all or some of the 12 numbers (real parts $R(f)$ and imaginary parts $X(f)$) as input and tried to train the neural network to learn the diagnosis in the children patients.

The resulting diagnostic system is, however, not yet perfect. For adult patients, if we use spirometry results as well as IOS, we get an almost perfect separation of asthma from healthy: its accuracy is 98–99% [3, 4]. However, when we only use IOS data, the current system's testing-data accuracy in distinguishing lung dysfunctions such as asthma, SAI, and PSAI from patients who do not have any of these diseases is only close to 60% [5].

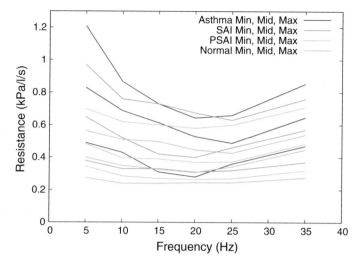

Fig. 1 Resistance maximum, middle and minimum patients' curves per class

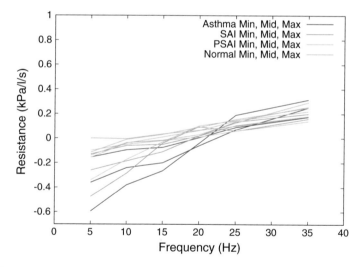

Fig. 2 Reactance maximum, middle and minimum patients' curves per class

Comment. Similar imperfect results were obtained for a related problem: predicting asthma deterioration one week ahead. For this problem, neural networks and other machine learning techniques result, at best, in 70–75% prediction accuracy; see, e.g., [14].

Resulting problem and what we do in this paper. It is therefore desirable to come up with better diagnostic techniques. Our approach is to help a neural network by providing an appropriate pre-processing of the input data. It turns out that an appropriate pre-processing indeed drastically improves the diagnostic results: for diagnosing the presence of a lung dysfunction, we have 100% accuracy.

2 First Pre-processing Stage: Scale-Invariant Smoothing

Need for further de-noising. The above-described filtering out of noisy high-frequency components eliminates some noise, but some noise remains.

Smoothing as a way to de-noise. To further decrease the noise level, it is desirable to take into account that in real life, almost all dependencies (including the dependence of the signal intensity on frequency) are smooth—in the sense that a small change in frequency leads to a small change in intensity.

How to smooth a signal. Thus, instead of considering the original (noisy) six complex numbers $Z(f)$ corresponding to $f = 5, 10, \ldots$, it makes sense to approximate these values by a smooth dependence, i.e., by a function of the type $\sum_{j=1}^{k} c_j \cdot e_j(f)$, for some smooth functions $e_1(f), \ldots, e_k(f)$.

Which level of smoothness to choose. Usually, real-life processes are very smooth—with few exceptions like phase transitions. It is therefore desirable to select approximating functions $e_j(f)$ which are as smooth as possible.

In general, for functions, there are several different degrees of smoothness. The simplest case is when a function is one time differentiable. The next—more smooth—case is when a function is two times differentiable, etc. Then, we have functions which are infinitely many time differentiable, and finally, the smoothest of all—analytical functions, functions that can be expanded in Taylor series. Thus, to achieve maximal smoothness, we will use analytical functions $e_j(x)$.

Which analytical functions should we choose: the idea of scale-invariance. There are many different analytical functions, which ones should we choose?

A natural requirement for this choice comes from the fact that we are approximating a function from numbers (f) to numbers (Z). These numbers represent the values of the corresponding physical quantities. However, the numerical value of each physical quantity depends not only on the quantity itself, it also depends on the measuring unit that we have selected for this quantity. If we replace the original measuring unit with a new one which is λ times smaller, all the numerical values will multiply by λ. In other words, for each frequency, instead of the original numerical value f, we will have a new numerical value $\widetilde{f} = \lambda \cdot f$ that describe the exact same physical quantity.

There is no physical reason why some measuring units would be preferable to others. Therefore, it makes sense to require that the selection of the resulting class C of linear combinations $\sum_{j=1}^{k} c_j \cdot e_j(f)$ should not change if we simply re-scale all the values by changing the measuring unit for frequencies.

The idea of scale-invariance is actively and successfully used in physics; see, e.g., [7, 20]. It is therefore reasonable to apply it to our problem as well.

In our case, scale-invariance means that the class C should be equal to the class $\widetilde{C}$ of all linear combinations $\sum_{j=1}^{k} c_j \cdot e_j(\widehat{f})$. In particular, this means that, for every i, the function $e_i(\lambda \cdot f)$ from the class $\widetilde{C}$ can be described as a linear combination $\sum_{j=1}^{k} c_{ji}(\lambda) \cdot e_j(f)$ of the original functions, with coefficients $c_{ji}(\lambda)$ depending on i and λ:

$$e_i(\lambda \cdot f) = \sum_{j=1}^{k} c_{ji}(\lambda) \cdot e_j(f). \tag{1}$$

Thus, we have a system of equations for the unknown functions $e_i(f)$. Let us solve this system.

Solving the corresponding system of equations. We know that the functions $e_i(f)$ are smooth. Let us show that the functions $c_{ji}(\lambda)$ are smooth as well. Indeed, for every i, by taking k different values $f_1, \ldots, f_k$ in the equation (1), we get a system of k linear equations with k unknowns $c_{1i}(\lambda), \ldots, c_{kj}(\lambda)$:

$$\sum_{j=1}^{k} c_{ji}(\lambda) \cdot e_j(f_1) = e_i(\lambda \cdot f_1);$$

$$\sum_{j=1}^{k} c_{ji}(\lambda) \cdot e_j(f_2) = e_i(\lambda \cdot f_2);$$

$$\cdots$$

$$\sum_{j=1}^{k} c_{ji}(\lambda) \cdot e_j(f_k) = e_i(\lambda \cdot f_k).$$

It is known that, in general, the solution to a system of linear equations can be described by Cramer's rule, as a ratio of two polynomials—and thus, a smooth function—depending on the coefficients and on the right-hand sides. In our case, the coefficients $e_j(f_1), \ldots, e_j(f_k)$ are constants—and thus, do not depend on λ at all, while the right-hand sides $e_i(\lambda \cdot f_1), \ldots, e_i(\lambda \cdot f_k)$ are smooth functions of λ. Thus, the solutions to this system of linear equations – i.e., the coefficients $c_{ji}(\lambda)$—are obtained by plugging smooth functions

$$e_j(f_1), \ldots, e_j(f_k), e_i(\lambda \cdot f_1), \ldots, e_i(\lambda \cdot f_k)$$

into a smooth expression (Cramer's rule) and are, thus, also smooth functions of λ.

Now, we can take the system of equations (1) corresponding to $i = 1, \ldots, k$:

$$e_1(\lambda \cdot f) = \sum_{j=1}^{k} c_{j1}(\lambda) \cdot e_j(f);$$

$$e_2(\lambda \cdot f) = \sum_{j=1}^{k} c_{j2}(\lambda) \cdot e_j(f);$$

$$\cdots$$

$$e_k(\lambda \cdot f) = \sum_{j=1}^{k} c_{jk}(\lambda) \cdot e_j(f).$$

All the expressions in this system are differentiable functions of λ. Thus, we can differentiate both sides of each equation by λ. As a result, we get the following system of equations:

$$f \cdot e_1'(\lambda \cdot f) = \sum_{j=1}^{k} c_{j1}'(\lambda) \cdot e_j(f);$$

$$f \cdot e_2'(\lambda \cdot f) = \sum_{j=1}^{k} c_{j2}'(\lambda) \cdot e_j(f);$$

$$\cdots$$

$$f \cdot e_k'(\lambda \cdot f) = \sum_{j=1}^{k} c_{jk}'(\lambda) \cdot e_j(f),$$

where e_i', as usual, denotes the derivative of the function e_i. If we substitute $\lambda = 1$ into these formulas, we get the following system of ordinary differential equations:

$$f \cdot e_1'(f) = \sum_{j=1}^{k} C_{j1} \cdot e_j(f);$$

$$f \cdot e_2'(f) = \sum_{j=1}^{k} C_{j2} \cdot e_j(f);$$

$$\cdots$$

$$f \cdot e_k'(f) = \sum_{j=1}^{k} C_{jk} \cdot e_j(f),$$

where we denoted $C_{ji} \stackrel{\text{def}}{=} c_{ji}'(1)$.

For each i, the expression $f \cdot e_i' = f \cdot \dfrac{de_i}{df}$ can be equivalently reformulated

as $\dfrac{de_i}{df/f}$. Here, $df/f = d(\ln(f))$. So, if we introduce a new variable $F = \ln(f)$

for which $f = \exp(F)$, then for the new functions $E_i(F) \overset{\text{def}}{=} e_i(\exp(F))$, we get

$f \cdot e_i'(f) = \dfrac{dE_i}{dF} = E_i'(F)$. Thus, in terms of the new functions, we get the following
system of equations:

$$E_1'(F) = \sum_{j=1}^{k} C_{j1} \cdot E_j(F);$$

$$E_2'(F) = \sum_{j=1}^{k} C_{j2} \cdot E_j(F);$$

$$\ldots$$

$$E_k'(F) = \sum_{j=1}^{k} C_{jk} \cdot E_j(F).$$

This is a system of linear differential equations with constant coefficients. It is known that a general solution to this system is a linear combination of terms $F^p \cdot \exp(\lambda \cdot F)$, where:

- λ is an eigenvalue of the matrix $\|C_{ji}\|$ (which is, in general, complex $\lambda = a + b \cdot i$), and
- the value $p \neq 0$ appears when we have a duplicate eigenvalue—in this case, p is a non-negative integer smaller than the dimension of the corresponding eigenspace.

In terms of real values, we get

$$F^p \cdot \exp(\lambda \cdot F) = F^p \cdot \exp((a + b \cdot i) \cdot F)$$
$$= F^p \cdot \exp(a \cdot F) \cdot (\cos(b \cdot F) + i \cdot \sin(b \cdot F)).$$

Substituting $F = \ln(f)$ into this expression, we conclude that the functions $e_i(f)$ are linear combinations of the expressions

$$(\ln(f))^p \cdot \exp(a \cdot \ln(f)) \cdot (\cos(b \cdot \ln(f)) + i \cdot \sin(b \cdot \ln(f))).$$

Here, $\exp(a \cdot \ln(f)) = (\exp(\ln(f)))^a = f^a$, so the above expression has the form

$$(\ln(f))^p \cdot f^a \cdot (\cos(b \cdot \ln(f)) + i \cdot \sin(b \cdot \ln(f))).$$

Let us take into account that the functions $e_i(f)$ should be analytical. Now, we can take into account that the functions $e_i(f)$ should be analytical, i.e., they should be expandable in Taylor series for $f = 0$. This requirement excludes possible logarithmic terms $(\ln(f))^p$, as well as cosines and sines of these logarithms, which leaves us with linear combinations of the powers f^a. Due to analyticity, all the powers should be natural numbers, so we conclude that all the functions $e_i(f)$ are linear combinations of expressions $f^0 = 1, f^1 = f, f^2, \dots$. In other words, due to scale-invariance, all the functions $e_i(f)$ should be polynomials.

We want to approximate the function $Z(f)$ by a linear combination of the functions $e_i(f)$. A linear combination of polynomials is also a polynomial. Thus, we arrive at the following conclusion.

General conclusion of this section. Due to the natural requirement of scale-invariance, we should approximate the impedance function $Z(f)$ by a polynomial.

3   Which Order Polynomials Should We Use?

Formulation of the problem. We want to find the polynomial that fits the observations. Of course, if we take a polynomial of a sufficiently large degree, we can always find a polynomial that fits all observed data exactly—this is a well-known Lagrange interpolation polynomial.

However, the whole purpose of the polynomial smoothing is to de-noise the signal, and if we keep all the values intact, we will retain all the noise. Thus, we should not use polynomials of too high order.

On the other hand, if we use polynomials of too low order—e.g., constants or linear functions—we get a smoothing, but the approximation is too crude, and we lose the information contained in the original signal. How can we find the adequate degree of approximating polynomials?

Natural idea: general case. A natural idea is to take into account the general monotonicity of IOS curves—as described, e.g., in [11]—and select the higher order of approximating polynomials that preserve this monotonicity.

Let us apply this general idea to our problem. According to [11], for all three diseases (asthma, SAI, and PSAI):

- for the real part $R(f)$ of the impedance $Z(f)$, the corresponding value first decreases with frequency f, and then increases;
- for the imaginary part $X(f)$ of the impedance $Z(f)$, the corresponding value increases with frequency f.

So:

- for each degree, we use the usual Least Squares techniques (see, e.g., [18]) to find the polynomial of this degree that best approximates the observed values, and then

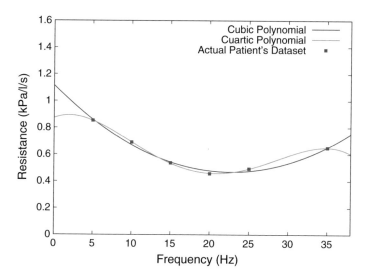

Fig. 3 Cubic versus cuartic versus a patient's dataset

- we select the largest degree for which, on the corresponding interval of values of frequency, the resulting best-approximation polynomials follow the same monotonicity pattern.

It turns out that:

- for quadratic and cubic polynomials, we have this feature, but
- for 4th order polynomials, we no longer have the desired monotonicity: for example, for the resistance $R(f)$, the corresponding 4th order polynomial first decreases, then increases, but then decreases again; see, e.g., Fig. 3.

Because of this, in this paper, we approximate the functions $R(f)$ and $X(f)$ by *cubic polynomials*. Let us denote the corresponding approximating polynomials by $R^a(f)$ and $X^a(f)$.

4 Second Pre-processing Stage: Using the Approximating Polynomials to Distinguish Between Different Diseases

Formulation of the problem. For each diagnosis d, we have several observations corresponding to patients with these diagnosis. Based on these observations, for each patient i with this diagnosis ($i \in d$), we find the smoothed functions $R_i^a(f)$ and $X_i^a(f)$.

Now, when we have a new patient with the corresponding functions $R^a(f)$ and $X^a(f)$, we would like to diagnose this patient, i.e., to classify this patient to one of the groups d. How can we do it?

How to separate different groups: general idea. How do we distinguish groups in general? How do we distinguish cats from dogs? Usually, in such situations:

- we have a mental picture of a typical cat,
- we have a mental picture of a typical dog, and
- we make our decision based on how similar the observed object is to one of these two typical ones.

A natural idea is thus:

- to form, for each group corresponding to a given diagnosis d, "typical" function $R_d(f)$ and $X_d(f)$ corresponding to this diagnosis, and then
- to base our diagnosis of new, yet-undiagnosed patients based on how similar their functions $R^a(f)$ and $X^a(f)$ are to the typical functions corresponding to each diagnosis.

A natural way to form a typical function. A natural way to form the typical function $R_d(f)$ corresponding to a given diagnosis d is to take all the IOS values $R_i(f)$ corresponding to patients i with this diagnosis ($i \in d$), and use Least Squares to find the cubic function $R_d(f)$ that best approximate all these measurement results, i.e., for which the sum $\sum_{i \in d} \sum_f (R_d(f) - R_i(f))^2$ attains its smallest possible value (here, the summation is over the IOS frequencies $f = 5, 10, 15, 20, 25, 35\,\text{Hz}$).

Similarly, a natural way to form the typical function $X_d(f)$ corresponding to a given diagnosis d is to take all the IOS values $X_i(f)$ corresponding to patients i with this diagnosis, and use Least Squares to find the cubic function $X_d(f)$ that best approximate all these measurement results, i.e., for which, the sum

$$\sum_{i \in d} \sum_f (X_d(f) - X_i(f))^2$$

attains its smallest possible value.

5   Third Pre-processing Stage: Scale-Invariant Similarity/Dissimilarity Measures

How to describe the similarity of functions: preliminary analysis. To use the above general idea for diagnosing a patient, we need to select a numerical measure of similarity/dissimilarity between the function $R^a(f)$ describing the new patient and the function $R_d(f)$ describing a typical patient with diagnosis d.

One thing to take into account is that usually the measurement record produced by the IOS device contains an initial spike, when the measured value jumps from the original 0 value to a non-zero value corresponding to the actual measurement. This initial impulse-type spike affect the Fourier transform values $R(f)$ and $X(f)$ produced by the measuring device. Since the Fourier transform of an impulse is a constant function, this means that to all the measured values $R_i(f)$ and $X_i(f)$, a constant is added that corresponds to the Fourier transform of this original impulse. Because of this added constant, the same constant gets added to the approximating cubic curves $R_i^a(f)$ and $X_i^a(f)$. For the same patient with the same disease, in different measurements, the initial impulse may be slightly different. Thus, the corresponding added constants may be different for the two measurements of the same patient. In other words, for the same patient, in two consequent measurements, we may get two functions differing by a constant.

So, to properly match, e.g., the patient's function $R^a(f)$ with the function $R_d(f)$ describing a typical patient with diagnosis d, we need to take this possible constant difference into account.

How do we estimate the corresponding constant difference? As we have mentioned earlier, the most informative part of the IOS results correspond to the 5–15 Hz range. The central point of this range is the 10 Hz. It is therefore reasonable to take the difference $\Delta R_{id} \overset{\text{def}}{=} R_d(10) - R_i^a(10)$ of the values corresponding to this central frequency as an estimate for the constant difference. Thus, we should compare the typical function $R_d(f)$ not with the actual patient's function $R_i^a(f)$, but with the "shifted" function $R_i^a(f) + \Delta R_{id}$, shifted so as to provide the best match between the two functions.

In other words, to diagnose a patient i, we need to describe the similarity/dissimilarity between the functions $R_i^a(f) + \Delta R_{id}$ and $R_d(f)$.

Similarly, for reactance, we need to describe the similarity/dissimilarity between the functions $X_i^a(f) + \Delta X_{id}$ and $X_d(f)$, where $\Delta X_{id} \overset{\text{def}}{=} X_d(10) - X_i^a(10)$.

How to describe the similarity/dissimilarity of functions: general idea. In general, if we have two functions $F(f)$ and $G(f)$, how can we describe their similarity/dissimilarity? For each frequency f, the larger the absolute value $|F(f) - G(f)|$ of the difference, the less similar are the corresponding values. Thus, it makes sense to assume that the degree of dissimilarity between these two values is a monotonic function of this absolute value: $m(|F(f) - G(f)|)$, for some increasing function $m(x)$.

The overall degree of dissimilarity we can then estimate by simply adding the degree corresponding to different frequencies f, i.e., by considering an integral

$$\int m(|F(f) - G(f)|)\, df.$$

Remaining question. Which similarity/dissimilaroty measures—i.e., which functions $m(x)$—should we use?

Main idea: use scale-invariance. To select an appropriate similarity/dissimilarity measure, let us use the same scale-invariance idea that we used to select an approximating family of functions.

Scale-invariance: from idea to formulas. If, for measuring real and imaginary components of the impedance, we select a new unit which is λ times smaller than the original one, then all the corresponding numerical values $F(f)$ and $G(f)$ get multiplied by λ:

- instead of $F(f)$, we get $\widetilde{F}(f) = \lambda \cdot F(f)$, and
- instead of $G(f)$, we get $\widetilde{G}(f) = \lambda \cdot G(f)$.

In this case, the absolute value of the difference $x = |F(f) - G(f)|$ also gets multiplied by λ:

$$\widetilde{x} = |\widetilde{F}(f) - \widetilde{G}(f)| = |\lambda \cdot F(f) - \lambda \cdot G(f)| = \lambda \cdot |F(f) - G(f)| = \lambda \cdot x.$$

We want to select the function $m(x)$ (that describes degree of similarity/dissimilarity) in such a way that if we change a measuring unit for $F(f)$ and $G(f)$, the resulting value of closeness will not change—provided, of course, that we appropriately change a unit for measuring dissimilarity.

Such an appropriate re-scaling is often necessary; see, e.g., [7, 20]. For example, a simple physical formula—like the fact $D = v \cdot t$ that the distance D is equal to velocity v times time t—does not change if we change the unit of time (e.g., from hours to seconds), but we need to appropriately change the related unit of velocity—from km/h to km/sec.

For each λ, let $\mu(\lambda)$ denote an appropriate re-scaling of the dissimilarity value $m(x)$. This means that if we replace x with $\widetilde{x} = \lambda \cdot x$, then the resulting dissimilarity value $m(\widetilde{x}) = m(\lambda \cdot x)$ differs from the original value $m(x)$ only by a re-scaling factor $\mu(\lambda)$:

$$m(\lambda \cdot x) = \mu(\lambda) \cdot m(x).$$

It is known (see, e.g., [1]) that all measurable solutions of this functional equations have the form $m(x) = C \cdot x^{\alpha}$.

By selecting an appropriate unit for measuring dissimilarity, we can make the coefficient C equal to 1. Thus, we arrive at the following conclusion.

Conclusion of this section. Due to scale-invariance, we measure dissimilarity between two functions as

$$\int |F(f) - G(f)|^{\alpha} \cdot df.$$

Remaining question. What value α should we use?

6   How to Select α: Need to Have Efficient and Robust Estimates

General idea. In the computer, we can only represent finitely many different values $f_1, f_2, \ldots$ So, an integral, in effect, means a (weighted) sum $\sum |F_i - G_i|^\alpha$, where we denoted $F_i \stackrel{\text{def}}{=} F(f_i)$ and $G_i \stackrel{\text{def}}{=} G(f_i)$.

From this viewpoint, which value α should we choose?

Need for efficient estimates. We want to be able to have an efficient algorithm that finds the closest approximation, i.e., an approximation for which the dissimilarity degree $\sum |F_i - G_i|^\alpha$ is the smallest possible.

It is known (see, e.g., [13, 21]) that, in general, feasible algorithms exist for minimizing convex objective functions, while in many non-convex cases, optimization is NP-hard (i.e., crudely speaking, not feasible). Moreover, it has been proven [10] that, in general, minimization is feasible *only* for convex objective functions. Thus, it makes sense to select the value α in such a way that the objective function $\sum |F_i - G_i|^\alpha$ be convex.

In general, according to calculus, a function is convex if its second derivative is non-negative. For $x > 0$, the first derivative of the function x^α is $\alpha \cdot x^{\alpha-1}$, and the second derivative is equal to $\alpha \cdot (\alpha - 1) \cdot x^{\alpha-2}$. Here, $\alpha > 0$—the larger the difference, the less similar are the functions, and the value $x^{\alpha-2}$ is also always positive. Thus, the second derivative is non-negative if and only $\alpha - 1 \geq 0$, i.e., if and only if $\alpha \geq 1$.

Thus, to make sure that the corresponding optimization problems can be efficiently solved, we need to select $\alpha \geq 1$.

Need for robustness. Another important requirement for selecting α is to make sure that the resulting estimates are the least affected by noise, i.e., are the most *robust*.

It is known (see, e.g., [9]), that among all the methods based on the objective function $\sum |F_i - G_i|^\alpha$ with $\alpha \geq 1$, the most robust is the method corresponding to $\alpha = 1$. Thus, to guarantee the desired robustness, we will use $\alpha = 1$. So, we arrive at the following conclusion.

Conclusion of this section. Among all computationally efficient scale-invariant dissimilarity measures, the most robust (i.e., the most resistant to noise) is the dissimilarity measure

$$\int |F(f) - G(f)| \cdot df.$$

In our case, $F(f) = R_d(f)$ and $G(f) = R_i^a(f) + \Delta R_{id}$, so we need to use the dissimilarity measure

$$\int |R_d(f) - (R_i^a(f) + \Delta R_{id})| \, df.$$

When this dissimilarity measure is close to 0, this means that the function $R_i^a(f)$ corresponding to the i-th patient is very similar to the typical function $R_d(f)$ corresponding to diagnosis d. The more dissimilar these two functions, the larger the value of this dissimilarity measure.

Similarly for reactance, we use the dissimilarity measure

$$\int |X_d(f) - (X_i^a(f) + \Delta X_{id})| \, df.$$

7 Scale-Invariance Helps to Take into Account That Signal Informativeness Decreases with Time

For IOS, the starting part of the signal is more informative. In the previous sections, we implicitly assumed that the values of the signal $y(t)$ at different moments of time are equally informative. However, a typical IOS measurement lasts for 30-45 s—a reasonable time to be tied in to a strange apparatus. As a result, children's level of stress somewhat increases as the measurement process continues. This stress level affects the breathing process—and thus, the measurement results.

So, we can conclude that values $y(t)$ corresponding to earlier time are more informative than values corresponding to later moments of time t.

How can we take this phenomenon into account. A reasonable idea of taking the above phenomenon into account is to consider not the original signals $y(t)$, but the signals weighted with some weight $w(t)$ which decreases with time. In other words, instead of the original signals $y(t)$, we consider weighted signals $w(t) \cdot y(t)$.

Which weight function should we choose: let us again apply scale-invariance. Which weight function $w(t)$ should we choose? A reasonable idea is to again use scale-invariance. In other words, we assume that if we change the unit of time to a one which is λ times smaller—which means changing all numerical values of time from t to $\widetilde{t} = \lambda \cdot t$, then the formula for the weight remains the same—once we appropriately re-scale the weight function w as well, from w to $\widetilde{w} = \mu(\lambda) \cdot w$, for some function $\mu(\lambda)$.

This means that if in the original units, we have $f(t) = w$, then in the new units, we will have $f(\widetilde{t}) = \widetilde{w}$. Substituting the expressions for $\widetilde{w}$ and $\widetilde{t}$ into this formula, we conclude that $f(\lambda \cdot t) = \mu(\lambda) \cdot w$ and thus,

$$f(\lambda \cdot t) = \mu(\lambda) \cdot f(t).$$

We have already mentioned that all measurable (in particular, all monotonic) solutions of this functional equation have the form $w = A \cdot t^\alpha$ [1]. Since we assume that the weight decreases with time, we must have $\alpha < 0$.

Which value α should we choose? Whether we use the original signal or its weighted form, what we will do next is apply Fourier transform. The original IOS device already returns the Fourier coefficients of the original signal $y(t)$. Thus, from the computational viewpoint, it is desirable to select α for which the Fourier transform of the weighted function $y(t) \cdot t^{\alpha}$ can be described in terms of the Fourier transform of the original function $y(t)$.

It is known that such a description is possible only for integer values α; namely:

- the Fourier transform of $y(t)/t$ is proportional to the integral of the Fourier transform of $y(t)$;
- the Fourier transform of $y(t)/t^2$ is proportional to the second integral (integral of an integral) of the Fourier transform of $y(t)$; etc.

The simplest of these cases is the case $\alpha = -1$, which corresponds to the integral.

Thus, in addition to the original Fourier transform values, we should consider integrals of these values.

Details. Of course, when computing these integrals, we should take into account the smoothing that we have applied to the original signal. In other words, we should integrate not the original values $Z(f)$, but the corresponding smoothing polynomial approximations.

Integration should be considered over the most informative part of the spectrum—from 5 15 Hz. Thus, we arrive at the following conclusion.

Conclusion of this section. In addition to the smoothed signals $R^a(f)$ and $X^a(f)$, we should also consider their integrals $I_R(f) = \int_5^f R^a(x)\, dx$ and $X_R(f) = \int_5^f X^a(x)\, dx$.

8 Pre-processing Summarized: What Information Serves as an Input to a Neural Network

Let us summarize. Let us summarize the scale-invariance-motivated pre-processing steps, and thus, describe what inputs are fed into a neural network.

This whole process consists of two stages:

- In the first, preliminary stage, we process data about known patients to find the "typical" functions $R_d(f)$ and $X_d(f)$ that correspond to each diagnosis d.
- On the working stage, we use these typical functions to diagnose a new patient.

Preliminary stage. First, for each diagnosis d, we process patients with known diagnoses d to find the typical functions $R_d(f)$ and $X_d(f)$ corresponding to each of these diagnoses. This is done as follows.

For each patient with the known diagnosis, we get the IOS values $R(f)$ and $X(f)$ corresponding to frequencies f equal to 5, 10, 15, 20, 25, 35 Hz.

Then, we use the Least Squares techniques to find the coefficients r_0, r_1, r_2, and r_3 of the cubic polynomial

$$R_d(f) = r_0 + r_1 \cdot f + r_2 \cdot f^2 + r_3 \cdot f^3$$

that best approximates the measured values $R_i(f)$ corresponding to the patients with this diagnosis d (of course, we only use patients from the training set, to be able to test our results of the patients from the testing set). In other words, we find the coefficients of the cubic polynomial for which the sum

$$\sum_{i \in d} \sum_f (R_d(f) - R_i(f))^2$$

is the smallest possible.

Similarly, we use the Least Squares techniques to find the coefficients x_0, x_1, x_2, and x_3 of the cubic polynomial

$$X_d(f) = x_0 + x_1 \cdot f + x_2 \cdot f^2 + x_3 \cdot f^3$$

that best approximates the measured values $X_i(f)$ corresponding to all the patients with this diagnosis d. In other words, we find the coefficients of the cubic polynomial for which the sum $\sum_{i \in d} \sum_f (X_d(f) - X_i(f))^2$ is the smallest possible.

For each of these functions, we then compute the integrals $I_{R,d}(f) = \int_5^f R_d(x)\,dx$ and $I_{X,d}(f) = \int_5^f X_d(x)\,dx$.

Working stage. For a new patient, we get the IOS values $R(f)$ and $X(f)$ corresponding to frequencies f equal to 5, 10, 15, 20, 25, 35 Hz. Then:

- We use the Least Squares techniques to find the coefficients r_0, r_1, r_2, and r_3 of the cubic polynomial

$$R^a(f) = r_0 + r_1 \cdot f + r_2 \cdot f^2 + r_3 \cdot f^3$$

that best approximates the measured values $R(f)$. We also compute the integral $I_R(f) = \int_5^f R^a(x)\,dx$.
- After that, we use the Least Squares techniques to find the coefficients x_0, x_1, x_2, and x_3 of the cubic polynomial

$$X^a(f) = x_0 + x_1 \cdot f + x_2 \cdot f^2 + x_3 \cdot f^3$$

that best approximates the measured values $X(f)$. We also compute the integral $I_X(f) = \int_5^f X^a(x)\,dx$.
- Finally, for each of the four diagnoses d, we compute the following values:

  - $\int_5^{15} |R_d(f) - (R^a(f) + \Delta R_d)|\,df$, where $\Delta R_d \overset{\text{def}}{=} R_d(10) - R^a(10)$;
  - $\int_5^{15} |I_{R,d}(f) - (I_R^a(f) + \Delta I_{R,d})|\,df$, where $\Delta I_{R,d} \overset{\text{def}}{=} I_{R,d}(10) - I_R^a(10)$;
  - $\int_5^{15} |X_d(f) - (X^a(f) + \Delta X_d)|\,df$, where $\Delta X_d \overset{\text{def}}{=} X_d(10) - X^a(10)$;
  - $\int_5^{15} |I_{X,d}(f) - (I_X^a(f) + \Delta I_{X,d})|\,df$, where $\Delta I_{X,d} \overset{\text{def}}{=} I_{X,d}(10) - I_X^a(10)$.

These four tuples of four values corresponding to four diagnoses—the total of 16 values—can then be used to train a neural networks to diagnose the patient.

Do we need all these 16 inputs? In data processing, it is known that if we use too many inputs, the prediction accuracy decreases. Indeed, if we use too many inputs, then, together with the most informative ones, we also add less informative ones. These additional inputs add noise to the result of data processing without providing us with any useful information.

Because of this, in data processing in general, it is a good idea not just to use all possible inputs, but also to check if selecting only some of these inputs will leads to more accurate results.

In our case, we tested whether we need both values corresponding to resistance R and values corresponding to reactance X. Interestingly, it turned out that the reactance-related values only decrease the prediction quality. As a result, our recommendation is to only use resistance-related values when diagnosing patients.

9 The Results of Training Neural Networks on These Pre-processed Data

Available data. In our research, we used the data collected by our colleague Erika Meraz [2]. This data consists of 288 IOS data sets from patients with known diagnoses.

Pre-processing: first stage. First, for each data set, we used Least Square to find the coefficients of cubic polynomials $R_i^a(f)$ that best fit the observed IOS values $R_i(f)$. Then, we computed the integral $I_{R,i}(f) = \int_5^f R_i^a(x)\, dx$.

Neural network: general idea. We trained a neural network to distinguish patients with lung dysfunctions from patients without lung dysfunction.

A neural networks consist of neurons. Each neuron takes several inputs $x_1, \ldots, x_n$ and transforms them into the signal

$$y = s_0(w_1 \cdot x_1 + \cdots + w_n \cdot x_n - w_0),$$

where $w_0, w_1, \ldots, w_n$ are coefficients that need to be determined during training, and $s_0(z)$ is a nonlinear function known as the *activation function*.

In this research, we used neural networks with sigmoid activation function $s_0(z) = 1/(1 + \exp(-z))$, the most widely used activation function. It is worth mentioning that this function can also be justified by invariance: this time, by shift-invariance (and not scale-invariance, as in the previous examples); see, e.g., [17].

Separation into training and validation data sets. Overall, we had 288 data sets, of which:

- 257 data sets correspond to patients with lung dysfunctions, and
- 31 data sets correspond to patients without lung dysfunctions.

To train a neural networks, we separated the corresponding data set into training data set (used for training) and validation data set (used for validation). In all three cases, we used approximately 75% of the data for training and approximately 25% for validation; see, e.g., [8, 18]. Specifically:

- we selected 214 data sets for training, among which 191 corresponded to patients with lung dysfunctions and 23 corresponded to patients without lung dysfunctions, and
- the remaining 74 data sets were used for validation, among which 66 corresponded to patients with lung dysfunctions, and 8 patients without lung dysfunctions.

We have four possible diagnoses: asthma (a), SAI (s), PSAI (p), and the absence of lung dysfunctions (n). Within the training set, for each of these four diagnoses d, we applied the Least Square method to all the values $R_i(f)$ corresponding to the data sets with this diagnosis to compute the typical values $R_d(f)$ corresponding to this diagnosis. We then computed the integral $I_{R,d}(f) = \int_5^f R_d(x)\,dx$.

Resulting typical functions $R_d(f)$. As a result of this analysis, we got the following typical functions corresponding to different diagnoses; see Fig. 4:

$$R_a(f) = 1.152 - 7.842 \cdot 10^{-2} \cdot f + 2.686 \cdot 10^{-3} \cdot f^2 - 2.443 \cdot 10^{-5} \cdot f^3;$$

$$R_s(f) = 8.960 \cdot 10^{-1} - 5.738 \cdot 10^{-2} \cdot f + 2.067 \cdot 10^{-3} \cdot f^2 - 2.024 \cdot 10^{-5} \cdot f^3;$$

$$R_p(f) = 6.076 \cdot 10^{-1} - 2.717 \cdot 10^{-2} \cdot f + 8.278 \cdot 10^{-4} \cdot f^2 - 4.888 \cdot 10^{-6} \cdot f^3;$$

$$R_n(f) = 4.612 \cdot 10^{-1} - 1.508 \cdot 10^{-2} \cdot f + 4.789 \cdot 10^{-4} \cdot f^2 - 2.424 \cdot 10^{-6} \cdot f^3.$$

Pre-processing: second stage. For each patient i, and for each of the four diagnoses d, we computed the following two similarity/dissimilarity measures:

- $\int |R_d(f) - (R_i^a(f) + \Delta R_{id})|\,df$, where $\Delta R_{id} \overset{\text{def}}{=} R_d(10) - R^a(10)$; and
- $\int |I_{R,d}(f) - (I_{R,i}^a(f) + \Delta I_{R,i,d})|\,df$, where $\Delta I_{R,i,d} \overset{\text{def}}{=} I_{R,d}(10) - I_{R,i}^a(10)$.

The resulting eight values serve as input to the neural network.

The results of training the neural network. The purpose of the neural network was to separate patients with lung dysfunctions from patients without lung dysfunction.

During the training, the network selected 50 neurons in the hidden layer. On the validation data set, the neural network achieved 100% accuracy on the validation set: all 74 cases were classified correctly.

This is much better that in the previous studies. The resulting classification accuracy is much better than the 60% accuracy achieved by neural networks without pre-processing.

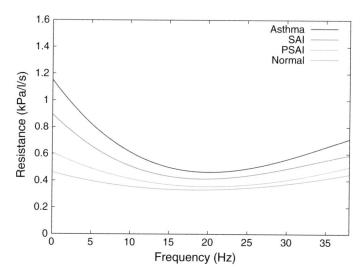

Fig. 4 Cubic resistance functions per class

Acknowledgements This research was partially supported:
• by the National Council of Science and Technology of México (CONACYT) via a doctoral fellowship for Nancy Avila,
• by the Research Program on Migration and Health (PIMSA), a program funded by the University of California, Berkeley, Health Initiative of the Americas (HAI), and
• by the US National Science Foundation via grant HRD-1242122 (Cyber-ShARE Center of Excellence).
The data for this research were acquired by Erika Meraz as part of the NIH grant #1 S11 ES013339-01A1: "UTEP-UNM HSC ARCH Program on Border Asthma" (Pilot 1) granted to Dr. Homer Nazeran.

References

1. Aczél, J., Dhombres, J.: Functional Equations in Several Variables. Cambridge University Press (2008)
2. Avila, N., Nazeran, H., Meraz, E., Gordillo, N., Aguilar, C.: Characterization of impulse oscillometric measures of respiratory small airway function in children. In: Advances in Electrical and Electronic Engineering
3. Badnjevic, A., Cifrek, M.: Classification of asthma utilizing integrated software suite. In: Proceedings of the 6th European Conference of the International Federation for Mechanical and Biological Engineering IFMBE'2015, Dubrovnik, Croatia, pp. 415–418 (2015)
4. Badnjevic, A., Gurbeta, L., Cifrek, M., Marjanovic, D.: Classification of asthma using artificial neural networks. In: Proceedings of the 39th IEEE International Convention on Information and Communication Technology, Electronics, and Microelectronics MIPRO'2016, Opatija, Croatia, pp. 387–390 (2016)

5. Barua, M., Nazeran, H., Nava, P., Granda, V., Diong, B.: Classification of pulmonary diseases based on impulse oscillometric measurements of lung function using neural networks. In: Proceedings of the 26th Annual International Conference of the IEEE Engineering in Medicine and Biology Society, San Francisco, California, September 1–5, pp. 3848–3851 (2004)
6. Bickel, S., Popler, J., Lesnick, B., Eid, N.: Impulse oscillometry: interpretation and practical applications. Chest **146**(3), 841–847 (2014)
7. Feynman, R., Leighton, R., Sands, M.: The Feynman Lectures on Physics. Addison Wesley, Boston, Massachusetts (2005)
8. Gholamy, A., Kreinovich, V., Kosheleva, O.: Why 70/30 or 80/20 relation between training and testing sets: a pedagogical explanation. Int. J. Intel. Technol. Appl. Stat. **11**(2), 105–111 (2018)
9. Huber, P.J., Ronchetti, E.M.: Robust Statistics. Wiley, Hoboken (2009)
10. Kearfott, R.B., Kreinovich, V.: Beyond convex? Global optimization is feasible only for convex objective functions: a theorem. J. Glob. Optim. **33**(4), 617–624 (2005)
11. Komarow, H.D., Myles, I.A., Uzzaman, A., Metcalfe, D.D.: Impulse oscillometry in the evaluation of diseases of the airways in children. Ann. Aller. Asthma Immunol. **106**(3), 191–199 (2011). https://doi.org/10.1016/j.anai.2010.11.011
12. Komarow, H.D., Skinner, J., Young, M., Gaskins, D., Nelson, C., Gergen, P.J., Metcalfe, D.D.: A study of the use of impulse oscillometry in the evaluation of children with asthma: analysis of lung parameters, order effect, and utility compared with spirometry. Pediatr. Pulmonol. **47**(1), 18–26 (2012)
13. Kreinovich, V., Lakeyev, A., Rohn, J., Kahl, P.: Computational Complexity and Feasibility of Data Processing and Interval Computations. Kluwer, Dordrecht (1998)
14. Luo, G., Stone, B.L., Maloney, C.G., Gesteland, P.H., Yerram, S.R., Nkoy, F.L.: Predicting asthma control deterioration in children. In: BMC Medical Informatics and Decision Making, vol. 15, Paper 84 (2015)
15. Marotta, A., Klinnert, M.D., Price, M.R., Larsen, G.L., Liu, A.H.: Impulse oscillometry provides an effective measure of lung dysfunction in 4-year-old children at risk for persistent asthma. J. Aller. Clin. Immunol. **112**(2), 317–322 (2003). https://doi.org/10.1067/mai.2003.1627
16. Mochizuki, H., Hirai, K., Tabata, H.: Forced oscillation techniques and childhood asthma. Allergol. Int. **61**(3), 373–383 (2012)
17. Nguyen, H.T., Kreinovich, V.: Applications of Continuous Mathematics to Computer Science. Kluwer, Dordrecht (1997)
18. Sheskin, D.J.: Handbook of Parametric and Nonparametric Statistical Procedures. Chapman and Hall/CRC, Boca Raton (2011)
19. Shi, Y., Aledia, A.S., Tatavoosian, A.V., Vijayalakshmi, S., Galant, S.P., George, S.C.: Relating small airways to asthma control by using impulse oscillometry in children. J. Aller. Clin. Immunol. **129**(3), 671–678 (2012)
20. Thorne, K.S., Blandford, R.D.: Modern Classical Physics: Optics, Fluids, Plasmas, Elasticity, Relativity, and Statistical Physics. Princeton University Press, Princeton, New Jersey (2017)
21. Vavasis, S.A.: Nonlinear Optimization: Complexity Issues. Oxford University Press, New York (1991)
22. World Health Organization (WHO): Asthma. http://www.who.int/mediacentre/factsheets/fs307/en/, downloaded on 31 Jan 2019

Comparative Analysis of Parametric Optimization Techniques: Fuzzy DSS for Medical Diagnostics

Oleksiy Kozlov, Yuriy Kondratenko, and Oleksandr Skakodub

Abstract This paper addresses the issues of the research and comparative analysis of the advanced bio-inspired techniques for parametric optimization of fuzzy systems (FS). Several modifications of particle swarm optimization (PSO) and grey wolf optimization (GWO) multi-agent algorithms, adapted for optimization of FS parameters, are compared with the conventional search methods. The studies and comparative analysis are conducted on a specific example, namely, at parametric optimization of the fuzzy decision support system (DSS) of Takagi-Sugeno type for the medical diagnostics. The obtained simulation results confirm the high efficiency of the presented bio-inspired optimization techniques, taking into account the achieved performance of the DSS and the spent computing costs.

Keywords Fuzzy decision support system · Parametric optimization · Bio-inspired techniques · Particle swarm optimization · Grey wolf optimization · Medical diagnostics

1 Introduction

At designing complicated systems and devices in various fields of human activity (technology, medicine, economics, agriculture etc.) the tasks of search for rational solutions in the multidimensional space of alternatives often arise [1–3]. In many cases, application of conventional optimization techniques for solving these tasks becomes ineffective, due to the complexity of the relief and the multimodality of the

O. Kozlov (✉) · Y. Kondratenko · O. Skakodub
Department of Intelligent Information Systems, Petro Mohyla Black Sea National University, 10, 68Th Desantnykiv Str, Mykolaiv 54003, Ukraine
e-mail: kozlov_ov@ukr.net

Y. Kondratenko
e-mail: y_kondrat2002@yahoo.com

O. Skakodub
e-mail: aleksandrskakodub1996@gmail.com

studied functions, which describe the nonlinear dependencies of the quality indicators of solutions on the desired parameters [4–6]. Currently, approximate methods
of global optimization, become enough popular that make it possible finding high
quality solutions using acceptable (from a practical point of view) computational
and time costs [7–9]. Among them, bioinspired multi-agent and evolutionary optimization techniques are enough promising [10–12]. Unlike classical methods of local
search, bioinspired intelligent algorithms can be applied in cases where there is almost
no information about the nature and peculiarities of the studied objects and processes.
They have a strong ability of local minima avoiding and can be easily adapted
for solving completely different real-life optimization problems. Moreover, these
algorithms implement enough simple computational procedures, that mimic social
animals' behavior, evolutionary concepts of natural selection, physical phenomena,
etc. Finally, bioinspired intelligent techniques can be effectively hybridized with the
different local search methods, which makes it possible to significantly intensify the
optimization process through the rational use of the global and local search strategies
[13].

One of the promising fields of the bioinspired multi-agent and evolutionary techniques' application is the design and optimization of fuzzy decision support and
control systems of different types and purposes [14]. In turn, the results of numerous
cutting edge studies show that FSs, developed by means of bioinspired optimization techniques, allow solving complicated problems of decision support and control
in different spheres with a sufficiently high efficiency [15]. Thus, research, aimed
at the development, enhancement and implementation of bioinspired multi-agent
approaches and techniques for the synthesis and optimization of fuzzy systems of
various types, is undoubtedly relevant and essential.

This paper is devoted to the research and comparative analysis of the bio-inspired
multi-agent techniques at solving the task of parametric optimization of the hierarchical fuzzy decision support system of Takagi-Sugeno type for the medical diagnostics. In particular, the comparative analysis is conducted for several modifications
(basic, improved, hybrid) of particle swarm optimization and grey wolf optimization
multi-agent methods, that are previously adapted for parametric optimization of FS
parameters.

2 Hierarchical Fuzzy DSS for Medical Diagnostics

The problem of improving the quality of the health care system is one of the most
important in the modern society. Herewith, the task of early diagnosis of different
diseases in patients play a special role [16–18]. In turn, fuzzy DSSs give the opportunity to significantly increase the level of information support for the doctor, as well
as the effectiveness and objectification of the diagnosis establishing [16, 18].

In this paper, for a comparative analysis of bio-inspired optimization techniques,
we study the hierarchical fuzzy decision support system of Takagi–Sugeno type
(Fig. 1) for diagnosing heart disease [19]. This system has 13 input variables: x_1;

$x_2; \ldots; x_{13}$ are age of the studied patient (in full years); patient sex (2 sexes: 1—male, 0—female); chest pain type (4 types: 1, 2, 3 or 4); resting blood pressure (from 0 to 200 mmHg); serum cholesterol (from 0 to 600 mg/dl); fasting blood sugar (2 levels: 1—more than 120 mg/dl, 0—less than 120 mg/dl); resting electrocardiographic results (2 levels: 0 or 2); maximum heart rate achieved (from 0 to 200 bpm); exercise induced angina (2 levels: 0—absent, 1—present); ST depression induced by exercise relative to rest (from 0 to 10); the slope of the peak exercise ST segment (from 1 to 3 degrees); number of major vessels colored by fluoroscopy (4 levels: 0, 1, 2, 3); variable that determines the treatment status of heart disease (3 levels: 3—normal, 6—fixed defect, 7—reversible defect). Output variable y_5 is a variable of discrete type, the current value of which can be attributed to one of the two available classes of solutions: A1—absence of heart disease, A2—presence of heart disease.

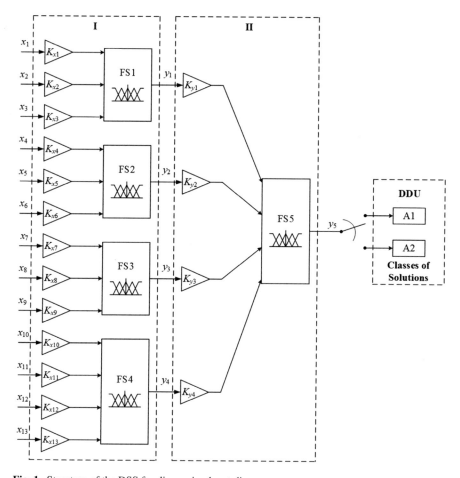

Fig. 1 Structure of the DSS for diagnosing heart disease

In turn, the signal 1 at the output of the system corresponds to the choice of solution A1, the signal 2—the solution A2. If at the output of the subsystem FS5 a signal with a value in the range from 0 to 1 is formed, the decision derivation unit DDU rounds it to 1. When a signal appears with a value in the range from 1 to 2, it is rounded up to 2. The developed DSS consists of 5 subsystems (FS1, FS2, ..., FS5) and has 2 hierarchical levels (I, II) of their organization. The coefficients $K_{x1}, K_{x2}, ..., K_{x13}$ are used for normalization of input variables of subsystems of the 1st hierarchical level $x_1, x_2, ..., x_{13}$. In turn, the coefficients $K_{y1}, ..., K_{y4}$—for normalization of input variables of the subsystem of the 2nd level of hierarchy $y_1, ..., y_4$.

In the presented in Fig. 1 hierarchical DSS for medical diagnostics, fuzzy subsystems FS1, FS2,..., FS5 formalize the relationship of input and output (main and intermediate) variables based on the nonlinear dependencies (1).

For input signals of FS1, FS2, ..., FS5 the following numbers and types of linguistic terms (LT) are chosen: for $x_1, x_4, x_5, x_8, x_{10}, x_{11}, x_{13}, y_1, y_2, y_3, y_4$—3 triangular-type LTs for each; for x_2, x_6, x_7, x_9—2 terms of trapezoidal type for each; for x_3, x_{12}—4 LTs of triangular type.

$$\begin{cases} y_1 = f_{FS1}(K_{x1}x_1, K_{x2}x_2, K_{x3}x_3); \\ y_2 = f_{FS2}(K_{x4}x_4, K_{x5}x_5, K_{x6}x_6); \\ y_3 = f_{FS3}(K_{x7}x_7, K_{x8}x_8, K_{x9}x_9); \\ y_4 = f_{FS4}(K_{x10}x_{10}, K_{x11}x_{11}, K_{x12}x_{12}, K_{x13}x_{13}); \\ y_5 = f_{FS5}(K_{y1}y_1, K_{y2}y_2, K_{y3}y_3, K_{y4}y_4). \end{cases} \qquad (1)$$

In turn, all these terms are evenly distributed over their operating ranges (from 0 to 1). The parameters of these LTs were not optimized. Thus, the vector of unknown parameters $\mathbf{X}$ is given by expression (2)

$$\mathbf{X} = \{\mathbf{K}_i, \mathbf{P}_{C1}, \mathbf{P}_{C2}, \mathbf{P}_{C3}, \mathbf{P}_{C4}, \mathbf{P}_{C5}\}, \qquad (2)$$

where $\mathbf{K}_i$ is the vector of normalizing coefficients, consisting of $K_{x1}, K_{x2}, ..., K_{x13}, K_{y1}, ..., K_{y4}$; $\mathbf{P}_{C1}, \mathbf{P}_{C2}, \mathbf{P}_{C3}, \mathbf{P}_{C4}$ and $\mathbf{P}_{C5}$ are the vectors of weight gains of consequents of rule bases (RB) for FS1, FS2,..., FS5, respectively.

The RBs of FS1, FS2,..., FS5 consist of 24, 18, 12, 108 and 81 rules, respectively. In turn, the rth rule of RB for each lth subsystem ($l = 1, 2, ..., 5$) in general is presented by the expression (3)

IF "$x_{1l} = LT_1$" AND "$x_{2l} = LT_2$" AND ...
AND "$x_{il} = LT_3$" AND... AND "$x_{nl} = LT_4$"
THEN "$y_l = k_{1rl}(x_{1l}) + k_{2rl}(x_{2l}) + \cdots + k_{irl}(x_{il}) + \cdots + k_{nrl}(x_{nl})$",   (3)

where x_{il} is the i_lth input signal of lth subsystem ($i_l = 1, ..., n_l, l = 1, ..., 5$); y_l is the output signal of lth subsystem; $k_{1rl}, k_{2rl}, ..., k_{irl}, ..., k_{nrl}$ are weight gains of consequents of rth rule of the lth subsystem RB; $LT_1, ..., LT_4$ are certain linguistic terms.

Vectors of weight gains of consequents $\mathbf{P}_{C1}$, $\mathbf{P}_{C2}$, $\mathbf{P}_{C3}$, $\mathbf{P}_{C4}$ and $\mathbf{P}_{C5}$ contain respectively 72, 54, 36, 432 and 324 gains that need to be optimized.

Thus, the vector of optimized parameters $\mathbf{X}$ in total, in this case, consists of 935 parameters. Synthesis and optimization of these parameters is carried out on the basis of a training sample [19]. In turn, this sample consists of 270 complete data rows. Herewith, 250 data rows are used directly for the optimization (training) process, and the remaining 20—for further testing of the already optimized system. The objective function is the number of incorrect decisions (as a percentage of 250 data rows used for optimization).

In this paper, it is proposed to carry out synthesis and parametric optimization of the given fuzzy DSS using such effective and well-proven bio-inspired techniques as particle swarm optimization and grey wolf optimization [13, 20, 21]. Also, several modifications of PSO and GWO algorithms should be applied with hybridization with local search methods in order to speed up their convergence. In particular, hybrid modifications of the PSO-method can be used based on the elite strategy with the gradient descent (GD) algorithm and with the extended Kalman filter (EKF) algorithm [12]. Moreover, it is proposed to apply an improved GWO method [21] as well as to hybridize it with techniques of GD and EKF. And finally, for a complete comparison, it is advisable to optimize the fuzzy DSS parameters using separately taken methods of local search (GD and EKF).

3 Parametric Optimization of the Fuzzy DSS for Heart Disease Diagnostics

The basic principles of the PSO method, as well as the features of its application for the synthesis and parametric optimization of FSs, are presented in detail in [13]. In turn, the hybridization of the PSO with GD and EKF based on the elite strategy to improve the processes of FS optimization was proposed by the authors in the paper [12]. The main idea of these modifications is to conduct an independent parallel search by the best swarm particle using GD or EKF, which can significantly speed up the convergence and reduce the computational costs of these methods. The main principles of the basic and improved GWO algorithms are discussed in [20, 21]. The improved method uses an additional dimension learning-based hunting (DLH) strategy for increasing the population diversity of the basic GWO algorithm and prevent the possibility of its premature convergence to suboptimal solutions [21]. In this paper, the authors propose to hybridize an improved version of GWO with local search techniques of GD and EKF. To do this, in addition to the strategies of group hunting and DLH, by analogy with hybrid PSO methods, alpha, beta, and delta wolves of the flock must also conduct a local search in their surrounding areas using GD or EKF.

The following parameters were used for PSO methods: number of particles in the swarm $Z_{max} = 30$; maximum particles velocity $V_{max} = 10$; value of accelerations C_1

Table 1 The best results of experiments obtained in the process of optimizing the hierarchical fuzzy DSS for medical diagnostics

Optimization technique	N_{Jopt}	υ_{Jopt}	J_{min}	N_{Jmin}	υ_{Jmin}
Basic PSO *Gbest*	179	5221	15.6	179	5221
Hybrid PSO with GD	93	2820	12.4	114	3450
Hybrid PSO with EKF	89	2700	11.6	109	3300
Basic GWO	165	4485	15.2	165	4485
Improved GWO (IGWO)	74	4026	13.6	94	5106
Hybrid IGWO with GD	44	2538	8.8	54	3108
Hybrid IGWO with EKF	41	2367	7.2	48	2766
GD	-	-	48	93	93
EKF	-	-	44.8	86	86

$= C_2 = 0.1$. For all modifications of GWO-methods, the number of wolves in the flock $Z_{max} = 30$. The optimal value of the objective function is $J_{opt} = 16\%$, the execution of the maximum number of iterations $N_{max} = 200$ is chosen as the criterion for the end of optimization. Parametric optimization of the vector **X** was performed 5 times using each of the studied methods, followed by selection of the best results. To evaluate the efficiency of the optimization techniques used in this paper, it is advisable to compare the found best values of the objective function J_{min} as well as the computational costs spent on this [12]. Also the computational costs necessary to find the given optimal value of the objective function J_{opt} are used for the assessment. Herewith, the computational costs of the considered algorithms are mainly determined by the total number of times υ_J of the objective function calculation required to achieve its certain value (υ_{Jopt}—for achievement of the optimal value J_{opt}, υ_{Jmin}—for achievement of the best value J_{min}).

The best experimental results obtained in the process of optimizing the vector **X** using each of the methods are presented in Table 1.

The Fig. 2 shows the convergence curves for the best values of the objective function in the process of optimizing the vector **X** based on the considered methods: 1) basic PSO *Gbest*, 2) hybrid PSO based on the elite strategy with GD, 3) hybrid PSO based on the elite strategy with EKF; 4) basic GWO, 5) improved GWO (IGWO), 6) hybrid IGWO with GD, 7) hybrid IGWO with EKF, 8) GD; 9) EKF.

As can be seen from Table 1 and Fig. 2, the hybrid IGWO-methods are generally more efficient than hybrid PSO-methods in the parametric optimization of the hierarchical fuzzy DSS. Thus, to find the optimal value J_{opt} using hybrid IGWO methods with EKF and GD, at best, it took 333 and 282 fewer calculations υ_{Jopt}, respectively, than when using hybrid PSO methods based on the elite strategy with EKF and GD.

In addition, the implementation of hybrid IGWO methods, on average, ensured the achievement of a smaller minimum value J_{min} compared to hybrid PSO methods.

To solve this specific problem, the most effective is the hybrid IGWO-method with EKF, at the implementation of which it was possible to achieve the optimal value J_{opt} of fuzzy DSS ($J \leq 16\%$) for the least number of calculations of the objective

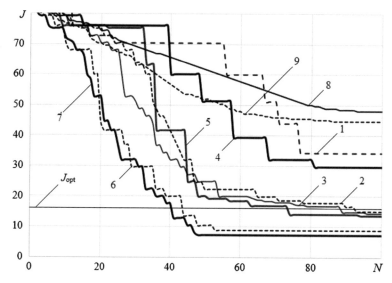

Fig. 2 Convergence curves for the best values of the objective function J in the process of optimization of the hierarchical fuzzy DSS (in the range $N = 0...100$)

function ($\upsilon_{J\text{opt}} = 2367$). Also, when implementing this technique, the lowest value of the objective function ($J_{\min} = 7.2\%$) was achieved on the 48th iteration (Fig. 2, curve 7).

Separate application of the GD and EKF algorithms, in this case, did not allow to reach the optimal value of the objective function ($J \leq 16\%$). In turn, the basic GWO showed slightly better results compared to the basic PSO (by the number of calculations of the objective function $\upsilon_{J\text{opt}}$ and by its minimum value found $J_{\min}$).

When testing fuzzy systems, optimized by various methods, on twenty (previously selected) data rows of the sample, which were not used in the optimization process, the lowest value of the objective function $J_T = 15\%$ was also shown by the DSS synthesized with the help of hybrid IGWO-method with EKF.

4   Conclusions

The research and comparative analysis of the advanced bio-inspired techniques for parametric optimization of fuzzy systems is presented in this paper. In particular, several modifications of PSO (basic, hybrid based on the elite strategy with GD and EKF) and GWO (basic, improved, hybrid improved with GD and EKF) multi-agent algorithms, adapted for optimization of FS parameters, are compared with each other and with the conventional search methods (GD and EKF).

The studies and comparative analysis are conducted on a specific example, namely, at parametric optimization of the hierarchical fuzzy decision support system of

Takagi–Sugeno type for the medical diagnostics. The obtained simulation results show that the hybrid IGWO-methods are generally more efficient than hybrid PSO-methods in the parametric optimization of the hierarchical fuzzy DSS for heart disease diagnosing. To solve this specific problem, the most effective is the hybrid IGWO-method with EKF. The developed fuzzy DSS by this hybrid bio-inspired technique has a fairly high accuracy in diagnosing heart disease: 7.2% of incorrect decisions—for the data set from the training sample; 15% of incorrect decisions—for the data set from the test sample. In addition, finding the best variant of the vector $\mathbf{X}_{best}$ using the hybrid IGWO-method with EKF did not require significant computational and time costs ($\upsilon_{Jmin} = 2367$), which generally confirms its high efficiency and expediency of application for parametric optimization of fuzzy decision support and control systems of different types and configurations.

References

1. Gogna, A., Tayal, A.: Metaheuristics: review and application. J. Exp. Theor. Artif. Intell. **25**, 503–526 (2013)
2. Kondratenko, Y.P., Korobko, O.V., Kozlov, O.V.: Synthesis and optimization of fuzzy controller for thermoacoustic plant, recent developments and new direction in soft-computing foundations and applications, studies in fuzziness and soft. Computing **342**, 453–467 (2016). https://doi.org/10.1007/978-3-319-32229-2_31
3. Kondratenko, Y.P., Klymenko, L.P., Al Zu'bi E.Y.M.: Structural optimization of fuzzy systems' rules base and aggregation models. Kybernetes **42**(5), 831–843 (2013). https://doi.org/10.1108/K-03-2013-0053
4. Land, A.H., Doig, A.G.: An automatic method for solving discrete programming problems. In: 50 Years of Integer Programming 1958–2008, pp. 105–132. Springer (2010)
5. Cornuéjols, G.: Valid inequalities for mixed integer linear programs. Math. Program. **112**, 3–44 (2008)
6. Kondratenko, Y.P., Kozlov, O.V.: Mathematic Modeling of reactor's temperature mode of multi-loop pyrolysis plant. In: Modeling and simulation in engineering, economics and management. Lecture Notes in Business Information Processing, vol. 115, pp. 178–187. Springer, Berlin, Heidelberg (2012)
7. Floudas, C.A., Pardalos, P.M. (eds.): Encyclopedia of Optimization. Springer, Berlin (2009)
8. Boussaïd, I., Lepagnot, J., Siarry, P.: A survey on optimization metaheuristics. Inf. Sci. **237**, 82–117 (2013)
9. Kondratenko, Y.P., Simon, D.: Structural and parametric optimization of fuzzy control and decision making systems. In: Zadeh L., Yager R., Shahbazova S., Reformat M., Kreinovich V. (eds.) Recent Developments and the New Direction in Soft-Computing Foundations and Applications, Studies in Fuzziness and Soft Computing, vol. 361, pp. 273–289. Springer, Cham (2018). https://doi.org/10.1007/978-3-319-75408-6_22
10. Blum, C., Puchinger, J., Raidl, G.R., Roli, A.: Hybrid metaheuristics in combinatorial optimization: a survey. Appl. Soft Comput. **11**(6), 4135–4151 (2011)
11. Simon, D.: Evolutionary Optimization Algorithms: Biologically Inspired and Population-Based Approaches to Computer Intelligence. Wiley (2013)
12. Kondratenko, Y.P., Kozlov, A.V.: Parametric optimization of fuzzy control systems based on hybrid particle swarm algorithms with elite strategy. J. Autom. Inf. Sci. **51**(12), 25–45 (2019). (Begel House Inc., New York)
13. Muthukaruppan, S., Er, M.J.: A hybrid particle swarm optimization based fuzzy expert system for the diagnosis of coronary artery disease. Expert Syst. Appl. **39**(14), 11657–11665 (2012)

14. Pedrycz, W., Li, K., Reformat, M.: Evolutionary reduction of fuzzy rule-based models. Fifty Years of Fuzzy Logic and its Applications, STUDFUZ, vol. 326, pp. 459–481. Springer, Cham (2015)
15. Talbi, E.-G.: Metaheuristics: From Design to Implementation, vol. 74. Wiley (2009)
16. Sheremet, A., et al.: Diagnosis of lung disease based on medical images using artificial neural networks. In: Proceedings of the 2021 IEEE 3rd Conference on Electrical and Computer Engineering (UKRCON), Lviv, Ukraine, IEEE, pp. 561–566 (2021). https://doi.org/10.1109/UKR CON53503.2021.9575961.
17. Mendel, J.M.: Uncertain rule-based fuzzy systems. Introduction and New Directions, 2nd edn. Springer International Publishing, 684 p (2017)
18. Atamanyuk, I., et al.: Calculation method for a computer's diagnostics of cardiovascular diseases based on canonical decompositions of random sequences. In: Batsakis, S., et al. (eds.) Proceedings of the 11th International Conference on ICTERI-2015, CEUR-WS, Vol. 1356, Lviv, Ukraine, 108–120 (2015)
19. Dua, D., Graff, C.: UCI machine learning repository. http://archive.ics.uci.edu/ml. University of California, School of Information and Computer Science, Irvine, CA (2019)
20. Mirjalili, S., Mirjalili, S.M., Lewis, A.: Grey wolf optimizer. Adv. Eng. Softw. **69**, 46–61 (2014)
21. Nadimi-Shahraki, M.H., Taghian, S., Mirjalili, S.: An improved grey wolf optimizer for solving engineering problems. J. Expert Syst. Appl. **166**, 113917 (2021)

What About Generic "And"- And "Or"-Operations: From Levels of Certainty (Philosophical–Physical–Mathematical) to a Natural Interpretation of Quantum-Like Negative Degrees of Certainty

Miroslav Svitek, Olga Kosheleva, Vladik Kreinovich, and Shahnaz N. Shahbazova ⓘ

Abstract Fuzzy techniques—techniques designed to convert imprecise human knowledge into precise computer-understandable terms—have many successful applications. Traditional applications of fuzzy techniques use only important general features of human reasoning and, to make an implement more efficient, ignore subtle details, details which are not important for the corresponding application. But from the more fundamental viewpoint, it is desirable to understand, in all the detail, how people actually reason. In this paper, we use general ideas of fuzzy approach to answer this question. Interestingly—and somewhat unexpectedly—the resulting analysis leads to a natural explanation of existence of several distinct levels of certainty and to a natural appearance of quantum-like negative degrees of certainty.

Keywords Fuzzy logic · Generic "And"-operations · Levels of certainty · Quantum computing

M. Svitek
Faculty of Transportation Sciences, Czech Technical University in Prague Konviktska 20, 110 00 Prague 1, Czech Republic
e-mail: svitek@fd.cvut.cz

O. Kosheleva · V. Kreinovich (✉)
University of Texas at El Paso, 500 W. University, El Paso, TX 79968, USA
e-mail: vladik@utep.edu

O. Kosheleva
e-mail: olgak@utep.edu

Sh. N. Shahbazova
Department of Digital Technologies and Applied Informatics, Azerbaijan State University of Economics, UNEC, Baku AZ1001, Azerbaijan
e-mail: shahbazova@gmail.com; shahnaz_shahbazova@unec.edu.az; shahbazova@cyber.az

Ministry of Science and Education of the Republic of Azerbaijan, Institute of Control Systems, 68 Bakhtiyar Bahabzadeh Street, Baku AZ1141, Azerbaijan

© The Author(s), under exclusive license to Springer Nature Switzerland AG 2023
Sh. N. Shahbazova et al. (eds.), *Recent Developments and the New Directions of Research, Foundations, and Applications*, Studies in Fuzziness and Soft Computing 423, https://doi.org/10.1007/978-3-031-23476-7_18

1 Why Study Generic "And"- And "Or"-Operations

"And"- and "or"-operations. Expert knowledge is often described by using imprecise ("fuzzy") words from natural languages such as "small'. To be able to use this knowledge in computer-based decision making, it is necessary to describe this knowledge in computer-understandable language. Computers have been originally designed to process numbers, this is still what they do most efficiently. So, a natural idea is to describe imprecise expert knowledge by numbers.

The main idea of such description comes from fuzzy logic (see, e.g., [3, 11, 14, 15, 17, 22]), where for each imprecise property like "small" and for each possible value x of the corresponding quantity, we ask the expert to mark, on scale of 0 to 1, his/her degree of certainty that x is small. (Alternatively, we can use another scale, e.g., a scale from 0 to 10, and then divide the result by 10 to re-scale it to the interval [0, 1].)

The problem is that many expert rules use logical connectives "and", "or", and "not" in their formulation. For example, a medical doctor can say that if a patient has a strong fever *and* a heavy headache then a certain medicine should be prescribed. To describe such rules in numerical form, we need to provide numerical degree of certainty not just to individual statements like "a patient has strong fever" and "a patient has a heavy headache", but also to logical combinations of these statements.

In the ideal world, we should ask the expert to estimate his/her degree of confidence in all possible logical combinations. However, the number of all such possible combinations is astronomical, in millions and billions, and there is no way that we can ask experts about all such combinations. For all other logical combinations, for which we cannot ask the expert directly, we need to be able to estimate the expert's degree of confidence in these statements based on the only information that we have, i.e., based on the expert-provided degree of confidence about combined statement.

For "and", this means that we should have an algorithm that, given the expert's degree of confidence a and b in statements A and B, will provide an estimate for the expert's degree of confidence in the "and"-combination $A \& B$ of these statements. This algorithm is called an *"and"-operation* or, for historical reasons, a *t-norm*. We will denote the result of applying this algorithm by $f_\&(a, b)$.

Similarly, for "or", this means that we should have an algorithm that, given the expert's degree of confidence a and b in statements A and B, will provide an estimate for the expert's degree of confidence in the "or"-combination $A \vee B$ of these statements. This algorithm is called an *"or"-operation* or, for historical reasons, a *t-conorm*. We will denote the result of this algorithm by $f_\vee(a, b)$.

What "and"- and "or"-operations are currently used in computer-based decision making. In the first papers that used fuzzy techniques, researchers tried to elicit the "and"- and "or'-operations that best describe actual human reasoning. However, it turned out that, taking into account that we are talking about very imprecise expert rules, adding more details to "and"- and "or"-operations did not add much to the quality of the resulting decisions and controls—while requiring a lot of time to elicit these operations and additional computation time to implement them in the actual decision-making procedures. As a result, researchers realized that from the application viewpoint, it is usually sufficient to limit ourselves to simple operations like $f_\&(a, b) = \min(a, b)$ or $f_\&(a, b) = a \cdot b$, and leave the study of the actual "and"- and "or"-operations to psychologists and linguists.

So what about the actual "and"- and "or"-operations: formulation of the problem. In this paper, we go back to the original idea of eliciting the actual "and"- and "or"-operations from the experts, with the purpose of analyzing what the analysis of such actual operations can tell us about human knowledge processing.

2   Analysis of the Problem

What are actual "and"- and "or"-operations? Empirical study of the actual human reasoning consistently shows that this reasoning is very complicated and very individualized. In particular, it is known that the actual "and"- and "or"-operations are complicated, and differ from one application area to another—and even from one person to another. This was shown, e.g., when the designers of one of the world's first expert systems MYCIN (see, e.g., [9]), who spent a lot of time and effort to find the "and"- and "or"-operations most appropriate for medical reasoning tried to apply the same operations to geophysics. It was immediately discovered that geophysicists think differently, and this difference makes perfect sense:

- a medical doctor has to be very cautious, and only recommend a surgery or other major intervention when he/she is reasonably confident that this intervention will help, while
- a company prospecting for oil usually starts exploring even if there is a remaining doubt that there is oil in this area: if they wait for additional experiments that would enable them to get a perfect confidence, competitors would have gone ahead of them.

In all these cases, simple—or even more complex—"and"- and "or"-operations are only an approximation to the actual, even more complex ones. It seems like the more data we gather about the actual human decision making, the more complex the resulting model becomes. In other words, in the set of all possible "and"- and "or"-operations, the actual "and"- and "or"-operations that describe human reasoning are not some simplified elements. In terms of complexity, they are *generic* elements of this set.

So, let us study what we can say about such generic operations.

In general, what do we know about generic elements. What do we mean by saying that some element is a generic element of a set—be it set of numbers, set of functions, or set of some other objects? In a large set, we usually have a few elements with unusual properties. For example, in the set of all people, we have people who are very short, and we have people who are very tall. Both very short and very tall people are rare, so such people are clearly not generic. Generic usually means that the corresponding object only has important properties that most elements of this set have. When we say "most", we mean that there is a natural probability measure on the corresponding set, and that a object is generic if it has the same properties as random elements in the sense of this measure.

There are several general results about random elements. In particular, there is a result by B. Tsirel'son [21] that under reasonable conditions, random elements are location on the boundary of the corresponding set—or at least close to this boundary. This general fact requires a complex proof, but it can be explained on a simple example. Suppose that you are testing a hypothesis that observations $x_1, \ldots, x_n$ are independent and normally distributed, with 0 mean and standard deviation 1. In this case, one of the most widely used tests is the t-test (see, e.g., [19]), i.e., checking whether the average $\frac{1}{n} \cdot \sum_{i=1}^{n} x_i^2$ is smaller than a certain threshold $t_{\alpha,n} \approx 1$, where α is the corresponding level of confidence. In such a situation, the set of all observations consistent with this hypothesis has the form

$$\left\{ (x_1, \ldots, x_n) : \frac{1}{n} \cdot \sum_{i=1}^{n} x_i^2 \leq t_{\alpha,n} \right\}.$$

However, according to the Law of Large Numbers, this mean tends to the variance—i.e., in this case, to 1. This means, informally speaking, that for large n, most observations are concentrated on the border of the above set, i.e., in the set

$$\left\{ (x_1, \ldots, x_n) : \frac{1}{n} \cdot \sum_{i=1}^{n} x_i^2 \approx 1 \right\},$$

for some relevant meaning of approximate equality.

Moreover, if the boundary itself has a kind of a boundary; e.g., the boundary of a square consists of four sides, and each of the sides has boundaries—vertices—then random elements are located on this "boundary of boundary", etc.

What does this general property of generic objects mean for the case of "and"-operations. In general, an "and"-operation $f_\&(a, b)$ is a function. There is a lot of research about random functions. This research started with Wiener measure (a mathematical description of Brownian motion), a description of functions of one variable. Still, most of the research now is about random functions of one variable—what is called random processes. There is research on random functions of two or more variables—known as random fields—but much less is done for them. Since an "and"-operation is a function of two variables, a natural idea is to look for related functions of one variable—so that more known results will be available.

A natural way to form a function of one variables is to consider the case when $a = b$, in this case we have a function $f(a) \stackrel{\text{def}}{=} f_\&(a, a)$. In general, the natural limitations on possible "and"-operations $f_\&(a, b)$ are:

- that this operation coincides with the usual "and"-operation when we know for sure whether each of the statements is true (1) or false (0), i.e., $f_\&(0, 0) = f_\&(0, 1) = f_\&(1, 0) = 0$ and $f_\&(1, 1) = 1$;
- that a person's degree of confidence $f_\&(a, b)$ in $A \& B$ is the same as this person's degree of confidence $f_\&(b, a)$ in $B \& A$: $f_\&(a, b) = f_\&(b, a)$;
- that a person's degree of confidence $f_\&(a, b)$ in $A \& B$ cannot exceed his/her degree of confidence in A or in B: $f_\&(a, b) \leq a$ and $f_\&(a, b) \leq b$; and
- that if the person becomes more confident in A and/or in B, then his/her degree of confidence in $A \& B$ will either increase or remain the same, but it cannot decrease: if $a \leq a'$ and $b \leq b'$, then $f_\&(a, b) \leq f_\&(a', b')$.

In relation to our function $F(a) = f_\&(a, a)$, this means that $F(0) = 0$, $F(1) = 1$, $F(a) \leq a$, and that if $a < a'$, then $F(a) \leq F(a')$.

In general, for a set determined by inequalities, its boundary is where some of these inequalities become equalities. For example, for an interval

$$[0, 1] = \{x : 0 \leq x \& x \leq 1\}$$

its boundary consists of the points in which one of these two inequalities $0 \leq x$ or $x \leq 1$ becomes equalities, i.e., when either $x = 0$ or $x = 1$.

Similarly, for the square

$$\{(x, y) : 0 \leq x \& x \leq 1 \& 0 \leq y \& y \leq 1\},$$

the boundary consists of four sides: the side where $x = 0$, the side where $x = 1$, the side where $y = 0$, and the side where $y = 1$.

In our case, this means that on the boundary, some of the inequalities become equalities, i.e., that we must have $F(a) = a$ for some $a \in (0, 1)$ and/or that we must have $F(a) = F(a')$ for some $a < a'$.

From this viewpoint, "boundary of boundary" corresponds to a situation when many inequalities become equalities, i.e., in particular, that we have $F(a) = a$ for several values $a \in (0, 1)$.

Which "and"-operations have these properties? To answer this question, let us recall that there is a known general description of all possible "and"-operations; see, e.g., [3, 11, 14, 15, 17]. This description starts with describing so-called *Archimedean* "and"-operations, i.e., operations for which either $f_\&(a, b) = f^{-1}(f(a) \cdot f(b))$ for some strictly monotonic function $f(a)$ or $f_\&(a, b) = f^{-1}(\max(f(a) + f(b) - 1, 0))$. For each of these operations, for all $a \in (0, 1)$, we have $f_\&(a, a) < a$.

To describe a general "and"-operation, on the interval $[0, 1]$, we select several intervals $I_i = [\underline{a}_i, \overline{a}_i]$ whose interiors do not intersect. On each of these intervals, the "and"-operation is isomorphic to an Archimedean one, and in all other cases—i.e., when a and b do not belong to the same interval I_i—we have $f_\&(a, b) = \min(a, b)$. In this arrangement, the equality $F(a) = a$ is satisfied if and only if the point a is one of the following two types:

- it is either an endpoint $\underline{a}_i$ or $\overline{a}_i$ of one of the selected intervals,
- or this point a does not belong to any of the selected intervals I_i at all.

So, we can conclude that in the actual generic "and"-operation, there are most probably several such intervals I_i and maybe an area that does not belong to any of these intervals.

This may sound as a somewhat unexpected conclusion, since in the usual pragmatic approach to fuzzy logic, such weird general "and"-operations are viewed as practically useless purely mathematical constructions—to the extent that many textbooks on fuzzy techniques do not even mention such operations or mention them in passing. What we showed is that these seemingly exotic over-complicated operations are the ones that describe the actual human reasoning!

OK, interesting mathematics, but so what? At first glance, our excitement sounds unjustified: OK, we have some interesting mathematical construction, but the goal of fuzzy techniques is to describe human reasoning. What can this mathematical construction tell us about human reasoning? Let us try to answer this question.

3   First Consequence: Levels of Certainty Naturally Appear

Usual continuous approach to certainty vs. the existence of levels of certainty: a seeming contradiction. In the traditional fuzzy logic, certainty is described by a number from the interval [0, 1]. Alternative, we can use a probabilistic approach to describing uncertainty, in which case certainty of a statement can be described by the probability that this statement is true—and this probability is also a number from the same interval [0, 1]. In both approaches, we have a continuous scale and, thus, a *continuous* transition from less certain statements to more certain one. In general, if we use, e.g., the usual product "and"-operation $f_\&(a, b) = a \cdot b$, then, even if we start with several highly confidence statements with $a = b = \ldots = c = 0.99$, then, by combining these statements, we can get statement $A \& B \& \ldots \& C$ whose degree of certainty is as close to 0 as we want.

In reality, however, practical uncertainties fall into several clearly distinct levels. On the top, there is mathematical level, where we only consider absolutely correct, rigorously proved statements. Next is the level of reasoning in physics, where we can often make conclusions without rigorously proving existence, continuity, or differentiability—as long as the resulting predictions make physical sense. This is not just because of lack of mathematical ingenuity and the resulting inability to prove the corresponding existence—the situation is much more complex: e.g., in general quantum field theory, there is still no consistent precise mathematical description (not to mention quantum gravity and other similar phenomena); see, e.g., [10, 20]. There are some mathematical formulas, but it is known that in some cases, they lead to physically useless infinite values for appropriate quantities.

Then there is a level of other natural sciences like geo- or biosciences, where there is even fewer rigor—and thus, fewer certainty, etc. Once we assign a numerical value describing our degree of certainty in each statement, each level can be described by the range $[\underline{d}, \overline{d}]$ of acceptable degrees of uncertainty. Everything mathematician do should have to satisfy some minimal certainty requirements, and the same is true for physicists—mathematicians may make fun of the physicists' non-perfectly-strict conclusions, but there is a certain high level of rigor that a physical paper must satisfy.

We just mentioned the main levels of certainty, but within each level, there are sub-levels. For example, in mathematics, there is a level of constructive mathematics (seem, e.g., [1, 2, 4, 5, 7, 8, 12, 13, 18]) when the statement $\exists x\, P(x)$ that there exists an object x satisfying a property $P(x)$ is considered proven only if we have an algorithm that effectively constructs such an object. Another sublevel is when, in principle, we allow non-algorithmic constructions but do not allow Axiom of Choice, etc.

The interesting thing about each level (and sub-level) of certainty is that, in contrast to the above example of the product "and"-operation on the interval [0, 1], each level is self-sufficient: mathematicians only use mathematical level of certainty, physicists only use physical level of certainty, etc. In other words, any "and"-combination of statements from the same level belongs to this same level. If we have statement A, B, ..., C from some level, for which the degrees of certainty are at least as high as the minimum $\underline{d}$ acceptable at this level, then our degree of certainty in the combines statement $A \& B \& \ldots \& C$ cannot get lower than this level $\underline{d} > 0$.

So, at first glance, there seems to be contradiction between the usual fuzzy description of uncertainty and the observed existence of clearly distinct levels of certainty.

But is there a contradiction? Let us show that the seeming contradiction comes not from the essence of fuzzy logic, but from the fact that we are using a simplified example of an "and"-operation—and that for the actual, more complex "and"-operation, not only there is no contradiction, but the existence of layers naturally follows.

Indeed, according to the general description of "and"-operations, for any two values $v < v'$ for which $F(v) = f_\&(v, v) = v$ and $F(v') = f_\&(v', v') = v'$—and, as we mentioned, such levels exist for the actual "and"-operation—if we have any two statements A and B with degrees of certainty a and b from the interval $[v, v']$, then the degree of certainty $f_\&(a, b)$ in the statement $A \& B$ also belongs to the same interval. So, for the actual "and"-operation, the values v for which $F(v) = v$ naturally divide the whole range [0, 1] of possible values of degree of certainty into sub-ranges $[\underline{d}, \overline{d}]$ which are self-sufficient—in the sense that:

- if we have statement A, B, ..., C from this sub-range,
- then our degree of certainty in the combines statement $A \& B \& \ldots \& C$ also belongs to this sub-range.

In other words, levels of certainty indeed naturally appear.

4   Below 0 and Above 1?

What about the future? In the previous section, we talked about the current situation. A natural question is: how will it evolve? To answer this question, let us recall what we usually interpret as levels 0 and 1, and what happened with the corresponding levels of certainty in the past.

What do we choose as levels 0 and 1? Naturally, we identify 1 with the level which, at present, corresponds to the largest degree of certainty, and 0 with the level that corresponds to the lowest possible degree of certainty.

Resulting problem. This sounds reasonable but the problem is that, historically, new levels appear all the time—and will probably appear again. For example, the current mathematical level of rigor appeared, it is final form, only in the 19 century, with the "revolution of rigor" that lead to formal definitions of continuity, limits, etc.; see, e.g., [6]. Constructive mathematics—one of the highest level of rigor—appeared only in the 20th century.

Negative and larger-than-1 degrees naturally appear. It is natural to expect that new even higher levels of certainty will appear in the future. How will we be able to describe them—taking into account that the degree 1 is already taken by some level which is lower than this forthcoming one? Naturally, we will end up with degree of certainty which are higher than 1—and, similarly, degree of certainty which are lower than 0. In the probabilistic approach, this would mean having subjective probabilities smaller than 0 or larger than 1.

Can this be related to quantum computing? Such negative and larger-than-1 subjective probabilities may provide a new interpretation for negative value in quantum physics [10, 20]—and especially in quantum computing [16], where, in contrast to the general quantum physics that uses general complex numbers $a + b \cdot i$, only real numbers are used. In this case, real numbers from the interval [0, 1] can be, in principle, interpreted in the usual probabilistic sense, so all that remains is to interpret numbers which are smaller than 0 and numbers which are larger than 1—and this is exactly what actual fuzzy operations provide.

On the qualitative level, this interpretation makes sense. Indeed, as we have mentioned, both negative values and values which are larger than 1 correspond to future knowledge. Naturally, if we use some elements of future knowledge in our computations, this can make computations faster—this may explain why many quantum computing algorithms are indeed faster and more efficient than the corresponding non-quantum ones: e.g., the well-known Grover's quatum algorithm can find an element with the desired property in a n-element list in $\sqrt{n}$ computations steps, which a similar non-quantum algorithm requires at least n steps which is much longer: e.g., for $n = 1,000,000$, it is 1,000 times longer.

Instead of a conclusion: reminder. And all this—explaining levels of certainty, relating to quantum algorithms—naturally follows from the analysis of the actual "and"-operations!

Acknowledgments This work was supported in part by the National Science Foundation grants 1623190 (A Model of Change for Preparing a New Generation for Professional Practice in Computer Science), and HRD-1834620 and HRD-2034030 (CAHSI Includes), and by the AT&T Fellowship in Information Technology.
It was also supported by the program of the development of the Scientific-Educational Mathematical Center of Volga Federal District No. 075-02-2020-1478, and by a grant from the Hungarian National Research, Development and Innovation Office (NRDI).

References

1. Aberth, O.: Precise Numerical Analysis Using C++. Academic Press, New York (1998)
2. Beeson, M.J.: Foundations of Constructive Mathematics. Springer, New York (1985)
3. Belohlavek, R., Dauben, J.W., Klir, G.J.: Fuzzy Logic and Mathematics: A Historical Perspective. Oxford University Press, New York (2017)
4. Bishop, E.: Foundations of Constructive Analysis. McGraw-Hill (1967)
5. Bishop, E., Bridges, D.S.: Constructive Analysis. Springer, New York (1985)
6. Boyer, C.B., Merzbach, U.C.: A History of Mathematics. Wiley, New York (1991)
7. Bridges, D.S.: Constructive Functional Analysis. Pitman, London (1979)
8. Bridges, D.S., Vita, S.L.: Techniques of Constructive Analysis. Springer, New York (2006)
9. Buchanan, B.G., Shortliffe, E.H.: Rule Based Expert Systems: The MYCIN Experiments of the Stanford Heuristic Programming Project. Addison-Wesley, Reading, Massachusetts (1984)
10. Feynman, R., Leighton, R., Sands, M.: The Feynman Lectures on Physics. Addison Wesley, Boston, Massachusetts (2005)
11. Klir, G., Yuan, B.: Fuzzy Sets and Fuzzy Logic. Prentice Hall, Upper Saddle River, New Jersey (1995)
12. Kreinovich, V., Lakeyev, A., Rohn, J., Kahl, P.: Computational Complexity and Feasibility of Data Processing and Interval Computations. Kluwer, Dordrecht (1998)
13. Kushner, B.A.: Lectures on Constructive Mathematical Analysis. American Mathematical Society, Providence, Rhode Island (1984)
14. Mendel, J.M.: Uncertain Rule-Based Fuzzy Systems: Introduction and New Directions. Springer, Cham, Switzerland (2017)
15. Nguyen, H.T., Walker, C.L., Walker, E.A.: A First Course in Fuzzy Logic. Chapman and Hall/CRC, Boca Raton, Florida (2019)
16. Nielsen, M.A., Chuang, I.L.: Quantum Computation and Quantum Information. Cambridge University Press, Cambridge, U.K. (2000)
17. Novák, V., Perfilieva, I., Močkoř, J.: Mathematical Principles of Fuzzy Logic. Kluwer, Boston, Dordrecht (1999)
18. Pour-El, M., Richards, J.: Computability in Analysis and Physics. Springer, New York (1989)
19. Sheskin, D.J.: Handbook of Parametric and Nonparametric Statistical Procedures. Chapman and Hall/CRC, Boca Raton, Florida (2011)
20. Thorne, K.S., Blandford, R.D.: Modern Classical Physics: Optics, Fluids, Plasmas, Elasticity, Relativity, and Statistical Physics. Princeton University Press, Princeton, New Jersey (2017)
21. Tsirel'son, B.S.: A geometrical approach to maximum likelihood estimation for infinite dimensional Gaussian location. I. Theor. Probab. Its Appl. **27**, 411–418 (1982)
22. Zadeh, L.A.: Fuzzy sets. Inf. Control **8**, 338–353 (1965)

Fuzzy Control Systems

A Survey of Empathetic Generative Chatbots

Carolina Martín-del-Campo-Rodríguez, Grigori Sidorov, and Ildar Batyrshin

Abstract The purpose of Chatbots also known as Dialogue Systems or Conversational Agents is to have natural conversations indistinguishable from human ones. Perception and expression of emotion are key factors to the success of Dialogue systems or conversational agents. In this review article, we focus on the literature on Empathetic Chatbots. The main goal of this review is to serve as a comprehensive guide to research and development on Empathetic Generative Chatbot and to suggest future directions in this domain.

Keywords Chatbots · Emotion and perception · Dialog systems

1 Introduction

The purpose of a Chatbot is to have natural conversations indistinguishable from human ones. Many companies are working in the development of Dialog Systems, such as Google (Google Assistant,[1] LaMDA[2]), Apple (Siri[3]), OpenAI (GPT series[4]), Amazon (Alexa[5]), among others.

There are some common challenges when building a Dialogue System, most of which are active research areas [2], such as: incorporating context (responses need to incorporate both linguistic context and physical context), coherent person-

---

[1] https://assistant.google.com/.
[2] https://blog.google/technology/ai/lamda/.
[3] https://www.apple.com/mx/siri/.
[4] https://openai.com/blog/openai-api/
[5] https://developer.amazon.com/en-US/alexa.

C. Martín-del-Campo-Rodríguez · G. Sidorov · I. Batyrshin (✉)
Centro de Investigación en Computación (CIC), Instituto Politécnico Nacional (IPN),
Mexico City, Mexico
e-mail: batyr1@gmail.com

G. Sidorov
e-mail: sidorov@cic.ipn.mx

Sh. N. Shahbazova et al. (eds.), *Recent Developments and the New Directions of Research,
Foundations, and Applications*, Studies in Fuzziness and Soft Computing 423,
https://doi.org/10.1007/978-3-031-23476-7_19

ality (responses should ideally produce consistent answers to semantically identical inputs), intention and diversity (avoid the use of generic responses); [5] proactivity (capability of autonomously acting on the user's behalf), damage control (to deal with either conflict or failure situations); [31] emotional intelligence (to perceive, integrate, and understand emotions).

In recent years, research on the perception and expression of emotions by Chatbots has grown, since showing empathy leads to a more natural conversation.

In this paper were summarize research which focuses on including emotional information into Chatbots. Section 2 describes the classification of Chatbots as well as evaluation for them. Section 3 discusses research that focuses on a Seq2Seq architecture based on Recurrent Neural Networks (RNNs) and their variants to create Emotional-Aware Chatbots. Section 4 compares approaches taken in the state of the art using Transformers [28]. A conclusion is given in Sect. 5 and possible future research areas.

2   Theorical Framework

In this section, a definition of Chatbots is given, as well as an introduction to the different types of evaluations.

2.1   Chatbots Classification

Chatbots are a type of Dialogue System.[6] There are two categories for Dialogue Systems: Task-Oriented and Non-Task-Oriented.

1. **Task-Oriented Dialogue Systems**: single-purpose programs that focus on performing one function. Using rules, NLP, and very little ML, they generate automated but conversational responses to user inquiries. Interactions with these Dialogue Systems are highly specific and structured and are most applicable to support and service functions-think robust, interactive FAQs [7].

2. **Non-Task-Oriented Dialogue Systems (Chatbots)**: these systems focus on conversing with humans on open domains. This type of Dialogue System is used mainly for entertainment, however it can also for practical purposes like making task-oriented agents more natural [12]. There are two models for non-task-oriented Dialogue Systems [2]:

   – Retrieval-Based Model: this model uses a repository of predefined responses and some kind of heuristic to pick an appropriate response based on the input and context. The heuristic could be as simple as a rule-based expression match,

---

[6] Programs that communicate with users in natural language (text, speech, or both) [12].

or as complex as an ensemble of Machine Learning classifiers. These systems don't generate any new text, they just pick a response from a fixed set.

– Generative Model: unlike the retrieval-based model, this model does not rely on pre-defined responses. They generate new responses from scratch. Generative models are typically based on Machine Translation techniques, but instead of translating from one language to another, they "translate" from input to output (response).

This article focuses on the investigation of Generative Chatbots that center on carrying out an empathetic conversation.

2.2   Chatbot Evaluation

An important aspect of Generative Chatbots is how to evaluate the quality of the generated response [17]. There are two methods to evaluate Chatbots: automatic measures and human evaluation.

– **Automatic Measures**: For Dialogue Systems (usually close-domain) measures for translation and summarizing have been used, like BLEU [20] (Bilingual Evaluation Understudy), METEOR [13] (Metric for Evaluation of Translation with Explicit ORdering), and ROUGE [15] (Recall-Oriented Understudy for Gisting Evaluation), these metrics measure the word overlap between the generated responses and a reference. Many researchers argue that these metrics are not appropriate for the evaluation of open-domain Dialogue Systems since there are many possible correct responses to the same input, while the number of reference responses in a test set is always limited. In [17], Liu et al. showed that neither of the word-overlap-based scores has any correlation to human judgments.

   In the state of the art also is used the automatic measure perplexity [4] to evaluate the model at a context level. In [24], the authors demonstrated that perplexity is strongly negatively correlated with the Sensibleness and Specificity Average (SSA) score, a human evaluation metric for open-domain Dialogue Systems.

– **Human Evaluations**: While automatic evaluation measures dimensions of dialogue objectively, human evaluation captures the subjective assessment from the user's point of view. All human evaluations involve gathering external annotators who answer questions regarding the dialogues resulting from a Dialogue System. There are two categories for human evaluation of open-domain Dialogue Systems: static evaluation and interactive evaluation [9]:

   • Static Evaluation: is an offline procedure where the evaluators never directly interact with the Dialogue Systems under review; instead, they are provided with dialogue excerpts.

   • Interactive Evaluations: unlike static evaluation, interactive evaluation has the same person playing the role of both the user (one who interacts with the system) and the evaluator. In this setup, the evaluator has a conversation with the dialogue system and assesses at the end of the conversation.

Authors typically use both types of evaluations to determine the performance of Chatbots and have a point of comparison with the state of the art. For automatic evaluation, BLEU and perplexity are the most used measures. On the other hand, there are many approaches to human evaluations to determine not only how well generated sentences are (grammatically) and how aware of the context they are, but also to determine how empathetic Chatbots are. Some of the human evaluations used are:

A/B Test A/B Test consists of showing the evaluator a prompt and two possible responses from models which are being compared [25]. The prompt can consist of a single utterance or a series of utterances. The user picks the better response or specifies a tie. When both model responses are the same, a tie is automatically recorded.

Empathy, Relevance, and Fluency In [23] authors used Empathy, Relevance, and Fluency to evaluate their Chatbot. They collected 100 ratings and asked about three aspects of performance, all rated on a Likert scale[7] (1: not at all, 3: somewhat, 5: very much)

- Empathy/Sympathy: did the responses show understanding of the feelings of the person talking about their experience?
- Relevance: did the responses seem appropriate to the conversation? Were they on-topic?
- Fluency: could you understand the responses? Did the language seem accurate?

ACUTE-EVAL In [14] authors introduced ACUTE-EVAL, a pairwise relative comparison setup for multi-turn dialogues. In each trial, the authors show the annotator two whole conversations (conv_A and conv_B), with the second speaker (sp_2) in each conversation highlighted, as the judgment should be independent of the quality of the first speaker. Then show a question with two choices: sp_2 of conv_A or sp_2 of conv_B, where the question measures the desired quality such as which speaker is more engaging, interesting, or knowledgeable.

Sensibleness and Specificity Average In 2020, Adiwardana et al. [1] proposed Sensibleness and Specificity Average (SSA), a human evaluation metric that combines two aspects of a human-like Chatbot: making sense and being specific. Authors crowd-source free-from conversations. For each utterance, the evaluator answers two questions using common sense:

- does it make sense?
- is it specific?

---

[7] The Likert scale is a five (or seven) point scale which is used to allow the individual to express how much they agree or disagree with a particular statement. It assumes that the strength/intensity of an attitude is linear [19].

Despite research efforts to create a standard for Chatbots evaluation, there is no agreement on which one should be followed.

3   Emotional-Aware Generative Chatbots

Recent research focuses not only on generating semantically appropriate responses, but also considering the emotions involved to generate empathy in responses.

The first work to address the emotion factor in large-scale conversation generation was the one proposed by Zhou et al. in 2017 [31]. The authors proposed an Emotional Chatting Machine (ECM). For training, an emotion classifier is used; it receives a corpus of post-response pairs and generates the emotion label of each response; then ECM is trained on the data of triples: posts, responses, and emotion labels. In the inference process, a post is fed to ECM to generate emotional responses conditioned on different emotion categories (angry, disgust, sad, happy, and like). The ECM model is composed of an encoder-decoder GRU-based [6] with external memory (to model emotion expressions explicitly). Authors represent each emotion as a real-valued, low dimensional vector (learned during the training process), this vector is passed to the decoder along with the context vector and the embedding vector to update the decoder state. For the evaluation of the Chatbot, the authors compared with two baselines, a general Seq2Seq model and an emotion category embedding model (proposed by the authors). The authors used the following evaluations:

- The automatic evaluations perplexity and accuracy (due to the dataset used, Emotional STC[8] (ESTC, based on STC, a one-round conversation data collected from a microblogging service), specify the expected emotion for the answer)
- For human evaluations: given a post and an emotion category, responses generated from all the models were randomized and presented to three human annotators that qualified content (using a rating scale 0, 1, 2; annotators were asked to qualify whether the response is appropriate and natural to a post and could plausibly have been produced by a human) and emotion(using a rating scale 0, 1; annotator qualify whether the emotional expression of a response agrees with the given emotion category).

ECM obtained the best performance considering accuracy, but not in perplexity. According to the human evaluation, ECM obtained the best performance: 27.2% of the responses generated by ECM have a Content score of 2 and an Emotion score of 1, while only 22.8% for Embeddings and 9.0% for Seq2Seq.

In 2018, Sun et al. [27] proposed the use of Reinforcement Learning (RL) and Generative Adversarial Networks (GANs [10]) in the architecture to recognize the emotions and generate responses according, using a dataset constructed from Weibo[9]

---

[8] Emotional STC: Short-Text Conversation Dataset [26] annotated with six emotion categories (Angry, Disgust, Happy, Like, Sad and Other).

[9] Weibo: is a Chinese social networking website.

posts and replies/comments. Authors use 'emotional tags' for posts and responses, that represent the emotion in the sentences. The emotional labels (happiness, anger, disgust, sadness, like, and other) are concatenated at the end of the sentences (post and responses). For the generator a complete Seq2Seq model is used (LSTM-based [11]), using RL (due to the limitations of the GANs with discrete tokens); a policy gradient method reward conversation sequences that seem more human, and in addition the emotion tags represent the response, which they use as a rewarding part of it so that the emotions of the real responses can be closer to the specified emotions. The discriminator is a network that distinguishes between human and machine-generated dialogues. The network gives a probability p base on the input pair post, response; p represents the probability that the input pair is generated from humans. The authors evaluated their results based in two aspects:

1. Emotional Consistency: Whether the emotional category of a generated response is the same as the pre-specified category
2. Coherence: Whether the response is appropriate in terms of both logic and content

Compare with related work, the authors obtain an improvement concerning the emotion consistency, but the coherence of the sentence generated had a bad performance.

In [30] (2018), Zhong et al. proposed a model to generate natural responses and rich in affect; based on an extension of the Seq2Seq model (LSTM-based) and adopted VAD[10] [29] (Valence, Arousal and Dominance) affective notations to embed each word with affects. The authors assigned VAD values to words based in their lemmas (the VAD value was concatenated to the embedding of each word). Authors used an attention mechanism to solve the problem of the limited representation power output of the encoder and, under the assumption that humans pay extra attention to affect-rich words during conversations, the authors bias the attention towards affect-rich words (affective attention) and proposed an affective loss which introduces a probability bias into the decoder language model towards affect-rich words. To evaluate the model the authors used perplexity and for human evaluation, authors ask five annotators to evaluate two aspects, content and emotion:

- +2: (content) The response has correct grammar and is relevant and natural/(emotion) The response has adequate and appropriate emotions conveyed
- +1: (content) The response has correct grammar but is too universal/(emotion) The response has inadequate but appropriate emotions conveyed
- 0: (content) The response has either grammar errors or is completely irrelevant/(emotion) The response has either no or inappropriate emotions conveyed

The authors demonstrated that their approach (the use of affective attention and affective loss) had an improvement concerning models with similar architectures.

---

[10] Three components of emotions are traditionally distinguished: valence (the pleasantness of a stimulus), arousal (the intensity of emotion provoked by a stimulus), and dominance (the degree of control exerted by a stimulus).

4   The Use of Transformers to Build Emotional Chatbots

Transformers were introduced by Vaswani et al. [28] in 2017. Transformers are language models based solely on attention mechanisms, dispensing with recurrence. They provide training parallelization that permits training on larger datasets. This led to the development of pre-trained systems such as BERT (Bidirectional Encoder Representations from Transformers) [8] and GPT series (Generative Pre-trained Transformer) [3, 21, 22], which were trained using huge datasets and may be fine-tuned for specific applications, such as dialogue generation. The rest of the section will explain some research related to empathetic generative Chatbots based on Transformers.

In 2019 Rashkin et al. proposed the dataset Empathetic Dialogues [23], a benchmark for empathetic dialogue generation, grounded in emotional situations. The EmpatheticDialogues dataset consists of 25 thousand single-turn and multi-turn empathetic conversations, where each conversation contains a label that describes the overall emotion of the conversation. They also proposed a Transformer-based baseline, pre-trained in Reddit conversations and fine-tuned with EmpatheticDialogues. The authors use perplexity and BLEU to evaluate their proposal, as well as human evaluation Empathy, Relevance, and Fluency; showing that fine-tuning conversation models with their dataset, leads to responses that are evaluated as more empathetic. Table 1 shows the results obtained.

In 2019, Lin et al. proposed in [16] the model MoEL (Mixture of Empathetic Listeners). MoEL is composed of three components: an emotion tracker, emotion-aware listeners, and a meta listener (all composed of standard Transformers [28]).

- The emotion tracker uses the standard Transformer encoder. First, all dialogue turns are concatenated and map each token into its vectorial representation using the context embedding (summing up the word embedding, dialogue state embedding, and positional embedding), then the encoder encodes the context sequence into a context representation. The authors added a query token QRY at the beginning of each input sequence to compute the weighted sum of the output tensor. This is used as the query for generating the emotion distribution
- The emotion-aware listeners consist of a shared listener that learns shared information for all emotions and n independently parameterized Transformer decoders that learn how to appropriately react given to a particular emotional state
- The meta Listener is another transformer decoder layer, which further transforms the representation of the listeners and generates the final response. The intuition is that each listener specializes to a certain emotion and the Meta Listener gathers the opinions generated by multiple listeners to produce the final response

The result of the work is shown in Table 1.

Majumder et al. [18], proposed in 2020 MIME based on the assumption that empathetic responses often mimic the emotion of the speaker. The authors used the same approach that the one used in MoEL for emotion tracker, enforcing emotion understanding in the context representation by classifying user emotion during training. For the response emotion, the authors group the 32 emotions used into two groups

Table 1 Comparison of evaluation of generative Chatbots based in transformers

Approach	BLEU	Empathy	Relevance	Fluency
Rashkin et al. [23]	6.27	3.25	3.33	**4.30**
Lin et al. [16]	2.90	3.44	**3.70**	3.47
Majumder et al. [18]	2.98	**3.87**	3.60	4.28

Best results are shown in bold

containing positive and negative emotions, resulting in 13 positive and 19 negative emotions (this split was made by the intuition of the authors). The authors introduce stochasticity in the response-emotion determination that results in emotionally varied responses. To evaluate the model the authors used BLEU and Empathy, Relevance, and Fluency (using three human annotators). Table 1 show the results.

5 Discussion

Despite the current advances in the fields of Deep Learning and Natural Language Processing, Empathetic Chatbot still has some shortcomings, such as: resorting to generic and repetitive answers; generating responses that are not relevant (out of context); generating responses that are grammatically wrong; poor recognition of the user's emotions, therefore generating some responses that are not considered empathetic. With the use of transformers, some of these problems have been mitigated however, there is still no model that can solve these problems.

According to the results of the state of the art, transformer-based models seem to be the right direction to try to build Empathetic Chatbots. To take advantage of the state of the art, research can focus on fine-tuning pre-trained transformer-based models so that, together with a complementary architecture, they can correctly represent and process emotions.

Regarding the evaluation of Empathetic Chatbot, although some automatic evaluations are widely used in the state of the art, there is no specific metric or set of metrics that are used as a reference, which limits testing and comparison with other models. In addition, there is no automatic evaluation measure to specifically evaluate the emotional aspect. For this reason, many models are based on human evaluation which can evaluate emotions aspects of the content but, like automatic evaluations, lack a standardized procedure, which makes it difficult to reproduce the results and compare the models directly.

The costs, time consumption, and bias of human-based evaluations will lead to research for scalable and robust procedures that provide a performance evaluation of Empathetic Chatbots in the context of coherence, correctness, and emotion of their dialogues.

Acknowledgement The work was done with support from the Mexican Government through the grant of the CONACYT, Mexico and grant 20220857 of the Secretaría de Investigación y Posgrado of the Instituto Politécnico Nacional, Mexico.

References

1. Adiwardana, D., Luong, M.T., So, D.R., Hall, J., Fiedel, N., Thoppilan, R., Yang, Z., Kulshreshtha, A., Nemade, G., Lu, Y., Le, Q.V.: Towards a Human-Like Open-Domain Chatbot (2020)
2. Britz, D.: Deep Learning for Chatbots, Part 1—Introduction. http://www.wildml.com/2016/04/deep-learning-for-chatbots-part-1-introduction/
3. Brown, T.B., Mann, B., Ryder, N., Subbiah, M., Kaplan, J., Dhariwal, P., Neelakantan, A., Shyam, P., Sastry, G., Askell, A., Agarwal, S., Herbert-Voss, A., Krueger, G., Henighan, T., Child, R., Ramesh, A., Ziegler, D.M., Wu, J., Winter, C., Hesse, C., Chen, M., Sigler, E., Litwin, M., Gray, S., Chess, B., Clark, J., Berner, C., McCandlish, S., Radford, A., Sutskever, I., Amodei, D.: Language Models Are Few-Shot Learners (2020)
4. Campagnola, C.: Perplexity in Language Models (2020). https://chiaracampagnola.io/2020/05/17/perplexity-in-language-models/
5. Chaves, A.P., Gerosa, M.A.: How should my chatbot interact? A survey on social characteristics in human-chatbot interaction design. Int. J. Hum.-Comput. Interact. 1–30 (2020). https://doi.org/10.1080/10447318.2020.1841438
6. Cho, K., van Merrienboer, B., Gulcehre, C., Bahdanau, D., Bougares, F., Schwenk, H., Bengio, Y.: Learning Phrase Representations Using RNN Encoder-Decoder for Statistical Machine Translation (2014)
7. Corporation, O.: What Is a Chatbot? https://www.oracle.com/middleeast/chatbots/what-is-a-chatbot/
8. Devlin, J., Chang, M.W., Lee, K., Toutanova, K.: Bert: Pre-training of deep bidirectional transformers for language understanding (2019)
9. Finch, S.E., Choi, J.D.: Towards Unified Dialogue System Evaluation: A Comprehensive Analysis of Current Evaluation Protocols (2020)
10. Goodfellow, I.J., Pouget-Abadie, J., Mirza, M., Xu, B., Warde-Farley, D., Ozair, S., Courville, A., Bengio, Y.: Generative Adversarial Networks (2014)
11. Hochreiter, S., Schmidhuber, J.: Long short-term memory. Neural Comput. **9**, 1735–80 (12 1997). https://doi.org/10.1162/neco.1997.9.8.1735
12. Jurafsky, D.: Speech and Language Processing, 3rd edn. draft (2020). https://web.stanford.edu/~jurafsky/slp3/
13. Lavie, A., Agarwal, A.: Meteor: an automatic metric for mt evaluation with high levels of correlation with human judgments. In: Proceedings of the Second Workshop on Statistical Machine Translation, pp. 228–231. StatMT '07, Association for Computational Linguistics, USA (2007)
14. Li, M., Weston, J., Roller, S.: Acute-Eval: Improved Dialogue Evaluation with Optimized Questions and Multi-turn Comparisons (2019)
15. Lin, C.Y.: Rouge: a package for automatic evaluation of summaries. In: Proceedings of ACL Workshop on Text Summarization Branches Out, p. 10 (2004). http://research.microsoft.com/~cyl/download/papers/WAS2004.pdf
16. Lin, Z., Madotto, A., Shin, J., Xu, P., Fung, P.: Moel: Mixture of Empathetic Listeners (2019)

17. Liu, C.W., Lowe, R., Serban, I., Noseworthy, M., Charlin, L., Pineau, J.: How NOT to evaluate your dialogue system: an empirical study of unsupervised evaluation metrics for dialogue response generation. In: Proceedings of the 2016 Conference on Empirical Methods in Natural Language Processing. pp. 2122–2132. Association for Computational Linguistics, Austin, Texas (2016). https://doi.org/10.18653/v1/D16-1230, https://aclanthology.org/D16-1230

18. Majumder, N., Hong, P., Peng, S., Lu, J., Ghosal, D., Gelbukh, A., Mihalcea, R., Poria, S.: Mime: Mimicking Emotions for Empathetic Response Generation (2020)

19. McLeod, S.: Likert Scale Definition, Examples and Analysis (2019). https://www.simplypsychology.org/likert-scale.html

20. Papineni, K., Roukos, S., Ward, T., Zhu, W.J.: Bleu: a method for automatic evaluation of machine translation. In: Proceedings of the 40th Annual Meeting on Association for Computational Linguistics, pp. 311–318. ACL '02, Association for Computational Linguistics, USA (2002). https://doi.org/10.3115/1073083.1073135

21. Radford, A., Sutskever, I.: Improving Language Understanding by Generative Pre-training (2018)

22. Radford, A., Wu, J., Child, R., Luan, D., Amodei, D., Sutskever, I.: Language Models Are Unsupervised Multitask Learners (2019)

23. Rashkin, H., Smith, E.M., Li, M., Boureau, Y.L.: Towards Empathetic Open-Domain Conversation Models: A New Benchmark and Dataset. ACL (2019)

24. Roller, S., Dinan, E., Goyal, N., Ju, D., Williamson, M., Liu, Y., Xu, J., Ott, M., Shuster, K., Smith, E.M., Boureau, Y.L., Weston, J.: Recipes for Building an Open-Domain Chatbot (2020)

25. Sedoc, J., Ippolito, D., Kirubarajan, A., Thirani, J., Ungar, L., Callison-Burch, C.: ChatEval: a tool for chatbot evaluation. In: Proceedings of the 2019 Conference of the North American Chapter of the Association for Computational Linguistics (Demonstrations), pp. 60–65. Association for Computational Linguistics, Minneapolis, Minnesota (2019). https://doi.org/10.18653/v1/N19-4011, https://aclanthology.org/N19-4011

26. Shang, L., Lu, Z., Li, H.: Neural Responding Machine for Short-Text Conversation (2015)

27. Sun, X., Chen, X., Pei, Z., Ren, F.: Emotional human machine conversation generation based on seqgan. In: 2018 First Asian Conference on Affective Computing and Intelligent Interaction (ACII Asia), pp. 1–6 (2018). https://doi.org/10.1109/ACIIAsia.2018.8470388

28. Vaswani, A., Shazeer, N., Parmar, N., Uszkoreit, J., Jones, L., Gomez, A.N., Kaiser, L.U., Polosukhin, I.: Attention is all you need. In: Guyon, I., Luxburg, U.V., Bengio, S., Wallach, H., Fergus, R., Vishwanathan, S., Garnett, R. (eds.) Advances in Neural Information Processing Systems, vol. 30. Curran Associates Inc. (2017). https://proceedings.neurips.cc/paper/2017/file/3f5ee243547dee91fbd053c1c4a845aa-Paper.pdf

29. Warriner, A., Kuperman, V., Brysbaert, M.: Norms of valence, arousal, and dominance for 13,915 english lemmas. Behav. Res. Methods **45** (2013). https://doi.org/10.3758/s13428-012-0314-x

30. Zhong, P., Wang, D., Miao, C.: An Affect-Rich Neural Conversational Model with Biased Attention and Weighted Cross-Entropy Loss (2018)

31. Zhou, H., Huang, M., Zhang, T., Zhu, X., Liu, B.: Emotional chatting machine: emotional conversation generation with internal and external memory. CoRR abs/1704.01074 (2017). http://arxiv.org/abs/1704.01074

Two-Level Classifier for Detecting Categories of Offensive Language

Segun Taofeek Aroyehun⬤, Alexander Gelbukh⬤, and Grigori Sidorov⬤

Abstract We explore the task of offensive content classification on the HASOC 2021 shared task dataset. Our approach is based on two-level classification scheme (corresponding to Subtasks 1A and 1B respectively). The first level is a binary classification (offensive or not). The second level further classifies only the offensive instances. The classifier at each level is a fine-tuned transformer model. Our model on the English dataset achieves an overall best macro F1 score of 0.831 and 0.666 on Subtasks 1A and 1B, respectively. The model performance on Hindi and Marathi is competitive: macro F1 score of 0.778 for Hindi Subtask 1A and 0.553 for Hindi Subtask 1B (fourth place on the leaderboard), and a macro F1 score of 0.847 for Marathi Subtask 1A (13th on the leaderboard).

Keywords Offensive content identification · Deep learning · Text classification · Multilingual · Sentiment analysis · Social media

1 Introduction

Interactions online sometimes result in cases of uncivil behaviour which can negatively affect individuals, groups, and the society at large. As a result, online plat-

The authors thank CONACYT for the computer resources provided through the INAOE Supercomputing Laboratory's Deep Learning Platform for Language Technologies. We are grateful for the support from Microsoft and Google through the Latin America PhD Award and the Latin America Research Award, respectively.

S. T. Aroyehun (✉)
Graz University of Technology, Graz, Austria
e-mail: aroyehun.segun@gmail.com

A. Gelbukh · G. Sidorov
CIC, Instituto Politécnico Nacional Mexico City, Mexico City, Mexico
e-mail: gelbukh@cic.ipn.mx

G. Sidorov
e-mail: sidorov@cic.ipn.mx

forms have the responsibility of putting in place measures that will facilitate safety of its users. One of such measures is content moderation to ensure that users are not exposed to harmful content. However, the scale of user-generated content requires automatic approaches for moderation. It is a usual practice to flag harmful contents for automatic removal and/or review by human moderators.

Several approaches have been explored to automatically identify offensive content. Traditionally, feature engineering in combination with classical machine learning models such as support vector machines, logistic regression and Naive Bayes have shown competitive performance [12]. More recently, neural networks have been shown to perform better than traditional approaches by using architectures such as GRU, LSTM, and CNN in combination with word embeddings [2]. The introduction of contextual word embeddings based on the pre-trained language models [9] and the transformer architecture [20] has paved way for achieving superior results on several NLP tasks including offensive content identification [17]. Current approaches vary from fine-tuning using the downstream task [19], intermediate pre-training [16] to adapting to the domain of the target task [11].

Pre-trained models have different inductive biases with respect to the downstream task depending on the prior knowledge acquired during pre-training. This is usually a function of the model architecture, source corpora, and the task(s). We evaluate the downstream performance of different pre-trained transformer models on the task of identification and classification of offensive content on the data provided for the HASOC 2021 shared task [14] in English, Hindi, and Marathi as team NLP-CIC. In addition, we use a two-level classification scheme for our submission to Subtask 1B (only for the last submission on English data). For English, we fine-tune monolingual models with our best model achieving an overall best macro F1 score of 0.831 and 0.666 on Subtasks 1A and 1B respectively in the competition. For Hindi and Marathi, we use multilingual models on the combination of datasets for both languages resulting in competitive performance. On Hindi dataset, we achieve a macro F1 score of 0.778 and 0.553 (fourth on the leaderboard) respectively for Subtasks 1A and 1B. The F1 score on Subtask 1A for Marathi is 0.847 (13th on the leaderboard).

2   Methodology

Task. We tackle the problem of classifying offensive content in social media comments written in English [13], Hindi [13], and Marathi [10] spanning Subtasks 1A and 1B. Subtask 1A deals with binary classification of comments as containing hate, profane or offensive content (HOF) or not (NOT). Subtask B addresses multiclass classification problem, where a given comment is to be assigned a label out of the categories non-offensive (NONE), profane (PRFN), offensive (OFFN), and hate (HATE).

Data. The HASOC 2021 shared task [14] made available annotated twitter datasets for Subtasks 1 and 2. Subtask 1 focuses on comments without conversational context. Subtask 2 considers comments in their conversational thread that are written

Table 1 Details of the dataset for Subtasks 1A and 1B for each language

	1A		1B				Total
	NOT	HOF	NONE	OFFN	PRFN	HATE	
English	1342	2501	1342	1196	622	683	3843
Hindi	3161	1433	3161	654	566	213	4594
Marathi	1205	669	–	–	–	–	1874

Total is the number of labeled examples per language
HOF—hate, Offensive or profane, and PRFN—profane

using code-mixed languages. We only use the dataset for Subtask 1. Table 1 shows the number of examples per category for each language. We normalize the text by transforming occurrences of user mentions and links to @USER and URL. Consecutive whitespaces are replaced by a single one and punctuation marks are separated from words. The text is tokenized with the respective tokenizer for the transformer models.

Approach. Our aim is to evaluate different pre-trained transformer models with different inductive biases on the task of offensive content classification. For English, we consider the following pre-trained monolingual models, the identifier on the Huggingface model hub is in parenthesis:

- Muppet (facebook/muppet-roberta-base) [1]: derived from a large-scale multitask learning on several datasets and tasks starting from a RoBERTa-base model with the goal of learning a representation that can generalize better to many downstream tasks.
- HateBERT (GroNLP/hateBERT) [7]: a BERT model re-trained on a collection of Reddit comments in English gathered from communities known to be offensive. This model has been shown to outperform a generic BERT model on tasks related to offensive content identification.
- BERTweet-base (vinai/bertweet-base) [15]: a pre-trained language model for English tweets with the same architecture as BERT-base.
- BERTweet-base-sentiment (cardiffnlp/bertweet-base-sentiment): Same as BERTweet-base and with additional training on sentiment analysis data [18] (the split provided in Tweeteval [6]).
- BERTweet-large (vinai/bertweet-large) [15]: a pre-trained language model for English tweets with the same architecture as BERT-large.

For Hindi and Marathi, we use variants of the multilingual XLM-RoBERTa [8] as they have been shown to be competitive for offensive language identification [4], the respective model identifier on the Huggingface model hub is in parenthesis:

- XLM-twitter (cardiffnlp/twitter-xlm-roberta-base) [5]: XLM-RoBERTa language model trained on 200M tweets for over 30 languages.
- XLM-twitter-sentiment (cardiffnlp/twitter-xlm-roberta-base-sentiment): The same as XLM-twitter model, it has been fine-tuned on a multilingual sentiment analysis dataset for eight languages.

- XLM-large (xlm-roberta-large) [8]: multilingual version of RoBERTa covering 100 languages. The transformer architecture has 24 layers and a hidden dimension of 1024.

On Subtask 1B, our last submission for English employ a two-level classification scheme. The first level is binary classification and the second level is multiclass classification only on the HOF subset of the dataset. During inference, only examples that receive a HOF label in the fist stage are considered for prediction using the second level classifier. This approach is similar to the one in [3], where the model for the second level is initialized with weights from the first level. Here, we instead use the model prediction to selectively pass samples to the second level classifier during inference. Our experiments on Hindi and Marathi follow the same approach. In addition, we train the multilingual models on the combination of datasets for Hindi and Marathi. Our intuition is that the two languages are similar and are likely to benefit from the larger size of the combined dataset.

Experimental Settings. We split the HASOC 2021 labeled dataset using stratified sampling into train (90%) and development (10%) sets. We fine-tune the transformer models in addition to a dense layer during training. We use the Huggingface Transformers library [21]. The sequence length is set to 128 tokens in all experiments. We use a random seed of 42. The model training uses a maximum learning rate of 1e-5 with AdamW as optimizer (without bias correction), a drop out rate of 0.2 for Subtask 1A and 0.1 for Subtask 1B on the input to the dense layer, a batch size of 128 or 64, label smoothing factor of 0.1, weight decay of 0.01, warm up ratio of 0.1, and a maximum gradient norm of 1.0. We train for a maximum of 10 epochs or until 2 evaluations occur without any improvement of the model performance on the development set. Evaluation on the development set is done after every 20 batches. We run all experiments on a single Nvidia V100 GPU (32GB).

3   Results

Table 2 shows the scores received by our submissions for the English Subtask 1A. The BERTweet-large model achieved the best scores with an F1 score of 0.831. The second best model in terms of F1 score is the BERTweet-base model with an F1 score

Table 2 Macro-average scores on the development and test sets for English Subtask 1A

Model		Test set		
	Dev. F1	Precision	Recall	F1
Muppet	**0.816**	0.829	0.783	0.796
HateBERT	0.787	0.796	0.783	0.788
BERTweet-base	0.803	0.830	0.812	0.819
BERTweet-base-sentiment	0.807	0.813	0.800	0.805
BERTweet-large	0.787	**0.841**	**0.823**	**0.831**

Best scores are in bold

Table 3 Macro-average scores on the development and test sets for English Subtask 1B

Model	Dev. F1	Test set		
		Precision	Recall	F1
Muppet	0.668	0.623	0.632	0.622
HateBERT	0.654	0.620	0.627	0.620
BERTweet-base	0.663	0.648	0.628	0.622
BERTweet-base-sentiment	0.674	0.615	0.623	0.609
BERTweet-large-HOF	–	**0.669**	**0.668**	**0.666**

Best scores are in bold
BERTweet-large-HOF is a model trained only on the offensive, hate, and profane examples

Table 4 Macro-average scores on the development and test sets for Hindi (HI) and Marathi (MR) Subtask 1A

Model	Dev. F1 (HI+MR)	HI Test set			MR Test set		
		P	R	F1	P	R	F1
XLM-twitter	**0.811**	0.769	0.742	0.752	0.833	0.836	0.835
XLM-twitter-sentiment	0.797	0.766	0.770	0.766	**0.855**	0.841	**0.847**
XLM-large	0.809	**0.786**	**0.771**	**0.778**	0.836	**0.845**	0.840

P is Precision and R is RecallBest scores are in bold

of 0.819. These results show that the inductive bias derived from training on tweets is relevant for the task. For English Subtask 1B, the scores in Table 3 indicate that the BERTweet-large-HOF is the best performing model with an F1 score of 0.666. This model is only trained on the HOF subset of the training set consisting of the HATE, OFFN, and PRFN samples. We use the prediction made by the corresponding model for Subtask 1A if it is NOT as NONE for Subtask 1B. The performance of this approach shows that it is effective. The second best model which did not employ the prediction from Subtask 1A is about 4 F1 points lower at 0.622.

For Hindi and Marathi, the performance of the multilingual models for Subtask 1A is in Table 4. On Hindi test data, the model with a larger number of parameters, XLM-large, performs best out of our submissions with the XLM-twitter-sentiment model in second place. For Marathi, the model that has been further trained on Twitter data and multilingual sentiment analysis dataset achieves the best performance, slightly better than the XLM-large model by 0.7 F1 point. These results provide some evidence that a multilingual model that has been trained on multilingual tweets is competitive for offensive language identification on Hindi and Marathi datasets, even when compared with a larger model. Table 5 presents the performance scores for Hindi Subtask 1B. The XLM-twitter-sentiment model performs best with an F1 score of 0.553. To our surprise, the XLM-large model has the least performance. We think that the number of training examples in the HOF subset (about 1400) is a likely factor as large models tend to overfit on small training sets. While we combine the prediction from Subtask

Table 5 Macro-average scores on the development and test sets for Hindi Subtask 1B. Best scores in bold.

Model		Test set		
	Dev. F1 HOF subset	Precision	Recall	F1
XLM-twitter	0.625	**0.562**	0.547	0.548
XLM-twitter-sentiment	**0.637**	0.555	**0.555**	**0.553**
XLM-large	0.571	0.503	0.457	0.469

1A and the prediction of the same model trained on the HOF subset, it would be interesting to examine whether a combination from different models can improve final predictions on Subtask 1B.

4 Conclusion

We explore the task of offensive content identification by evaluating different pre-trained transformer models with varying inductive biases within a two-level classification scheme. We perform experiments on English, Hindi, and Marathi datasets. The macro F1 score for our best monolingual model on the English dataset rank our approach first on the leaderboard with 0.831 for Subtask 1A and 0.666 for Subtask 1B. We record competitive macro F1 scores on Hindi and Marathi using multilingual models: 0.778 and 0.553 (fourth on the leaderboard) for Hindi Subtasks 1A and 1B, respectively, and 0.847 (13th on the leaderboard) for Marathi Subtask 1A. As future work, we will explore effective and efficient approaches for combining the knowledge from different pre-trained models for offensive language identification.

References

1. Aghajanyan, A., Gupta, A., Shrivastava, A., Chen, X., Zettlemoyer, L., Gupta, S.: Muppet: Massive multi-task representations with pre-finetuning (2021). arXiv preprint arXiv:2101.11038
2. Aroyehun, S.T., Gelbukh, A.: Aggression detection in social media: using deep neural networks, data augmentation, and pseudo labeling. In: Proceedings of the First Workshop on Trolling, Aggression and Cyberbullying (TRAC-2018), pp. 90–97 (2018)
3. Aroyehun, S.T., Gelbukh, A.: NLP-CIC at HASOC 2020: multilingual offensive language detection using all-in-one model. In: FIRE (Working Notes), pp. 331–335 (2020). http://ceur-ws.org/Vol-2826/T2-31.pdf
4. Aroyehun, S.T., Gelbukh, A.: Evaluation of intermediate pre-training for the detection of offensive language. In: Proceedings of the Iberian Languages Evaluation Forum (IberLEF 2021), pp. 313–320 (2021, September). http://ceur-ws.org/Vol-2943/meoffendes_paper8.pdf

5. Barbieri, F., Anke, L.E., Camacho-Collados, J.: XLM-T: A multilingual language model toolkit for Twitter. arXiv preprint arXiv:2104.12250 (2021)
6. Barbieri, F., Camacho-Collados, J., Espinosa Anke, L., Neves, L.: TweetEval: Unified benchmark and comparative evaluation for tweet classification. In: Findings of the Association for Computational Linguistics: EMNLP 2020, pp. 1644–1650. Association for Computational Linguistics, Online (2020, November). https://doi.org/10.18653/v1/2020.findings-emnlp.148. https://www.aclweb.org/anthology/2020.findings-emnlp.148
7. Caselli, T., Basile, V., Mitrović, J., Granitzer, M.: HateBERT: Retraining BERT for abusive language detection in English. In: Proceedings of the 5th Workshop on Online Abuse and Harms (WOAH 2021), pp. 17–25. Association for Computational Linguistics, Online (2021, August). https://doi.org/10.18653/v1/2021.woah-1.3, https://aclanthology.org/2021.woah-1.3
8. Conneau, A., Khandelwal, K., Goyal, N., Chaudhary, V., Wenzek, G., Guzmán, F., Grave, E., Ott, M., Zettlemoyer, L., Stoyanov, V.: Unsupervised cross-lingual representation learning at scale. In: Proceedings of the 58th Annual Meeting of the Association for Computational Linguistics, pp. 8440–8451. Association for Computational Linguistics, Online (2020, July). https://doi.org/10.18653/v1/2020.acl-main.747. https://www.aclweb.org/anthology/2020.acl-main.747
9. Devlin, J., Chang, M.W., Lee, K., Toutanova, K.: BERT: Pre-training of deep bidirectional transformers for language understanding. In: Proceedings of the 2019 Conference of the North American Chapter of the Association for Computational Linguistics: Human Language Technologies, Volume 1 (Long and Short Papers), pp. 4171–4186. Association for Computational Linguistics, Minneapolis, Minnesota (2019, June). https://doi.org/10.18653/v1/N19-1423. https://www.aclweb.org/anthology/N19-1423
10. Gaikwad, S., Ranasinghe, T., Zampieri, M., Homan, C.M.: Cross-lingual offensive language identification for low resource languages: the case of marathi. In: Proceedings of RANLP (2021)
11. Gururangan, S., Marasović, A., Swayamdipta, S., Lo, K., Beltagy, I., Downey, D., Smith, N.A.: Don't stop pretraining: adapt language models to domains and tasks. In: Proceedings of the 58th Annual Meeting of the Association for Computational Linguistics, pp. 8342–8360. Association for Computational Linguistics, Online (2020, July). https://doi.org/10.18653/v1/2020.acl-main.740, https://www.aclweb.org/anthology/2020.acl-main.740
12. Malmasi, S., Zampieri, M.: Detecting hate speech in social media. In: Proceedings of Recent Advances in Natural Language Processing (RANLP), pp. 467–472. Varna, Bulgaria (2017)
13. Mandl, T., Modha, S., Shahi, G.K., Madhu, H., Satapara, S., Majumder, P., Schäfer, J., Ranasinghe, T., Zampieri, M., Nandini, D., Jaiswal, A.K.: Overview of the HASOC subtrack at FIRE 2021: hate speech and offensive content identification in English and Indo-Aryan languages. In: Working Notes of FIRE 2021—Forum for Information Retrieval Evaluation. CEUR (2021, December). http://ceur-ws.org/
14. Modha, S., Mandl, T., Shahi, G.K., Madhu, H., Satapara, S., Ranasinghe, T., Zampieri, M.: Overview of the HASOC subtrack at FIRE 2021: hate speech and offensive content identification in English and Indo-Aryan languages and conversational hate speech. In: FIRE 2021: Forum for Information Retrieval Evaluation, Virtual Event, 13th–17th December 2021. ACM (2021, December)
15. Nguyen, D.Q., Vu, T., Tuan Nguyen, A.: BERTweet: A pre-trained language model for English tweets. In: Proceedings of the 2020 Conference on Empirical Methods in Natural Language Processing: System Demonstrations, pp. 9–14. Association for Computational Linguistics, Online (2020, October). https://doi.org/10.18653/v1/2020.emnlp-demos.2. https://www.aclweb.org/anthology/2020.emnlp-demos.2
16. Pruksachatkun, Y., Phang, J., Liu, H., Htut, P.M., Zhang, X., Pang, R.Y., Vania, C., Kann, K., Bowman, S.R.: Intermediate-task transfer learning with pretrained language models: when and why does it work? In: Proceedings of the 58th Annual Meeting of the Association for Computational Linguistics, pp. 5231–5247. Association for Computational Linguistics, Online (2020, July). https://doi.org/10.18653/v1/2020.acl-main.467. https://www.aclweb.org/anthology/2020.acl-main.467

17. Risch, J., Krestel, R.: Bagging bert models for robust aggression identification. In: Proceedings of the Second Workshop on Trolling, Aggression and Cyberbullying, pp. 55–61 (2020)

18. Rosenthal, S., Farra, N., Nakov, P.: Semeval-2017 task 4: Sentiment analysis in twitter. In: Proceedings of the 11th International Workshop on Semantic Evaluation (SemEval-2017), pp. 502–518 (2017)

19. Sun, C., Qiu, X., Xu, Y., Huang, X.: How to fine-tune bert for text classification? In: China National Conference on Chinese Computational Linguistics, pp. 194–206. Springer, Berlin (2019)

20. Vaswani, A., Shazeer, N., Parmar, N., Uszkoreit, J., Jones, L., Gomez, A.N., Kaiser, L., Polosukhin, I.: Attention is all you need. In: Guyon, I., Luxburg, U.V., Bengio, S., Wallach, H., Fergus, R., Vishwanathan, S., Garnett, R. (eds.) Advances in Neural Information Processing Systems 30, pp. 5998–6008. Curran Associates, Inc. (2017). http://papers.nips.cc/paper/7181-attention-is-all-you-need.pdf

21. Wolf, T., Debut, L., Sanh, V., Chaumond, J., Delangue, C., Moi, A., Cistac, P., Rault, T., Louf, R., Funtowicz, M., Davison, J., Shleifer, S., von Platen, P., Ma, C., Jernite, Y., Plu, J., Xu, C., Le Scao, T., Gugger, S., Drame, M., Lhoest, Q., Rush, A.: Transformers: state-of-the-art natural language processing. In: Proceedings of the 2020 Conference on Empirical Methods in Natural Language Processing: System Demonstrations, pp. 38–45. Association for Computational Linguistics, Online (2020, October). https://doi.org/10.18653/v1/2020.emnlp-demos.6. https://www.aclweb.org/anthology/2020.emnlp-demos.6

Tuning of Fuzzy Control Systems by Artificial Bee Colony with Dynamic Parameter Values Algorithm for Traction Power System

Shahnaz N. Shahbazova[ID] **and Dursun Ekmekci**[ID]

Abstract Traction power supply systems are one of the main parameters for high-speed train transportation, which has become increasingly common recently. The system can be controlled by a combination of different techniques. Simulation modeling shows that traction power supply systems make it possible to improve the power quality and reduce power losses in traction substations. This study proposes a fuzzy control system (FCS) using the Sugeno inference system to control the traction power supply system. Artificial Bee Colony with Dynamic Parameter Values (ABC-DPV) algorithm was used in the parameter optimization of MFs and rule selection of the designed system. ABC-DPV is a version of ABC that dynamically updates the parameter values of the algorithm in the search process to improve the exploitation and exploration capability of the algorithm. The proposed FCS has been tested on samples obtained from the simulation results, and it produces reasonable solutions at negligible error levels.

Keywords Fuzzy control systems · Artificial bee colony with dynamic parameter values algorithm · Traction power system

Sh. N. Shahbazova
Ministry of Sciences and Education of the Republic of Azerbaijan, Institute of Control Systems, 68 Bakhtiyar Vahabzadeh Street, Baku AZ1141, Azerbaijan
e-mail: shahbazova@cyber.az; shahnaz_shahbazova@unec.edu.az

Sh. N. Shahbazova · D. Ekmekci (✉)
Department of Digital Technologies and Applied Informatics, Azerbaijan State University of Economics, UNEC, 6 Istiqlaliyat str, Baku AZ1001, Azerbaijan

D. Ekmekci
Karabuk University, Karabuk 78050, Turkey
e-mail: dekmekci@karabuk.edu.tr

1 Introduction

The fuzzy logic control (FLC) has proven to be a successful control approach to many nonlinear, complex systems, and even non-analytical systems. In the engineering field, it has been suggested as an alternative approach to conventional control techniques in many cases [1]. However, assigning the correct parameters for the FLC is a complex problem that consumes considerable time. To solve this difficulty, different methods have been proposed that can be integrated into FLC. In this context, the methods used to automate the design of the fuzzy control systems (FCSs) can be classified under the following main headings [2]: analytical methods, heuristic methods, optimization methods, and adaptive tuning methods.

Difficulties in tuning controller parameters are related to the complexity of the system. Systems that have nonlinear characteristics can have many inputs and outputs. Besides, disturbances and uncertainties regarding operational conditions and some real-world limitations make it difficult to assign appropriate controller parameters that achieve the desired performance. Moreover, control systems demand that several performance specifications are met at once, generally in conflict with each other. The aforementioned reasons make heuristic methods more attractive than neural systems [3, 4].

Commonly used heuristics generally mimic the lives of creatures living in colonies. These algorithms usually generate random solutions at the initialization, and then iteratively derive new solutions with their specific parameters. The Bat Algorithm (BA) [5], the Firefly Algorithm (FA) [6], the Lion Optimization (LO) [7] the Chicken Swarm Optimization (CSO) [8], and the Spider Monkey (SM) algorithm [9] can be cited as relatively recent algorithms compared to other swarm intelligence-based algorithms.

One of the other proposed swarm intelligence-based methods is the artificial bee colony (ABC) algorithm [10]. In many studies, ABC and its many versions and extensions were used in the design of control systems [2]. In several studies, ABC has been used for network training such as feed-forward neural network [11], modular neural network [12], Adaptive network fuzzy inference system [13] (ANFIS) [14, 15]. Simulations of function approximations show that ABC optimizes the ANFIS better than back-propagation algorithm [16]. In several studies, ABC was applied to design or parameter tuning for FCSs. In [17] the ABC algorithm was used for tuning the parameters of Membership Functions (MFs) in the Mamdani-type fuzzy controller to improve the performance of suspension systems. In the study [18], which proposed an artificial pancreas model that mimics the blood glucose concentration profile of the healthy person with exogenous insulin infusion, Mamdani-type control parameters were optimized with the ABC algorithm. In a different study [19], an ABC based rule generation method is proposed for automated fuzzy power system stabilizer design to improve power system stability and reduce the design effort. In [20] a new method using ABC to control the reinforcement fuzzy system was developed. In [21] ABC was used to tune Sugeno parameters on predicting wind speed, and the Interval Type-2 Fuzzy PID Controller designing [22]. In many studies

[23–25], unlike the control system, ABC has been applied for fuzzy clustering. In numerous studies, algorithm was used for optimal Proportional-Integral-Derivative (PID) controller designing [26–28]. With another control model using the ABC algorithm, optimal performance is aimed for the automatic voltage regulator [29]. At the end of the experiments, it has been proved that the automatic voltage regulator system performance can be improved by the ABC algorithm. The results obtained by ABC showed that ABC has a better tuning capability for this control application compared to other similar population-based optimization algorithms. In a similar studies [30, 31], ABC and other swarm intelligence-based meta-heuristic methods were compared to adjust PI and PID parameters in control generation systems. In different studies [32, 33], respectively ABC and modified ABC were applied to tune fractional-order PID (FOPID) parameters, which include derivative order, and integral order in addition to the conventional PID, for speed control in DC Motor Drive. The algorithm was also applied for the multi-objective optimal power flow problem that has many discrete and continuous variables [34], was implemented on the multi-objective optimization for the adaptive robust controller [35], and linear-quadratic optimal controller designing [36].

When the brief literature review is analyzed, the ABC algorithm can be applied to the design of control systems with different structures and can produce more successful solutions compared to similar heuristics. In this study, the power required for the train to travel at the desired speed in different track conditions is controlled. For the proposed control model, the Sugeno-type inference system has been applied. The MFs control parameters of the model have been optimized by the Artificial Bee Colony with Dynamic Parameter Values (ABC-DPV) algorithm, which is a new ABC version. The proposed method dynamically changes ABC parameters in each iteration during the searching process for more successful exploration and efficient exploitation. The ABC-DPV algorithm was first introduced in [37], and in the paper, it was analyzed that it produces better solutions than classical ABC for numerical optimization problems. The model was designed according to 70% of the data set obtained from real-time measurements and tested with 30%.

The remainder of the paper is designed as follows: In Sect. 2, the traction power system is introduced, the system parameters are explained, and equations that calculate the energy required for the movement of the train, are given. Section 3 describes the ABC algorithm and procedures and how the algorithm parameter values are calculated by ABC-DPV during the search process. The section also explains in detail how the proposed method is applied to the problem. In the fourth section, the results obtained by the model on the training and test data are discussed. In Sect. 5, the study is concluded.

2 Traction Power Systems

Electric traction is widely used around the world on high-speed lines that need electric traction to achieve the required speeds, especially on routes with heavy traffic such as urban and suburban railways, or inter-city travel. The primary function of the traction power supply system is to continuously supply the electrical energy required by traction motors to ensure an uninterrupted and high-quality train operation. One of the main components of the system is the substations where electrical energy is supplied. According to the characteristics of the rail system project, for the required power supply, AC or DC systems, substations of the gas-insulated switchgear or the air-insulated switchgear, substations with different catenary operating voltage and frequency values are preferred. The electrical energy required for the movement of the train is provided through these substations, which are established at appropriate intervals on the line.

2.1 Train Movement Calculation

Train movement calculations are based on Newton's laws of motion, taking into operation modes, account gradients, and speed restriction [38]. Also, factors such as train resistance, track gradient force, and so on are opposed parameters for the train movement, and cannot be ignored in the calculation. Indeed, the train can be modeled as a point following the track as formulated in (1) [39].

$$\sum_{i=1}^{n} F_i = m^* a \tag{1}$$

where F_i are the different forces acting on the train, m^* is the mass of the train, corrected to consider the inertia of the rotating mass ($m^* = \xi m$) and a is the acceleration of the train along the track.

The forces acting on the train running on a curved track are represented in Fig. 1.

As in Fig. 1, the different forces acting on a train can be grouped into two basic categories: Forces (F_{in}) generated by the train that act positively in traction and negatively in braking, and forces (F_{ex}) that mostly opposite the train movement. If these forces are considering together, Eq. (1) is updated as Eq. (2).

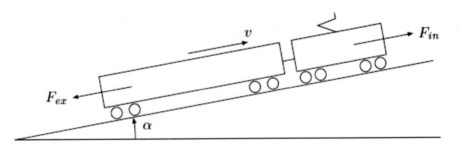

Fig. 1 A train running on a track (adapted from [40])

$$F_{in} - F_{ex} = m * a \qquad (2)$$

Let F_r is the train resistance to forwarding motion, F_c is the resistance due to the curves, and F_{gr} is the force due to gradients. F_{ex} as the sum of these forces against train movement can be formulated as (3).

$$F_{in} = F_r + F_c + F_{gr} \qquad (3)$$

The train resistance formula is according to EN 14,067–4 (European Standard for the requirements and test procedures for aerodynamics on open track) is given Eq. (4).

$$F_r = C1 + C2xv + C3xv^2 \qquad (4)$$

where the coefficients C1, C2, C3 are the train resistance factors, and v is speed. These values are all constants, referred to as the Davis coefficients [41].

The resisting force produced by the curve is formulated as in Eq. (5).

$$F_c = \frac{k_e}{r} 10^{-3} mg \qquad (5)$$

where ke is a coefficient depending on the track gauge, g is the gravitational acceleration, and r is the radius of the curve.

As the other parameter, the gradient force is simply defined in (6).

$$F_{gr} = m * g * \sin a \qquad (6)$$

where α is the gradient angle illustrated in Fig. 1.

2.2 Power Consumed by a Train

The power consumed by a train corresponding to tractive effort Fin and instantaneous speed v is given by the Eq. (7) [39].

$$P = \frac{F_{in} x v}{\eta} \qquad (7)$$

where η denotes the efficiency of conversion of electrical input power to the mechanical output power at the wheels.

3   Proposed Method and Implementation

In this section, ABC method and its parameters are explained, and mathematical formulations used in the algorithm steps are given. Then, it is explained how algorithm parameters are dynamically computed at each cycle for a more successful convergence with fewer operations. Finally, how the proposed method is applied to the problem is stated.

3.1   Artificial Bee Colony (ABC) Algorithm

The ABC algorithm is a swarm intelligence-based heuristic method that simulates the individual and collective behaviors of honeybees participating in foraging and food-collecting processes. The algorithm was introduced to the literature by Karaboga in 2005 [10]. In the social approach, the process begins with scout bees randomly searching for food around the hive. Then, employed bees are assigned to each of the food sources found. The employed bees, who visit these sources together with their neighbors, choose the more suitable source with a greedy approach. There are also as many onlooker bees as the employed bees in the hive. Therefore, there are twice as many honeybees as the initial solutions in the colony. Employed bees share information about sources with onlooker bees in the hive. Onlooker bees search by evaluating the information they have obtained from employed bees. In this context, they decide for their mind which source to straight to. When any food source is exhausted, employed or onlooker bees act as a scout bee and seek out a new food source in nature.

In the algorithmic model, food sources represent feasible solutions in the search space. So artificial bee procedures are used to find the best solution. Initial solutions are randomly generated in the scout bee phase. The connected activities of employed and onlooker bees affect the exploitation ability of the algorithm. A roulette wheel is used so that onlooker bees can evaluate more successful solutions. To redirect the searching activity of the algorithm to different search regions from the local optima trap, the "limit" parameter is used. In this context, if a more successful solution could not be derived with any solution in a "limit" number of cycles, a new random solution is produced instead of this solution. Iterative processes continue until the requirements are met.

The main steps of the ABC algorithm are as Fig. 2:

To produce the initial solutions Eq. (8) is used.

$$s_{m,i} = l_i + rand(0, 1) * (u_i - l_i) \qquad (8)$$

where S is the set of solutions, $s_{m,i}$ is the ith dimension value of the solution m. l_i and u_i represent the upper and lower bounds of dimension i respectively.

When deriving new solutions in the employed bees phase, (9) is used.

Fig. 2 Pseudo code of the ABC algorithm

Algorithm 1 General structure of the ABC Algorithm

Scout bees phase
repeat
    Employed bees phase
    Onlooker bees phase
    Scout bees phase
until requirements are met

$$t_{m,i} = s_{m,i} + \phi_{m,i}(0, 1) * \left(s_{m,i} - s_{r,i}\right) \tag{9}$$

where s_m is the next solution, sr is a random solution selected from the solution set (S), i is a dimension of the solution, selected randomly. $\phi_{m,i}$ is a coefficient value that selected randomly between $[-1, 1]$.

The fitness value $(fit(t_m))$ of the new solution is calculated and compared with the fitness value of the old solution, and the better one is preferred. The fitness value of the solutions is determined by using the value $(f(t_m))$ calculated according to the objective function by Eq. (10).

$$fit(t_m) = \begin{cases} \frac{1}{1+fit(t_m)} & \text{if}(f(t_m) \geq 0) \\ \frac{1}{abs(fit(t_m))} & \text{if}(f(t_m) < 0) \end{cases} \tag{10}$$

When the fitness values of the solutions are compared, if the fitness value of the new solution is greater, the old solution is deleted in the solution set, and the new solution is saved instead.

After the employed bees phase, to evaluate more successful solutions in the onlooker bees phase, all the current solutions are placed on the roulette wheel. In this context, the selecting probability (r_m) of the solution m is calculated by Eq. (11).

$$r_m = \frac{fit(s_m)}{\sum_{i=1}^{SN} s_i} \tag{11}$$

Onlooker bees, like employed bees, compare the new solution with the old solution and record the better one. After the onlooker bees phase is completed, the failure counters of the solutions are checked. Solutions whose failure counter reaches "limit" are deleted, and instead of these solutions, random solutions are produced as in the beginning.

3.2   Algorithm Artificial Bee Colony with Dynamic Parameter Values (ABC–DPV) Algorithm

One of the biggest problems for metaheuristic methods that can generate valid solutions for optimization problems in a reasonable time is to assign appropriate values to algorithm parameters. For the algorithm to adequately search the active regions of the solution space, and to produce better solutions from the obtained solutions, appropriate values should be assigned to the control parameters. Therefore, algorithm performance is directly related to parameter values. Researchers have recently developed many online [42, 43] and offline [44] methods that adjust parameter values for optimization algorithms to the optimum value. In the scope of the study, the ABC-DPV algorithm was preferred as an optimization method. ABC-DPV was developed to make the exploitation capability of the ABC algorithm more efficient and, on the other hand, to increase exploration performance. In the proposed method, ABC parameter values are dynamically updated in each cycle of the search process.

In the classical ABC model, only "colony size—CS" and "limit" parameters are used. Since the ABC algorithm uses the explorative process efficiently by providing enough diversity in the population for the specified CS, the algorithm does not need large CS to solve large-scale optimization problems [45].

If the initialization phase is ignored, CS represents the numbers of employed and onlooker bees. In the algorithm acceptance, the number of employed and onlooker bees is equal to CS. But ABC-DPV adopts an elitist strategy in the onlooker bees phase and uses fewer onlooker bees. In each cycle of ABC-DPV, first, the average fitness value is calculated with Eq. (12). Then, onlooker bees are used as many as the number of solutions with a higher fitness value than the calculated value, and onlooker bees are directed to these solutions. Thus, with less computing, all solutions that make the algorithm successful are visited without exception.

$$fit(avg) = \frac{\sum_{n=1}^{CS/2} fit(s_n)}{CS} \tag{12}$$

The limit parameter of the ABC algorithm is the parameter that determines the exploitation/exploration balance of the algorithm, and the limit is of equal value for all solutions. Detailed analysis [46] shows that the most efficient value for the limit parameter can be calculated with Eq. (13). (D represents the dimension of the solutions.)

$$limit_i = (CS^* D)/2 \tag{13}$$

Therefore, the total limit number ($limit_{sum}$) used by the ABC algorithm is the result of the value calculated by Eq. (6) multiplied by the number of solutions. In the ABC-DPV, the $limit_{sum}$ is shared in proportion to the fitness values of the solutions, considering the global best solution. In each cycle, after the onlooker bees phase, the total fitness value ($fit(sum)$) of the population (current solutions and the global best solution) is calculated with Eq. (14).

$$fit(sum) = fit(s_{best}) + \sum_{n=1}^{CS/2} fit(s_n) \qquad (14)$$

Then the $limit_{sum}$ is shared among the current solutions in the S. In this context, the limit value to be assigned to the solution m is calculated with Eq. (15). Pseudocode of the ABC-DPV is given in Fig. 3.

$$limit_m = limit_{sum} * \frac{fit(s_m)}{fit(sum)} \qquad (15)$$

3.3 FCS Design by ABC–DPV

In the paper, it is suggested to design traction power control with the ABC-DPV algorithm. When the traction power control system described in the previous section is analyzed, it is noticed that the parameters affecting the power consumed by train are the gradient of the road, the radius of the curve, the cant, and the instantaneous speed of the train. In this context, the structure of the control system is arranged as consist of four inputs and one output. Sugeno method was preferred for the inference system of FCS.

In the study, the ABC-DPV algorithm has been used for parameter optimization and rule selection of the system whose preliminary design was explained. In each cycle, the outputs of the system constructed with the results of the algorithm were compared with the targets in the data set, and the error was determined. In this context, the objective function of the problem is as given in (16).

$$\min \left\{ \sum_{i=1}^{n} \left(target_i^2 - output_i^2 \right) \right\} \qquad (16)$$

where n is the number of samples for training or testing.
    The proposed control model was designed as in Fig. 4.

4 Experimental Study

The proposed control model has been trained and its performance tested using the data set obtained from simulations. In this section, firstly, the prepared data set and experiment design are explained, then the results obtained by the system from the experiments are shared.

Fig. 3 Pseudo code of proposed method

Algorithm 1 Pseudo code of the ABC-DPV Algorithm

Assign the *food* value
// *Scout bees phase*
for $m = 1$ to *food* **do**
    Produce a random solution (s_m) with Eq.(1)
    $failure_m = 0$
end for
repeat
    // *Employed bees phase*
    $fit(sum) = 0$
    for $m = 1$ to *food* **do**
        Derive a new solution (t_m) with Eq.(2)
        Calculate the fitness value of t_m with Eq.(3)
        Apply the greedy approach for t_m and s_m
        $fit(sum) = fit(sum) + fit(s_m)$
    end for
    $fit(avg) = fit(sum)/food$
    // *Onlooker bees phase*
    for $m = 1$ to *food* **do**
        if $fit(s_m) \geq fit(avg)$ **then**
            Derive a new solution (t_m) with Eq.(2)
            Calculate the fitness value of t_m with Eq.(3)
            Apply the greedy approach for t_m and s_m
        end if
    end for
    Memorize the best solution (s_{best})
    // *Calculate the limit values of the solutions*
    $fit(sum) = fit(sum) + fit(s_{best})$
    $limit_{sum} = food * Dimension * food$
    for $m = 1$ to *food* **do**
        $limit_m = limit_{sum} * fit(s_m) / fit(sum)$
    end for
    // *Scout bees phase*
    for $m = 1$ to *food* **do**
        if $limit_m \leq faliure_m$ **then**
            Produce a random solution (s_m) with Eq.(1)
            $failure_m = 0$
        end if
    end for
until (requirements are met)

4.1 Prepared Dataset

You will need to determine whether or not your equation should First of all, information on the rise, gradient, radius of the curve, curvature, and cant of the track of approximately 126 km the train will travel on were saved. Then, the train traveling on this track was simulated. In the simulation, the necessary electrical energy was supplied to the train for the speed it had to reach under the current geophysical conditions, and the process saved at 0.5-s intervals.

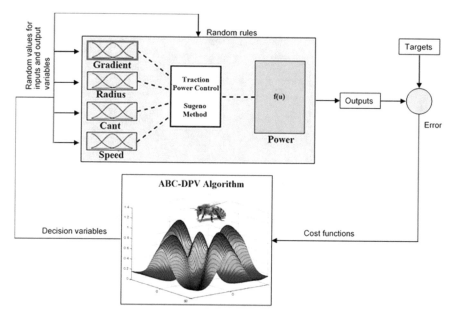

Fig. 4 Designing of the proposed model

When the simulation process was completed, a data set consisting of approximately 5000 samples was obtained. Normalization was applied for each column in the data set, and all results were reduced to the range [0–1]. A fair number of samples were selected to represent the entire data set, as using all samples would cause unnecessary processing load. In this context, the Statistical Package for the Social Sciences (SPSS) software was used for sample selection. All samples in the data set were divided into 50 clusters using the k-means clustering method and cluster centers were taken as reference.

4.2 Experiment Design

For the input variables of the proposed FCS, 5, 4, 3, and 1 MFs respectively, and 5 MFs for the output variable was used. All MFs are the type of Generalized Bell-Shaped Membership Function (gbellmf). The ABC-DPV method was run independently of each other 30 times. In each trial, 25 solutions were initially evaluated, and the maximum number of iterations was assigned as 10,000.

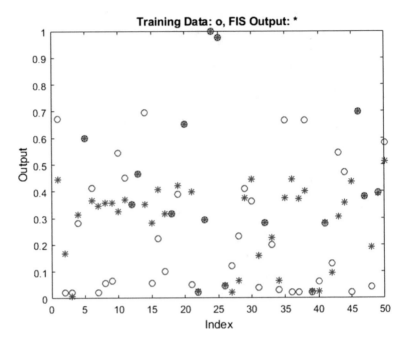

Fig. 5 Results of the proposed method for the data set

4.3 Results

The best result achieved by the proposed FCS for training data is 3.50. The mean for 30 trials is 3.79, the standard deviation is 0.17, and the worst result is 4.11. The results of the best solution for training data are illustrated in Fig. 5.

The results produced by the method for all the data obtained as a result of the simulation are plotted in Fig. 6.

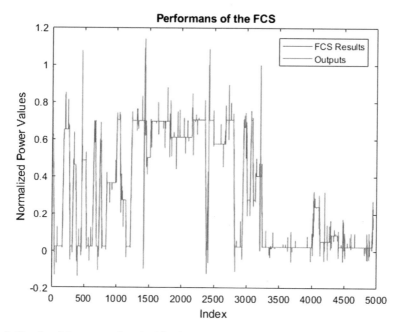

Fig. 6 Results of the proposed method for the whole data obtained by the simulation

5 Conclusion

Within the scope of the study, traction power supply systems were investigated, and FCS was proposed for the required power. The parameters for the proposed FCS have been optimized with the new version of the ABC algorithm, which is widely used for numerical optimization problems. The performance of the method has been tested with simulation data. The train movement on the track with different physical conditions was simulated, and the required power was saved. The obtained data set was sampled by the k-means clustering algorithm. The proposed method was trained with this data set and tested for whole simulation results. The comparison results have shown that the method can produce successful results and its usability for traction power supply systems.

References

1. Feng, G.: A survey on analysis and design of model-based fuzzy control systems. IEEE Trans. Fuzzy Syst. **14**, 676–697 (2006). https://doi.org/10.1109/TFUZZ.2006.883415
2. Rodríguez-Molina, A., Mezura-Montes, E., Villarreal-Cervantes, M.G., Aldape-Pérez, M.: Multi-objective meta-heuristic optimization in intelligent control: a survey on the controller tuning problem. Appl. Soft Comput. **93**, 106342 (2020). https://doi.org/10.1016/j.asoc.2020.106342

3. Viharos, Z.J., Kis, K.B.: Survey on Neuro-Fuzzy systems and their applications in technical diagnostics and measurement. Measurement **67**, 126–136 (2015). https://doi.org/10.1016/j.mea surement.2015.02.001
4. Shihabudheen, K.V., Pillai, G.N.: Recent advances in neuro-fuzzy system: a survey. Knowl.-Based Syst. **152**, 136–162 (2018). https://doi.org/10.1016/j.knosys.2018.04.014
5. Yang, X.-S.: A new metaheuristic bat-inspired algorithm. In: Gonzalez JR et al. (eds.) Studies in Computational Intelligence. Springer Berlin Heidelberg, pp 65–74 (2010)
6. Yang, X.-S.: Firefly algorithms for multimodal optimization. In: Watanabe, O, Zeugmann, T. (eds.) Stochastic Algorithms: Foundations and Applications. SAGA 2009, pp. 169–178. Springer Berlin Heidelberg (2009)
7. Rajakumar, B.R.: The Lion's algorithm: a new nature-inspired search algorithm. Procedia Technol. **6**, 126–135 (2012). https://doi.org/10.1016/j.protcy.2012.10.016
8. Meng, X., Liu, Y., Gao, X., Zhang, H.: A new bio-inspired algorithm: chicken swarm optimiza-tion. In: Lecture Notes in Computer Science (Including Subseries Lecture Notes in Artificial Intelligence and Lecture Notes in Bioinformatics), pp. 86–94 (2014)
9. Bansal, J.C., Sharma, H., Jadon, S.S., Clerc, M.: Spider monkey optimization algorithm for numerical optimization. Memetic Comput. **6**, 31–47 (2014). https://doi.org/10.1007/s12293-013-0128-0
10. Karaboga D (2005) An Idea Based on Honey Bee Swarm for Numerical Optimization. Kayseri, Turkey
11. Ozturk, C., Karaboga, D.: Hybrid artificial bee colony algorithm for neural network training. In: 2011 IEEE Congress of Evolutionary Computation (CEC). IEEE, pp. 84–88 (2011)
12. Ling, W., Wang, Y.: Using modular neural network with artificial bee colony algorithm for classification. In: Advances in Swarm Intelligence, pp. 396–403. Springer, Berlin, Heidelberg (2013)
13. Jang, J.R.: ANFIS: Adaptive network-based fuzzy inferenc system. IEEE Trans Syst Man Cybern **23**, 665– 685 (1993). https://doi.org/10.1109/21.256541
14. Karaboga, D., Kaya, E.: An adaptive and hybrid artificial bee colony algorithm (aABC) for ANFIS training. Appl. Soft Comput. J. **49**, 423–436 (2016). https://doi.org/10.1016/j.asoc.2016.07.039
15. Karaboga, D., Kaya, E.: Adaptive network based fuzzy inference system (ANFIS) training approaches: a comprehensive survey. Artif. Intell. Rev. **52**, 2263–2293 (2019). https://doi.org/10.1007/s10462-017-9610-2
16. Karaboğa, D., Kaya, E.: Training ANFIS by using the artificial bee colony algorithm. Turkish J. Electr. Eng. Comput. Sci. **25**, 1669–1679 (2017). https://doi.org/10.3906/elk-1601-240
17. Aldair, A.A., Alsaedee, E.B., Abdalla, T.Y.: Design of ABCF control scheme for full vehicle nonlinear active suspension system with passenger seat. Iran J. Sci. Technol. Trans. Electr. Eng. **43**, 289–302 (2019). https://doi.org/10.1007/s40998-018-0134-9
18. Soylu, S., Danişman, K.: Blood glucose control using an ABC algorithm-based fuzzy-PID controller. Turkish J. Electr. Eng. Comput. Sci. **26**:172–183 (2018). https://doi.org/10.3906/elk-1704-203
19. Abedinia, O., Wyns, B., Ghasemi, A.: Robust fuzzy PSS design using ABC. In: 2011 10th International Conference on Environment and Electrical Engineering. IEEE, pp. 1–4 (2011)
20. Saeed, S., Niknafs, A.: Artificial bee colony-fuzzy Q learning for reinforcement fuzzy control (truck backer-upper control problem). Int. J. Uncertainty Fuzziness Knowl.-Based Syst. **24**, 123–136 (2016). https://doi.org/10.1142/S0218488516500070
21. Ismail, F.H., Aziz, M.A., Hassanien, A.E.: Optimizing the parameters of Sugeno based adap-tive neuro fuzzy using artificial bee colony: a case study on predicting the wind speed. In: Proceedings of the 2016 Federated Conference on Computer Science and Information Systems, FedCSIS 2016. Polish Information Processing Society, pp 645–651 (2016)
22. Kumar, A., Kumar, V.: Artificial bee colony based design of the interval type-2 fuzzy PID controller for robot manipulator. In: TENCON 2017 - 2017 IEEE Region 10 Conference. IEEE, pp. 602–607 (2017)

23. Karaboga, D., Ozturk, C.: Fuzzy clustering with artificial bee colony algorithm. Sci. Res. Essays **5**, 1899–1902 (2010). https://doi.org/10.5897/SRE.9000517
24. Zhao, X., Zhang, S.: An improved KFCM algorithm based on artificial bee colony. In: Communications in Computer and Information Science, pp. 190–198 (2011)
25. Patel, V., Tiwari, A., Patel, A.: A comprehensive survey on hybridization of artificial bee colony with particle swarm optimization algorithm and ABC applications to data clustering. In: Proceedings of the International Conference on Informatics and Analytics—ICIA-16. ACM Press, New York, New York, USA, pp 1–9 (2016)
26. Karaboga, D., Akay, B.: Proportional—integral—derivative controller design by using artificial bee colony, harmony search, and the bees algorithms. Proc. Inst. Mech. Eng. Part I J. Syst. Control Eng. **224**, 869–883 (2010). https://doi.org/10.1243/09596518JSCE954
27. Chang, W.-D.: Nonlinear CSTR control system design using an artificial bee colony algorithm. Simul. Model. Pract. Theory **31**, 1–9 (2013). https://doi.org/10.1016/j.simpat.2012.11.002
28. Liao W, Hu Y, Wang H (2014) Optimization of PID control for DC motor based on artificial bee colony algorithm. In: Proceedings of the 2014 International Conference on Advanced Mechatronic Systems. IEEE, pp 23–27
29. Gozde, H., Taplamacioglu, M.C.: Comparative performance analysis of artificial bee colony algorithm for automatic voltage regulator (AVR) system. J. Franklin Inst. **348**, 1927–1946 (2011). https://doi.org/10.1016/j.jfranklin.2011.05.012
30. Turanoglu, E., Ozceylan, E., Kiran, M.S.: Particle swarm optimization and artificial bee colony approaches to optimize of single input-output fuzzy membership functions. In: Proceedings of the 41st International Conference on Computers & Industrial Engineering, pp. 542–547 (2011)
31. Gozde, H., Cengiz Taplamacioglu, M., Kocaarslan, İ: Comparative performance analysis of Artificial Bee Colony algorithm in automatic generation control for interconnected reheat thermal power system. Int. J. Electr. Power Energy Syst. **42**, 167–178 (2012). https://doi.org/10.1016/j.ijepes.2012.03.039
32. Rajasekhar, A., Das, S., Abraham, A.: Fractional order PID controller design for speed control of chopper fed DC motor drive using artificial bee colony algorithm. In: 2013 World Congress on Nature and Biologically Inspired Computing. IEEE, pp. 259–266 (2013)
33. Rajasekhar, A., Kunathi, P., Abraham, A., Pant, M.: Fractinal order speed control of DC motor using levy mutated artificial bee colony algorithm. In: 2011 World Congress on Information and Communication Technologies. IEEE, pp. 7–13 (2011)
34. Rezaei Adaryani, M., Karami, A.: Artificial bee colony algorithm for solving multi-objective optimal power flow problem. Int. J. Electr. Power Energy Syst. **53**, 219–230 (2013). https://doi.org/10.1016/j.ijepes.2013.04.021
35. Mahmoodabadi, M.J., Shahangian, M.M.: A new multi-objective artificial bee colony algorithm for optimal adaptive robust controller design. IETE J. Res. 1–14 (2019). https://doi.org/10.1080/03772063.2019.1644211
36. Ata, B., Coban, R.: Artificial bee colony algorithm based linear quadric optimal controller design for a nonlinear inverted pendulum. Int. J. Intell. Syst. Appl. Eng. **3**, 1–6 (2015)
37. Ekmekci, D.: Dinamik Parametre Değerli Yapay Arı Koloni Algoritması (DPD-YAK). Eur J. Sci. Technol. 407–415 (2020). https://doi.org/10.31590/ejosat.780659
38. Mingpruk, N., Leeton, U., Kulworawanichpong, T.: Modeling and simulation of voltage unbalance in AC electric railway systems using MATLAB/Simulink. In: 2016 IEEE/SICE International Symposium on System Integration (SII). IEEE, pp. 13–18 (2016)
39. Seimbille, D.: Design of power supply system in DC electrified transit railways—influence of the high voltage network. Stocholm (2014)
40. Allenbach, J.-M., Chapas, P., Comte, M., Roger, K.: Traction Electrique. Presses polytechniques et universitaires romandes (2008)
41. Rochard, B.P., Schmid, F.: A review of methods to measure and calculate train resistances. Proc. Inst. Mech. Eng. Part F J. Rail. Rapid Transit **214**, 185–199 (2000). https://doi.org/10.1243/0954409001531306
42. Korkmaz Tan, R., Bora, Ş: Adaptive parameter tuning for agent-based modeling and simulation. SIMULATION **95**, 771–796 (2019). https://doi.org/10.1177/0037549719846366

43. Xue, Y., Jiang, J., Zhao, B., Ma, T.: A self-adaptive artificial bee colony algorithm based on global best for global optimization. Soft Comput. **22**, 2935–2952 (2018). https://doi.org/10.1007/s00500-017-2547-1
44. Pellegrini, P., Stützle, T., Birattari, M.: A critical analysis of parameter adaptation in ant colony optimization. Swarm Intell. **6**, 23–48 (2012). https://doi.org/10.1007/s11721-011-0061-0
45. Akay, B., Karaboga, D.: Parameter Tuning for the Artificial Bee Colony Algorithm, pp 608–619 (2009)
46. Karaboga, D., Akay, B.: A comparative study of artificial bee colony algorithm. Appl. Math. Comput. **214**, 108–132 (2009). https://doi.org/10.1016/j.amc.2009.03.090

New Approach for Solving Fully Interval-Value Fuzzy Transportation Problems

Ali Ebrahimnejad, **Farhad Hosseinzadeh Lotfi**,
and **Tofigh Allahviranloo**

Abstract The transportation problem (TP) is one of the most broadly studied fields in the literature of linear programming problems used to a wide variety of practical applications. Conventional TPs generally assume that the values of transportation costs and the values of demand and supply are defined by real variables, though these values are unpredictable in TPs due to some uncontrollable factors. The present study formulates a TP when all parameters are interval-valued triangular fuzzy numbers and uses a novel optimization structure to find its optimal solution. The novelty of such an approach resides in requiring less computational effort compared to existing ones. The applicability of the proposed approach is illustrated through a numerical example.

Keywords Transportation problem · Interval-valued fuzzy numbers ·
Computational complexity

1 Introduction

The transportation problem (TP), seeks an optimal transportation plan to satisfy the demands of destinations via the supplies of sources with minimal transportation cost. Transportation problems are strongly exposed to the subject of uncertainty since it is an unpredictable framework in which components and connections are very much affected by that uncertainty. For example, due to the fluctuation of customers' needs, the demands for the products in the market are not known precisely, which are

A. Ebrahimnejad (✉)
Department of Mathematics, Qaemshahr Branch, Islamic Azad University, Qaemshahr, Iran
e-mail: a.ebrahimnejad@qaemiau.ac.ir

F. H. Lotfi · T. Allahviranloo
Department of Mathematics, Science and Research Branch, Islamic Azad University, Tehran, Iran
e-mail: farhad@hosseinzadeh.ir

T. Allahviranloo
Faculty of Engineering and Natural Sciences, Istinye University, Istanbul, Turkey

© The Author(s), under exclusive license to Springer Nature Switzerland AG 2023
Sh. N. Shahbazova et al. (eds.), *Recent Developments and the New Directions of Research, Foundations, and Applications*, Studies in Fuzziness and Soft Computing 423,
https://doi.org/10.1007/978-3-031-23476-7_22

uncertain. As the demand is uncertain, the supply of source should be also uncertain. Therefore, in order to design more practical and reliable modeling of TPs, the decision makers are confronted with various sources of uncertainty. To handle uncertainty, fuzzy numbers can be used for modeling of TPs [1, 2], where the membership grade of a fuzzy number is considered as a crisp value. However in many practical applications where it is not possible to address a crisp membership for few uncertainties, the membership grade of a fuzzy number is given by an interval number and the resulting problem is called interval-valued fuzzy TP (IVFTP). Very few researchers have studied this kind of uncertainty in TPs. Chiang [3] have used interval-valued triangular fuzzy numbers (IVTFNs) instead of normal ones and suggested an approach for solving a single objective TP with IVTFN demands and supplies. Gupta and Kumar [4] studied a linear multi-objective TP with IVTFN costs, supplies and demands. Bharati and Singh [5] studied interval-valued intuitionistic FTP based on expected values. Ebrahimnejad [6] proposed an approach based on fuzzy linear programming technique for solving fully IVFTP. In this book chapter, we formulate fully IVFTP and propose a novel approach for finding the optimal solution of such TP by representing all decision variables and parameters as IVTFNs.

2 Preliminaries

Here, some basic definitions of IVTFNs are reviewed [6].

Definition 1 A level λ triangular fuzzy number (TFN) $\tilde{A}$ on R, denoted by $\tilde{A} = (a_1, a_2, a_3; \lambda), 0 < \lambda \leq 1$, is defined by following membership function:

$$\mu_{\tilde{A}}(x) = \begin{cases} \lambda\left(\frac{x-a_1}{a_2-a_1}\right), & a_1 \leq x \leq a_2, \\ \lambda\left(\frac{a_3-x}{a_3-a_2}\right), & a_2 \leq x \leq a_3, \\ 0, & \text{otherwise.} \end{cases} \tag{1}$$

Definition 2 A level $(\underline{\lambda}, \overline{\lambda})$–interval–valued triangular fuzzy number $\tilde{\tilde{A}}$, denoted by $\tilde{\tilde{A}} = \left[\tilde{\underline{A}}, \tilde{\overline{A}}\right] = \langle(\underline{a}_1, a_2, \underline{a}_3; \underline{\lambda}), (\overline{a}_1, a_2, \overline{a}_3; \overline{\lambda})\rangle$ is an interval–valued fuzzy set on R where the lower TFN $\tilde{\underline{A}}$ and the upper TFN $\tilde{\overline{A}}$ are expressed as follows, respectively:

$$\mu_{\tilde{\underline{A}}}(x) = \begin{cases} \underline{\lambda}\left(\frac{x-\underline{a}_1}{a_2-\underline{a}_1}\right), & \underline{a}_1 \leq x \leq a_2, \\ \underline{\lambda}\left(\frac{\underline{a}_3-x}{a_2-\underline{a}_3}\right), & a_2 \leq x \leq \underline{a}_3, \\ 0, & \text{otherwise} \end{cases} \tag{2}$$

$$\mu_{\tilde{\tilde{A}}}(x) = \begin{cases} \bar{\lambda}\left(\frac{x-\bar{a}_1}{a_2-\bar{a}_1}\right), & \bar{a}_1 \leq x \leq a_2, \\ \bar{\lambda}\left(\frac{\bar{a}_3-x}{a_2-\bar{a}_3}\right), & a_2 \leq x \leq \bar{a}_3, \\ 0, & \text{otherwise.} \end{cases} \tag{3}$$

where $\bar{a}_1 \leq \underline{a}_1 \leq a_2 \leq \underline{a}_3 \leq \bar{a}_3, 0 < \underline{\lambda} \leq \bar{\lambda} \leq 1$. The family of all level $(\underline{\lambda}, \bar{\lambda})$–IVTFNs is denoted by $F_{IVTN}(\underline{\lambda}, \bar{\lambda})$.

Definition 3 Let $\tilde{\tilde{A}} = \left[\tilde{A}, \tilde{\bar{A}}\right] = \langle (\underline{a}_1, a_2, \underline{a}_3; \underline{\lambda}), (\bar{a}_1, a_2, \bar{a}_3; \bar{\lambda}) \rangle$ and $\tilde{\tilde{B}} = \left[\tilde{B}, \tilde{\bar{B}}\right] = \langle (\underline{b}_1, b_2, \underline{b}_3; \underline{\lambda}), (\bar{b}_1, b_2, \bar{b}_3; \bar{\lambda}) \rangle$ belong to $F_{IVTN}(\underline{\lambda}, \bar{\lambda})$ and $k \geq 0$ be a real number. Then, interval-valued fuzzy arithmetic operations are defined as follows:

$$\tilde{\tilde{A}} + \tilde{\tilde{B}} = \langle (\underline{a}_1 + \underline{b}_1, a_2 + b_2, \underline{a}_3 + \underline{b}_3; \underline{\lambda}), (\bar{a}_1 + \bar{b}_1, a_2 + b_2, \bar{a}_3 + \bar{b}_3; \bar{\lambda}) \rangle,$$

$$\tilde{\tilde{A}} \otimes \tilde{\tilde{B}} = \langle (\underline{a}_1\underline{b}_1, a_2 b_2, \underline{a}_3\underline{b}_3; \underline{\lambda}), (\bar{a}_1\bar{b}_1, a_2 b_2, \bar{a}_3\bar{b}_3; \bar{\lambda}) \rangle, \quad \bar{a}_1 \geq 0,$$

$$k\tilde{\tilde{A}} = \begin{cases} \langle (k\underline{a}_1, ka_2, k\underline{a}_3; \underline{\lambda}), (k\bar{a}_1, ka_2, k\bar{a}_3; \bar{\lambda}) \rangle, & k > 0, \\ \langle (k\underline{a}_3, ka_2, k\underline{a}_1; \underline{\lambda}), (k\bar{a}_3, ka_2, k\bar{a}_1; \bar{\lambda}) \rangle, & k < 0, \\ \langle (0, 0, 0; \underline{\lambda}), (0, 0, 0, ; \bar{\lambda}) \rangle = \tilde{\tilde{0}}, & k = 0. \end{cases} \tag{4}$$

Definition 4 Two IVTFNs $\tilde{\tilde{A}} = \left[\tilde{A}, \tilde{\bar{A}}\right] = \langle (\underline{a}_1, a_2, \underline{a}_3; \underline{\lambda}), (\bar{a}_1, a_2, \bar{a}_3; \bar{\lambda}) \rangle$ and $\tilde{\tilde{B}} = \left[\tilde{B}, \tilde{\bar{B}}\right] = \langle (\underline{b}_1, b_2, \underline{b}_3; \underline{\lambda}), (\bar{b}_1, b_2, \bar{b}_3; \bar{\lambda}) \rangle$ are said to be equal, i.e., $\tilde{\tilde{A}} = \tilde{\tilde{B}}$ if $\bar{a}_1 = \bar{b}_1, \underline{a}_1 = \underline{b}_1, a_2 = b_2, \underline{a}_3 = \underline{b}_3, \bar{a}_3 = \bar{b}_3$. Also, $\tilde{\tilde{A}} \prec \tilde{\tilde{B}}$ if $\bar{a}_1 \leq \bar{b}_1, \underline{a}_1 \leq \underline{b}_1, a_2 \leq b_2, \underline{a}_3 \leq \underline{b}_3, \bar{a}_3 \leq \bar{b}_3$.

3 Fully Interval-Valued Fuzzy Transportation Problems

In this section, the fully Interval-valued fuzzy TP (FIVFTP) is formulated mathematically.

The FIVFTP mathematically is formulated as below [6]:

$$\min \quad \tilde{\tilde{Z}} = \sum_{i=1}^{m} \sum_{j=1}^{m} \tilde{\tilde{c}}_{ij}\tilde{\tilde{x}}_{ij}$$

$$s.t. \quad \sum_{j=1}^{n} \tilde{\tilde{x}}_{ij} = \tilde{\tilde{s}}_i, \quad 1 \leq i \leq m,$$

$$\sum_{i=1}^{m} \tilde{\tilde{x}}_{ij} = \tilde{\tilde{d}}_j, \quad 1 \leq j \leq n,$$

$$\tilde{\tilde{x}}_{ij} \geq 0, \forall i, \forall j. \tag{5}$$

Assume that the parameters of model (5) are denoted as the following non-negative IVTFNs. The FIVFTP (5) can then be rewritten as below:

$$\min \quad \tilde{\tilde{Z}} = \sum_{i=1}^{m}\sum_{j=1}^{m}\left\langle(\underline{c}_{ij,1},c_{ij,2},\underline{c}_{ij,3};\underline{\lambda}),(\overline{c}_{ij,1},c_{ij,2},\overline{c}_{ij,3};\overline{\lambda})\right\rangle$$

$$\left\langle(\underline{x}_{ij,1},x_{ij,2},\underline{x}_{ij,3};\underline{\lambda}),(\overline{x}_{ij,1},x_{ij,2},\overline{x}_{ij,3};\overline{\lambda})\right\rangle$$

$$s.t. \quad \sum_{j=1}^{n}\left\langle(\underline{x}_{ij,1},x_{ij,2},\underline{x}_{ij,3};\underline{\lambda}),(\overline{x}_{ij,1},x_{ij,2},\overline{x}_{ij,3};\overline{\lambda})\right\rangle$$

$$= \left\langle(\underline{s}_{i,1},s_{i,2},\underline{s}_{i,3};\underline{\lambda}),(\overline{s}_{i,1},s_{i,2},\overline{s}_{i,3};\overline{\lambda})\right\rangle, \quad \forall i$$

$$\sum_{i=1}^{m}\left\langle(\underline{x}_{ij,1},x_{ij,2},\underline{x}_{ij,3};\underline{\lambda}),(\overline{x}_{ij,1},x_{ij,2},\overline{x}_{ij,3};\overline{\lambda})\right\rangle$$

$$= \left\langle(\underline{d}_{j,1},d_{j,2},\underline{d}_{j,3};\underline{\lambda}),(\overline{d}_{j,1},d_{j,2},\overline{d}_{j,3};\overline{\lambda})\right\rangle, \quad \forall j$$

$$\left\langle(\underline{x}_{ij,1},x_{ij,2},\underline{x}_{ij,3};\underline{\lambda}),(\overline{x}_{ij,1},x_{ij,2},\overline{x}_{ij,3};\overline{\lambda})\right\rangle \geq 0, \forall i, \forall j. \tag{6}$$

By definitions given in Sect. 2, the FIVFTP (6) is reformulated as the following multi-objective linear programming problem:

$$\min \quad \tilde{\tilde{Z}} = \left\langle\left(\sum_{i=1}^{m}\sum_{j=1}^{m}\underline{c}_{ij,1}\underline{x}_{ij,1},\sum_{i=1}^{m}\sum_{j=1}^{m}c_{ij,2}x_{ij,2},\sum_{i=1}^{m}\sum_{j=1}^{m}\underline{c}_{ij,3}\underline{x}_{ij,3};\underline{\lambda}\right),\right.$$

$$\left.\left(\sum_{i=1}^{m}\sum_{j=1}^{m}\overline{c}_{ij,1}\overline{x}_{ij,1},\sum_{i=1}^{m}\sum_{j=1}^{m}c_{ij,2}x_{ij,2},\sum_{i=1}^{m}\sum_{j=1}^{m}\overline{c}_{ij,3}\overline{x}_{ij,3};\overline{\lambda}\right)\right\rangle$$

$$s.t. \quad \sum_{j=1}^{n}\underline{x}_{ij,1}=\underline{s}_{i,1},\sum_{j=1}^{n}x_{ij,2}=s_{i,2},\sum_{j=1}^{n}\underline{x}_{ij,3}=\underline{s}_{i,3},\sum_{j=1}^{n}\overline{x}_{ij,1}=\overline{s}_{i,1},\sum_{j=1}^{n}\overline{x}_{ij,3}=\overline{s}_{i,3}, \quad \forall i,$$

$$\sum_{i=1}^{m}\underline{x}_{ij,1}=\underline{d}_{j,1},\sum_{i=1}^{m}x_{ij,2}=d_{j,2},\sum_{i=1}^{m}\underline{x}_{ij,3}=\underline{d}_{j,3},\sum_{i=1}^{m}\overline{x}_{ij,1}=\overline{d}_{j,1},\sum_{i=1}^{m}\overline{x}_{ij,3}=\overline{d}_{j,3}; \forall j$$

$$\overline{x}_{ij,1} \geq 0, \quad \forall i, \forall j,$$

$$\underline{x}_{ij,1} - \overline{x}_{ij,1} \geq 0, \quad \forall i, \forall j,$$

$$x_{ij,2} - \underline{x}_{ij,1} \geq 0, \quad \forall i, \forall j,$$

$$\underline{x}_{ij,3} - x_{ij,2} \geq 0, \quad \forall i, \forall j,$$

$$\overline{x}_{ij,1} - \underline{x}_{ij,3} \geq 0, \quad \forall i, \forall j. \tag{7}$$

In the next section, a new approach for solving the same problem is suggested to reduce the complexity of the existing method [6].

4 New Approach

Note that non-negativity and boundary constraints in model (7) are used to preserve the form of the optimal solutions as non–negative interval-valued triangular fuzzy numbers. By removing these constraints, the feasible space of model (7) is separable in terms of decision variables. The idea behind the proposed approach is that first the non-negativity and boundary constraints are removed from the feasible space and then model (7) is decomposed into five classical TPs. The integration of the optimal solution of the resulting five crisp TPs provides the optimal feasible solution of model (7). Here, the ranking technique given in Definition 4 is used to develop our approach for solving model (7). The proposed method is included the following seven steps:

Step 1: Solve the following crisp TP:

$$\overline{Z}_1^* = \min \overline{Z}_1 = \sum_{i=1}^{m} \sum_{j=1}^{m} \overline{c}_{ij,1} \overline{x}_{ij}$$

$s.t.$

$$\sum_{j=1}^{n} \overline{x}_{ij,1} = \overline{s}_{i,1}, \quad \forall i,$$

$$\sum_{i=1}^{m} \overline{x}_{ij,1} = \overline{d}_{j,1}, \forall j,$$

$$\overline{x}_{ij,1} \geq 0, \quad \forall i, \forall j. \tag{8}$$

Step 2: The following bounded TP is solved assuming that $\overline{x}_1^* = (\overline{x}_{ij,1}^*)_{nm \times 1}$ as the optimal solution of model (8):

$$\underline{Z}_1^* = \min \underline{Z}_1 = \sum_{i=1}^{m} \sum_{j=1}^{m} \underline{c}_{ij,1} \underline{x}_{ij,1}$$

$s.t.$

$$\sum_{j=1}^{n} \underline{x}_{ij,1} = \underline{s}_{i,1}, \quad \forall i,$$

$$\sum_{i=1}^{m} \underline{x}_{ij,1} = \underline{d}_{j,1}, \forall j,$$

$$\underline{x}_{ij,1} \geq \overline{x}_{ij,1}^*, \quad \forall i, \forall j. \tag{9}$$

Step 3: Solve the following bounded TP assuming that $\underline{x}_1^* = (\underline{x}_{ij,1}^*)_{nm \times 1}$ as the optimal solution of model (9):

$$Z_2^* = \min Z_2 = \sum_{i=1}^{m} \sum_{j=1}^{m} c_{ij,2} x_{ij,2}$$

$$s.t.$$

$$\sum_{j=1}^{n} x_{ij,2} = s_{i,2}, \forall i,$$

$$\sum_{i=1}^{m} x_{ij,2} = d_{j,2}, \forall j,$$

$$x_{ij,2} \geq \underline{x}_{ij,1}^*, \quad \forall i, \forall j. \tag{10}$$

Step 4: Solve the following bounded TP assuming that $x_2^* = (x_{ij,2}^*)_{nm \times 1}$ as the optimal solution of model (10):

$$\underline{Z}_3^* = \min \underline{Z}_3 = \sum_{i=1}^{m} \sum_{j=1}^{m} \underline{c}_{ij,3} \underline{x}_{ij,3}$$

$$s.t.$$

$$\sum_{j=1}^{n} \underline{x}_{ij,3} = \underline{s}_{i,3}, \forall i,$$

$$\sum_{i=1}^{m} \underline{x}_{ij,3} = \underline{d}_{j,3}, \forall j,$$

$$\underline{x}_{ij,3} \geq x_{ij,2}^*, \quad \forall i, \forall j. \tag{11}$$

Step 5: Solve the following bounded TP assuming that $\underline{x}_3^* = (\underline{x}_{ij,3}^*)_{nm \times 1}$ as the optimal solution of model (11):

$$\overline{Z}_3^* = \min \overline{Z}_3 = \sum_{i=1}^{m} \sum_{j=1}^{m} \overline{c}_{ij,3} \overline{x}_{ij,3}$$

$$s.t.$$

$$\sum_{j=1}^{n} \overline{x}_{ij,3} = \overline{a}_{i,3}, \forall i,$$

$$\sum_{i=1}^{m} \overline{x}_{ij,3} = \overline{b}_{j,3}, \forall j,$$

$$\overline{x}_{ij,3} \geq \underline{x}_{ij,3}^*, \quad \forall i, \forall j. \tag{12}$$

Step 6: Determine the interval-valued triangular fuzzy optimal solution $\tilde{\underline{x}}^*$ by substituting $\overline{x}^*_{ij,1}, \underline{x}^*_{ij,1}, x^*_{ij,2}, \underline{x}^*_{ij,3}, \overline{x}^*_{ij,3}$ in $\tilde{\underline{x}}^*_{ij} = \left\langle (\underline{x}^*_{ij,1}, x^*_{ij,2}, \underline{x}^*_{ij,3}; \underline{\lambda}), (\overline{x}^*_{ij,1}, x^*_{ij,2}, \overline{x}^*_{ij,3}; \overline{\lambda}) \right\rangle$. Then, determine the minimum interval-valued triangular fuzzy transportation cost by substituting $\overline{Z}^*_1, \underline{Z}^*_1, Z^*_2, Z^*_3, \overline{Z}^*_3$ in $\tilde{\underline{Z}}^* = \left\langle (\underline{Z}^*_1, Z^*_2, \underline{Z}^*_3; \underline{\lambda}), (\underline{Z}^*_1, Z^*_2, \underline{Z}^*_3; \overline{\lambda}) \right\rangle$.

Note that the optimal interval-valued triangular fuzzy transportation cost, $\tilde{\underline{Z}}^* = \left\langle (\underline{Z}^*_1, Z^*_2, \underline{Z}^*_3; \underline{\lambda}), (\underline{Z}^*_1, Z^*_2, \underline{Z}^*_3; \overline{\lambda}) \right\rangle$, keeps the form of a non-negative IVTFN. Also, the optimal interval-valued triangular fuzzy solution, $\tilde{\underline{x}}^*_{ij} = \left\langle (\underline{x}^*_{ij,1}, x^*_{ij,2}, \underline{x}^*_{ij,3}; \underline{\lambda}), (\overline{x}^*_{ij,1}, x^*_{ij,2}, \overline{x}^*_{ij,3}; \overline{\lambda}) \right\rangle$, keeps the form of a non-negative IVTFN.

5 Numerical Example

Here, one numerical example is illustrated in order to validate the formulated model. Consider the following FIVFTP:

$$\min \ \tilde{\underline{Z}} = \langle (10, 12, 13; 0.5), (9, 12, 15; 1) \rangle \langle (\underline{x}_{11,1}, x_{11,2}, \underline{x}_{11,3}; \underline{\lambda}), (\overline{x}_{11,1}, x_{11,2}, \overline{x}_{11,3}; \overline{\lambda}) \rangle$$

$$+ \langle (8, 10, 11; 0.5), (7, 10, 14; 1) \rangle \langle (\underline{x}_{12,1}, x_{12,2}, \underline{x}_{12,3}; \underline{\lambda}), (\overline{x}_{12,1}, x_{12,2}, \overline{x}_{12,3}; \overline{\lambda}) \rangle$$

$$+ \langle (10, 12, 13; 0.5), (9, 12, 16; 1) \rangle \langle (\underline{x}_{13,1}, x_{13,2}, \underline{x}_{13,3}; \underline{\lambda}), (\overline{x}_{13,1}, x_{13,2}, \overline{x}_{13,3}; \overline{\lambda}) \rangle$$

$$+ \langle (2, 3, 7; 0.5), (1, 3, 8; 1) \rangle \langle (\underline{x}_{21,1}, x_{21,2}, \underline{x}_{21,3}; \underline{\lambda}), (\overline{x}_{21,1}, x_{21,2}, \overline{x}_{21,3}; \overline{\lambda}) \rangle$$

$$+ \langle (4, 8, 10; 0.5), (3, 8, 12; 1) \rangle \langle (\underline{x}_{22,1}, x_{22,2}, \underline{x}_{22,3}; \underline{\lambda}), (\overline{x}_{22,1}, x_{22,2}, \overline{x}_{22,3}; \overline{\lambda}) \rangle$$

$$+ \langle (5, 7, 8; 0.5), (4, 7, 9; 1) \rangle \langle (\underline{x}_{23,1}, x_{23,2}, \underline{x}_{23,3}; \underline{\lambda}), (\overline{x}_{23,1}, x_{23,2}, \overline{x}_{23,3}; \overline{\lambda}) \rangle$$

$s.t.$

$$\left\langle (\underline{x}_{11,1}, x_{11,2}, \underline{x}_{11,3}; \underline{\lambda}), (\overline{x}_{11,1}, x_{11,2}, \overline{x}_{11,3}; \overline{\lambda}) \right\rangle + \left\langle (\underline{x}_{12,1}, x_{12,2}, \underline{x}_{12,3}; \underline{\lambda}), (\overline{x}_{12,1}, x_{12,2}, \overline{x}_{12,3}; \overline{\lambda}) \right\rangle$$

$$+ \left\langle (\underline{x}_{13,1}, x_{13,2}, \underline{x}_{13,3}; \underline{\lambda}), (\overline{x}_{13,1}, x_{13,2}, \overline{x}_{13,3}; \overline{\lambda}) \right\rangle = \langle (2000, 3000, 4000; 0.5), (1500, 3000, 5000; 1) \rangle$$

$$\left\langle (\underline{x}_{21,1}, x_{21,2}, \underline{x}_{21,3}; \underline{\lambda}), (\overline{x}_{21,1}, x_{21,2}, \overline{x}_{21,3}; \overline{\lambda}) \right\rangle + \left\langle (\underline{x}_{22,1}, x_{22,2}, \underline{x}_{22,3}; \underline{\lambda}), (\overline{x}_{22,1}, x_{22,2}, \overline{x}_{22,3}; \overline{\lambda}) \right\rangle$$

$$+ \left\langle (\underline{x}_{23,1}, x_{23,2}, \underline{x}_{23,3}; \underline{\lambda}), (\overline{x}_{23,1}, x_{23,2}, \overline{x}_{23,3}; \overline{\lambda}) \right\rangle = \langle (1200, 1500, 1800; 0.5), (1000, 1500, 2000; 1) \rangle,$$

$$\left\langle (\underline{x}_{11,1}, x_{11,2}, \underline{x}_{11,3}; \underline{\lambda}), (\overline{x}_{11,1}, x_{11,2}, \overline{x}_{11,3}; \overline{\lambda}) \right\rangle + \left\langle (\underline{x}_{21,1}, x_{21,2}, \underline{x}_{21,3}; \underline{\lambda}), (\overline{x}_{21,1}, x_{21,2}, \overline{x}_{21,3}; \overline{\lambda}) \right\rangle$$

$$= \langle (1500, 2000, 2500; 0.5), (950, 2000, 3000; 1) \rangle,$$

$$\left\langle (\underline{x}_{12,1}, x_{12,2}, \underline{x}_{12,3}; \underline{\lambda}), (\overline{x}_{12,1}, x_{12,2}, \overline{x}_{12,3}; \overline{\lambda}) \right\rangle + \left\langle (\underline{x}_{22,1}, x_{22,2}, \underline{x}_{22,3}; \underline{\lambda}), (\overline{x}_{22,1}, x_{22,2}, \overline{x}_{22,3}; \overline{\lambda}) \right\rangle$$

$$= \langle (900, 1500, 2000; 0.5), (850, 1500, 2500; 1) \rangle,$$

$$\left\langle (\underline{x}_{13,1}, x_{13,2}, \underline{x}_{13,3}; \underline{\lambda}), (\overline{x}_{13,1}, x_{13,2}, \overline{x}_{13,3}; \overline{\lambda}) \right\rangle + \left\langle (\underline{x}_{23,1}, x_{23,2}, \underline{x}_{23,3}; \underline{\lambda}), (\overline{x}_{23,1}, x_{23,2}, \overline{x}_{23,3}; \overline{\lambda}) \right\rangle$$

$$= \langle (800, 1000, 1300; 0.5), (700, 1000, 1500; 1) \rangle,$$

$$\left\langle (\underline{x}_{ij,1}, x_{ij,2}, \underline{x}_{ij,3}; \underline{\lambda}), (\overline{x}_{ij,1}, x_{ij,2}, \overline{x}_{ij,3}; \overline{\lambda}) \right\rangle \geq 0, i = 1, 2, j = 1, 2, 3. \tag{13}$$

Regarding model (7), the FIVFTP (13) is reformulated as follows:

$$\min \ \widetilde{\widetilde{Z}} = \left\langle \begin{pmatrix} 10\underline{x}_{11,1} + 8\underline{x}_{12,1} + 10\underline{x}_{13,1} + 2\underline{x}_{21,1} + 4\underline{x}_{22,1} + 5\underline{x}_{23,1}, \ 12\,x_{11,2} + 10x_{12,2} \\ +12x_{13,2} + 3x_{21,2} + 8x_{22,2} + 7x_{23,2} + 13\underline{x}_{11,3} + 11\underline{x}_{12,3} + 13\underline{x}_{13,3} + 7\underline{x}_{21,3} + 10\underline{x}_{22,3} + 8\underline{x}_{23,3}; 0.5 \end{pmatrix}, \\ \begin{pmatrix} 9\overline{x}_{11,1} + 7\overline{x}_{12,1} + 9\overline{x}_{13,1} + \overline{x}_{21,1} + 3\overline{x}_{22,1} + 4\overline{x}_{23,1}, \ 12\,x_{11,2} + 10x_{12,2} \\ +12x_{13,2} + 3x_{21,2} + 8x_{22,2} + 7x_{23,2}, +15\overline{x}_{11,3} + 14\overline{x}_{12,3} + 16\overline{x}_{13,3} + 8\overline{x}_{21,3} + 12\overline{x}_{22,3} + 9\overline{x}_{23,3}; 1 \end{pmatrix} \right\rangle$$

s.t.

$\underline{x}_{11,1} + \underline{x}_{12,1} + \underline{x}_{13,1} = 2000, \ x_{11,2} + x_{12,2} + x_{13,2} = 3000, \ \underline{x}_{11,3} + \underline{x}_{12,3}\underline{x}_{12,3} = 4000,$

$\overline{x}_{11,1} + \overline{x}_{12,1} + \overline{x}_{13,1} = 1500, \overline{x}_{11,3} + \overline{x}_{12,3} + \overline{x}_{13,3} = 5000,$

$\underline{x}_{21,1} + \underline{x}_{22,1} + \underline{x}_{23,1} = 1200, \ x_{21,2} + x_{22,2} + x_{23,2} = 1500, \ \underline{x}_{21,3} + \underline{x}_{22,3}\underline{x}_{22,3} = 1800,$

$\overline{x}_{21,1} + \overline{x}_{22,1} + \overline{x}_{23,1} = 1000, \overline{x}_{21,3} + \overline{x}_{22,3} + \overline{x}_{23,3} = 2000,$

$\underline{x}_{11,1} + \underline{x}_{21,1} = 1500, \ x_{11,2} + x_{21,2} = 2000, \underline{x}_{11,3} + \underline{x}_{21,3} = 2500,$

$\overline{x}_{11,1} + \overline{x}_{21,1} = 950, \overline{x}_{11,3} + \overline{x}_{21,3} = 3000,$

$\underline{x}_{12,1} + \underline{x}_{22,1} = 900, \ x_{12,2} + x_{22,2} = 1500, \underline{x}_{12,3} + \underline{x}_{22,3} = 2000,$

$\overline{x}_{12,1} + \overline{x}_{22,1} = 850, \overline{x}_{12,3} + \overline{x}_{22,3} = 2500$

$\underline{x}_{13,1} + \underline{x}_{23,1} = 800, \ x_{13,2} + x_{23,2} = 1000, \underline{x}_{13,3} + \underline{x}_{23,3} = 1300,$

$\overline{x}_{13,1} + \overline{x}_{23,1} = 700, \overline{x}_{13,3} + \overline{x}_{23,3} = 1500,$

$\overline{x}_{ij,1} \geq 0, \qquad i = 1, 2, j = 1, 2, 3,$

$\underline{x}_{ij,1} - \overline{x}_{ij,1} \geq 0, \quad i = 1, 2, j = 1, 2, 3,$

$x_{ij,2} - \underline{x}_{ij,1} \geq 0, \quad i = 1, 2, j = 1, 2, 3,$

$\underline{x}_{ij,3} - x_{ij,2} \geq 0, \quad i = 1, 2, j = 1, 2, 3,$

$\overline{x}_{ij,3} - \underline{x}_{ij,3} \geq 0, \quad i = 1, 2, j = 1, 2, 3.$ \hfill (14)

By implementing the proposed solution algorithm the optimal IVTFN solution of FIVFTP (14) is given as follows:

$$\widetilde{\underline{x}}_{11}^* = \langle (350, 550, 750; 0.5), (0, 550, 1250; 1) \rangle$$

$$\widetilde{\underline{x}}_{12}^* = \langle (900, 1500, 2000; 0.5), (850, 1500, 2500; 1) \rangle$$

$$\widetilde{\underline{x}}_{13}^* = \langle (750, 950, 1250; 0.5), (650, 950, 1250; 1) \rangle$$

$$\widetilde{\underline{x}}_{21}^* = \langle (1150, 1450, 1750; 0.5), (950, 1450, 1750; 1) \rangle$$

$$\widetilde{\underline{x}}_{22}^* = \langle (0, 0, 0; 0.5), (0, 0, 0; 1) \rangle$$

$$\widetilde{\underline{x}}_{23}^* = \langle (20750, 37700, 60650; 0.5), (12950, 37700, 90000; 1) \rangle \hfill (15)$$

The minimum interval-valued triangular fuzzy transportation cost is then given as $\widetilde{\widetilde{Z}}^* = \left\langle (\underline{Z}_1^*, Z_2^*, \underline{Z}_3^*; \underline{\lambda}), (\underline{Z}_1^*, Z_2^*, \underline{Z}_3^*; \overline{\lambda}) \right\rangle =$ $\langle (20750, 37700, 60650; 0.5), (12950, 37700, 90000; 1) \rangle.$

It should be noted that the method proposed by Ebrahimnejad [6] and the proposed method have the same results for solving the FIVFTP (13). However, our proposed method is preferred because according to the method proposed by Ebrahimnejad [6], the LP problem (14) is solved to obtain the solution of FIVFTP (13) which is not an standard TP, whereas based on proposed approach transportation structured problems are used for solving FIVFTP (14).

6 Conclusions

In this paper, we have analyzed a TP with IVTFNs for the transportation costs, demands and supplies. The FIVFTP under consideration has been decomposed into five crisp TPs. According to this new decomposition, a novel optimization process has been used to solve FIVFTP. The proposed method is preferable to the Ebrahimnejad's approach [6] because of requiring less constraints and variables in the decomposed sub-problems. There are a few other interesting topics for future research works of this study by considering the proposed approach for fuzzy and interval fuzzy shortest path problems [7, 8].

References

1. Baykasoğlu, A., Subulan, K.: A direct solution approach based on constrained fuzzy arithmetic and metaheuristic for fuzzy transportation problems. Soft. Comput. **23**, 1667–1698 (2019)
2. Mahmoodirad, A., Allahviranloo, T., Niroomand, S.: A new effective solution method for fully intuitionistic fuzzy transportation problem. Soft. Comput. **23**, 4521–4530 (2019)
3. Chiang, J.: The optimal solution of the transportation problem with fuzzy demand and fuzzy product. J. Inf. Sci. Eng. **21**, 439–451 (2005)
4. Gupta, A., Kumar, A.: A new method for solving linear multi-objective transportation problems with fuzzy parameters. Appl. Math. Model. **36**, 1421–1430 (2012)
5. Bharati, S.K., Singh, S.R.: Transportation problem under interval-valued intuitionistic fuzzy environment. Int. J. Fuzzy Syst. **20**, 1511–1522 (2018)
6. Ebrahimnejad, A.: Fuzzy linear programming approach for solving transportation problems with interval-valued trapezoidal fuzzy numbers. Sādhanā **41**, 299–316 (2016)
7. Di Caprio, D., Ebrahimnejad, A., Alrezaamiri, H., Santos-Arteaga, F.J.: A novel ant colony algorithm for solving shortest path problems with fuzzy arc weights. Alex. Eng. J. **61**(5), 3403–3415 (2022)
8. Ebrahimnejad, A., Enayattabr, M., Motameni, H., Garg, H.: Modified artificial bee colony algorithm for solving mixed interval-valued fuzzy shortest path problem. Complex Intell. Syst. **7**(3), 1527–1545 (2021)

Generalized Differentiability of Fuzzy-Valued Convex Functions and Applications

T. Allahviranloo⬛, M. R. Baloochshahryari⬛, and O. Sedaghatfar⬛

Abstract In this paper, the concept of generalized differentiability and level-wise generalized Hukuhara differentiability are extended for one-dimensional fuzzy-valued convex functions from $\mathbb{R}$ into E. In addition, the properties of generalized differentiability and characterization for fuzzy-valued convex functions in terms of generalized differentiability and the fundamental theorem of calculus generalized differential and fuzzy integral are presented in detail. Moreover, the concepts of generalized subgradient and generalized subdifferential in terms of level-wise generalized Hukuhara differentiability are extended for fuzzy-valued convex functions. Finally, by using their properties, the convex fuzzy optimization for the one-dimensional fuzzy-valued convex functions is discussed.

Keywords Fuzzy numbers · Fuzzy-valued convex function · L_{gH}-differentiability · g-differentiability · g-subgradient · g-subdifferential · Fuzzy optimization

1 Introduction

There are very important topics in fuzzy optimization, one of the most significant of them is fuzzy convex analysis. Hereupon, in 1992 the concept of fuzzy mapping convexity was proposed by Nada and Kar [1]. Accordingly, the convexity for fuzzy mapping and application in fuzzy optimization have been studied in [2–6]. The concepts of convexity and quasi-convexity of fuzzy-valued functions have been

T. Allahviranloo (✉)
Faculty of Engineering and Natural Sciences, Istinye University, Istanbul, Turkey
e-mail: tofigh.allahviranloo@istinye.edu.tr

M. R. Baloochshahryari
Department of Mathematics, Kerman Branch, Islamic Azad University, Kerman, Iran

O. Sedaghatfar
Department of Mathematics, Yadegar-E-Imam Khomeini (RAH) Shahre Rey Branch, Islamic Azad University, Tehran, Iran

© The Author(s), under exclusive license to Springer Nature Switzerland AG 2023
Sh. N. Shahbazova et al. (eds.), *Recent Developments and the New Directions of Research, Foundations, and Applications*, Studies in Fuzziness and Soft Computing 423,
https://doi.org/10.1007/978-3-031-23476-7_23

explored by Yan-Xu [5]. The concepts of quasi-convex and pseudo-convex multi-variable fuzzy functions have been investigated by Syau [7]. Furukawa has interested in convexity and Lipschitz continuity of fuzzy-valued functions [8]. In [2, 3, 9], we can see several definitions and properties for various kinds of convexity or generalized convexity of fuzzy mapping. Thence, the concept and properties of fuzzy preinvex functions in the $\mathbb{R}$ field have been presented by Noor [10]. The Hukuhara difference (H-difference) exists between two fuzzy numbers only under very restricted situations, then Stefanini proposed the generalized Hukuhara difference (gH-difference) [11]. In many cases, when H-difference does not exist, the gH-difference exists, but notice that the gH-difference does not always exist. Then, the generalized difference (g-difference) has been presented by Bede and Stefanini, which always exists [12]. It is prominent that this difference in some cases does not support the convexity condition of fuzzy numbers, and thus is not a fuzzy number. Now, this debate is resolved by considering the results of Gomes and Barros studies, which investigated the resulting set convex hull [13]. Founded on these two differences, the generalized Hukuhara differentiability (gH-differentiability), level-wise generalized Hukuhara differentiability (LgH-differentiability), and generalized differentiability (g-differentiability) have been acquired [12]. In this regard, this paper included the concepts of g-differentiability and the topological properties, such as limit, continuity, integrability, and differentiability for one-dimensional fuzzy-valued convex functions. In addition, we express the properties of g-differentiability as well as the characterizations for the fuzzy-valued convex functions in terms of g-differentiability and the fundamental theorem of calculus g-differential and fuzzy integral in detail. Below, the g-subdifferential concept is developed for fuzzy-valued convex functions in terms of g-differentiability and LgH-differentiability. Finally, using the properties of the fuzzy-valued convex functions, the fuzzy convex optimization is expressed. This paper is formed as follows; In Sect. 2 we study one-dimensional fuzzy-valued convex functions, and the generalized differentiability of one-dimensional fuzzy-valued convex functions in Sect. 3 is considered, then g-subdifferential of one-dimensional fuzzy-valued convex functions are considered and some outcomes are gained in Sect. 4. At the end of this paper, in Sect. 5 a one-dimensional fuzzy-valued convex function and optimizations are performed, and eventually, these concepts will be found precisely by some examples prepared in this article.

2 One-Dimensional Fuzzy-Valued Convex Functions

In this section, let $\phi : I \subseteq \mathbb{R} \to E$ be a fuzzy valued function on some interval of the real line $\mathbb{R}$, and I can be open, half-open, closed, finite, or infinite. The interior of I is denoted by I^0. Also, E has been explained, that the set of fuzzy numbers, that is normal, fuzzy convex, upper semi-continuous, and compactly supported fuzzy sets that are determined over the real line. Suppose that $X \in E$ be a fuzzy number; for $0 < r \leq 1$, the level-set (or r-cut) of X is described by $[X]_r = \{t \in \mathbb{R} | X(t) \geq r\}$, and for $r = 0$ by the closure of the support $[X]_0 = cl\{t \in \mathbb{R} | X(t) > 0\}$ is illustrated.

Moreover, we explain $[X]_r = [X_r^-, X_r^+]$, that the r-cut $[X]_r$ is a closed interval $\forall r \in [0, 1]$ $[0, 1]$.

Definition 1 [14] Suppose that $\phi : I \subseteq \mathbb{R} \to E$, then ϕ is a fuzzy-valued convex function if

$$\phi(\theta t + (1 - \theta)y) \preceq_\theta \odot \phi(t) \oplus (1 - \theta) \odot \phi(y), \; \forall t, y \in I, \; \forall 0 \leq \theta \leq 1.$$

Closely related to fuzzy convexity is the following concept.

Definition 2 Suppose that $\phi : I \subseteq \mathbb{R} \to E$. Then ϕ is midpoint fuzzy-valued convex function if

$$\phi\left(\frac{t + y}{2}\right) \preceq \frac{1}{2}[\phi(t) \oplus \phi(y)], \; \forall t, y \in I.$$

Note that, if ϕ is a fuzzy-valued convex function, so ϕ is the midpoint fuzzy-valued convex function.

Definition 3 Suppose that $\phi : I \subseteq \mathbb{R} \to E$, then ϕ is a strictly fuzzy-valued convex function if

$$\phi(\theta t + (1 - \theta)y) \underset{\not\approx}{\prec} \theta \odot \phi(t) \oplus (1 - \theta) \odot \phi(y), \; with \; t \neq y, \; \forall t, y \in I, \; \forall 0 \leq \theta \leq 1.$$

Theorem 1 [14] *Suppose that* $\phi : I \subseteq \mathbb{R} \to E$ *with* $[\phi(t)]_r = [\phi_r^-(t), \phi_r^+(t)]$, *then*

(1) *The function* ϕ *is a convex fuzzy-valued if and only if for any fixed* $r \in [0, 1]$, *the functions* $\phi_r^-(t)$ *and* $\phi_r^+(t)$ *both are real-valued convex functions of* t.
(2) *The function* ϕ *is a nondecreasing (nonincreasing) if and only if for any fixed* $r \in [0, 1]$, *the functions* $\phi_r^-(t)$ *and* $\phi_r^+(t)$ *both are nondecreasing (nonincreasing) of* t.

Lemma 1 *Suppose that* $\phi : I \to E$, *then* ϕ *is a fuzzy-valued convex function if and only if*

$$\phi(y) \preceq \frac{z - y}{z - t} \odot \varphi(t) \oplus \frac{y - t}{z - t} \odot \phi(z),$$
$$\forall t, y, z \in I \; with \; t < y < z.$$

Lemma 2 *Let* $\phi : I \subseteq \mathbb{R} \to E$ *be a fuzzy-valued convex function and* $t \in I^0$ *with* $t + \gamma$ *and* $t - \gamma \in I^0$, *then the fuzzy right and left quotient w.r.t.* γ

$$\Re_{+g}(t, \gamma) = \frac{\phi(t + \gamma) \ominus_g \phi(t)}{\gamma},$$
$$\Re_{-g}(t, \gamma) = \frac{\phi(t) \ominus_g \phi(t - \gamma)}{\gamma},$$

both are respectively nondecreasing and nonincreasing on I^0.

3   g-Differentiability of Fuzzy-Valued Convex Functions

In this section, the g-differentiability and the basic facts about the g-differentiability properties of one-dimensional fuzzy-valued convex functions are discussed that can be easily visualized.

Definition 4 Let $\phi : I \subseteq \mathbb{R} \to E$ be a fuzzy-valued convex function, $t \in I$, γ with $t + \gamma \in I$, and $t - \gamma \in I$, then

$$\phi'_{+g}(t) = \lim_{\gamma \to 0^+} \frac{\phi(t + \gamma) \ominus_g \phi(t)}{\gamma},$$

$$\phi'_{-g}(t) = \lim_{\gamma \to 0^+} \frac{\phi(t - \gamma) \ominus_g \phi(t)}{\gamma},$$

exists on I^0. We demonstrate that ϕ is a fuzzy-valued convex function that right and left g-differentiable on I^0.

Definition 5 Suppose that $\phi : I \subseteq \mathbb{R} \to E$ is a fuzzy-valued convex function, $t \in I$, γ with $t + \gamma \in I$, and $t - \gamma \in I$, the quotient $\Re$ is right and left L_{gH}-difference for any $r \in [0, 1]$, as below:

$$\Re_{+L_{gH}}(t, \gamma)_r = \frac{\left[\phi(t + \gamma)\right]_r \ominus_{gH} [\phi(t)]_r}{\gamma},$$

and

$$\Re_{-L_{gH}}(t, \gamma)_r = \frac{[\phi(t)]_r \ominus_{gH} \left[\phi(t - \gamma)\right]_r}{\gamma}.$$

Note that, $\Re_{+L_{gH}}(t, \gamma)_r$ and $\Re_{-L_{gH}}(t, \gamma)_r \in K_C \, \forall r \in [0, 1]$. The family of interval-valued $\{\Re_{+L_{gH}}(t, \gamma)_r : r \in [0, 1]\}$ and $\{\Re_{-L_{gH}}(t, \gamma)_r : r \in [0, 1]\}$ are right and left L_{gH}-quotient of ϕ as a function of γ at t, and showed by $\Re_{+L_{gH}}(t, \gamma)$ and $\Re_{-L_{gH}}(t, \gamma)$.

Theorem 2 *Suppose that $\phi : I \subseteq \mathbb{R} \to E$ is a fuzzy-valued convex function, $t \in I^0$, γ with $t + \gamma, t - \gamma \in I^0$, and ϕ is uniformly right and left L_{gH}-differentiable at t. Then ϕ is right and left g-differentiable at t and there exists $\phi'_{+g}(t)_r$ and $\phi'_{-g}(t)_r \subset \mathbb{R}$ for every $r \in [0, 1]$, the family of interval-valued functions*

$$\Re_{+g}(t, \gamma)_r = cl \left(conv \bigcup_{\beta \geq r} \frac{\left[\phi(t + \gamma)\right]_\beta \ominus_{gH} [\phi(t)]_\beta}{\gamma} \right),$$

$$\mathfrak{R}_{-g}(t, \gamma)_r = cl\left(conv \bigcup_{\beta \geq r} \frac{[\phi(t)]_\beta \ominus_{gH} [\phi(t - \gamma)]_\beta}{\gamma}\right),$$

Uniformly w.r.t. $r \in [0, 1]$ converge to $\phi'_{+g}(t)_r$ and $\phi'_{-g}(t)_r$, as $\gamma \rightarrow 0^+$, respectively.

Theorem 3 *Suppose that $\phi : I \subseteq \mathbb{R} \rightarrow E$ is a fuzzy-valued convex function, let $t \in I^0$, γ with $t + \gamma$ and $t - \gamma \in I^0$. ϕ is right and left uniformly L_{gH}-differentiable at t, then $\phi'_{+g}(t)$ and $\phi'_{-g}(t) \in E$ at every point of I^0 exists, also $\phi'_{+g}(t) = \inf_{\gamma>0} \mathfrak{R}_{+g}(t, \gamma), \phi'_{-g}(t) = \sup_{\gamma>0} \mathfrak{R}_{-g}(t, \gamma)$.*

Corollary 1 *For fuzzy-valued convex function $\phi : I \rightarrow E$, with $t + \gamma$ and $t - \gamma \in I^0$, if $\phi'_{+g}(t)$ and $\phi'_{-g}(t) \in E$ exists on I^0, then,*

$$\phi'_{+g}(t) = \inf_{\gamma>0} \mathfrak{R}_{+g}(t, \gamma) = \inf_{\gamma>0} \frac{\phi(t + \gamma) \ominus_g \phi(t)}{\gamma},$$

$$\phi'_{-g}(t) = \sup_{\gamma>0} \mathfrak{R}_{-g}(t, \gamma) = \sup_{\gamma>0} \frac{\phi(t) \ominus_g \phi(t - \gamma)}{\gamma}.$$

Theorem 4 *LeT $\phi : I \rightarrow E$ be a fuzzy-valued convex function, with $t + \gamma$ and $t - \gamma \in I^0$. Then*

(1) *These two functions $\phi'_{+g}(t)$ and $\phi'_{-g}(t)$ are increasing on I^0.*
(2) *If $c \in I^0$ and $\gamma > 0$ with $c + \gamma$ and $c - \gamma \in I^0$, then:*

$$\phi'_{+g}(c) = \inf_{\gamma>0} \mathfrak{R}^{c,c+\gamma}_{+g}(\gamma) \succeq \sup_{\gamma>0} \mathfrak{R}^{c-\gamma,c}_{-g}(\gamma) = \phi'_{-g}(c).$$

(3) *If $c, d \in I^0$ with $c < d$ and $\gamma > 0$ be such that $c + \gamma$ and $c - \gamma \in I^0$ $d + \gamma$ and $d - \gamma \in I^0$. Then*

$$\phi'_{-g}(c) = \sup_{\gamma>0} \mathfrak{R}^{c-\gamma,c}_{-g}(\gamma) \preceq \inf_{\gamma>0} \mathfrak{R}^{c,c+\gamma}_{+g}(\gamma) = \phi'_{+g}(c).$$

$$\phi'_{+g}(c) = \inf_{\gamma>0} \mathfrak{R}^{c,c+\gamma}_{+g}(\gamma) \preceq \sup_{\gamma>0} \mathfrak{R}^{d-\gamma,d}_{-g}(\gamma) = \phi'_{-g}(d).$$

Definition 6 Suppose that $\phi : I \rightarrow E$. Let $I_0 \subseteq I$ be a subinterval. It is said to be $\phi(t)$ is fuzzy Lipschitzian relative to I_0 if there exists $M > 0$ such that

$$D(\phi(t), \phi(y)) \leq M|t - y|, \forall t, y \in I_0.$$

Theorem 5 *Suppose that $\phi : I \rightarrow E$ is a fuzzy-valued convex function and $[a, b] \subset I^0$. Then*

(1) *$\phi(t)$ is fuzzy Lipschizian relative to $[a, b]$.*

(2) $\phi(t)$ is g-continuous on I^0.

Theorem 6 *Suppose that $\phi : I \subseteq \mathbb{R} \to E$ is a fuzzy-valued convex function. Then $\phi'_{+g}(t) \in E$ is the right g-continuous function and $\phi'_{-g}(t) \in E$ is left g-continuous function on I^0.*

Theorem 7 *Suppose that $\phi : I \to E$ is the midpoint fuzzy-valued convex function and g-continuous on I. Then $\phi(t)$ is a fuzzy-valued function.*

Below theorem, give a characterization for fuzzy-valued convex function in terms of twice g-differentiability.

Theorem 8 *Suppose that $\phi : (a, b) \to E$ is twice g-differentiable. Then $\phi(t)$ is a fuzzy-valued convex function if and only if $\phi''_g(t) \succeq 0$, $\forall t \in (a, b)$.*

Example 1 *Suppose that $\phi(t) := X \odot t^p$, $t > 0$, $p \geq 1$, and $X \in E$ be fixed. The fuzzy-valued convex function $\phi(t)$ is twice g-differentiable. Then*

$$\phi'_g(t) = X \odot pt^{p-1},$$
$$\phi''_g(t) = X \odot p(p-1)t^{p-2} \succeq 0, \ \forall t > 0,$$

therefore $\phi(t)$ is a fuzzy-valued convex function.

Theorem 9 *Suppose that $\phi : (a, b) \to E$. Then ϕ is a fuzzy-valued convex function with no switching on (a, b) if and only if there exists $g : (a, b) \to E$ such that g is an increasing right g-continuous function and*

$$\int_c^t g(x)dx = \phi(t) \ominus_g \phi(c), \ \forall c, t \in (a, b).$$

4 g-Subdifferential of the Fuzzy-Valued Convex Function

In the next definition, we introduce the g-subdifferential of a one-dimensional fuzzy-valued convex function.

Definition 7 Let $\phi : I \subseteq \mathbb{R} \to E$ be a fuzzy-valued convex function, and $t_0 \in I^0$. If there exists $\phi'_{-g}(t_0)$ and $\phi'_{+g}(t_0) \in E$, and

$$\begin{cases} \phi(t) \ominus_g \phi(t_0) \succeq \phi'_{+g}(t_0) \odot (t - t_0), \ \forall t \in I, \\ \phi(t) \ominus_g \phi(t_0) \succeq \phi'_{-g}(t_0) \odot (t - t_0), \ \forall t \in I. \end{cases} \tag{1}$$

Then it is said to be $\phi'_{+g}(t_0)$ and $\phi'_{-g}(t_0)$ both are right and left g-subgradient of ϕ at t_0 and the set of all g-subgradient of ϕ at t_0 is called g-subgradient of ϕ at t_0, and denoted by $\partial_g \phi(t_0)$. Moreover, ϕ is g-differentiable on I, then $\partial_g \phi(t_0) = \{\phi'_g(t_0)\}$, $\forall t_0 \in I. \partial_g \phi(t_0) = \{\phi'_g(t_0)\}$, $\forall t_0 \in I$.

Definition 8 Let ϕ is right and left uniformly L_{gH}-differentiable at t on I^0 then the representation of Eq. (1) for any $r \in [0, 1]$ is as below:

$$
\begin{cases}
cl\left(conv \bigcup_{\beta \geq r} [\phi(t)]_\beta \ominus_{gH} [\phi(t_0)]_\beta\right) \geq cl\left(conv \bigcup_{\beta \geq r} \phi'_{+L_{gH}}(t_0)_\beta\right)(t - t_0), \forall t \in I, \\
cl\left(conv \bigcup_{\beta \geq r} [\phi(t)]_\beta \ominus_{gH} [\phi(t_0)]_\beta\right) \geq cl\left(conv \bigcup_{\beta \geq r} \phi'_{-L_{gH}}(t_0)_\beta\right)(t - t_0), \forall t \in I.
\end{cases}
$$

Then it is said to be $\phi'_{+L_{gH}}(t_0)$ and $\phi'_{-L_{gH}}(t_0)$ are right and left L_{gH}-subgradient of ϕ at t_0 and the set of all right and left L_{gH} -subgradient of ϕ at t_0 is called L_{gH} -subgradient of ϕ at t, and denoted by $\partial_{L_{gH}} \phi(t_0)$.

Theorem 10 *Suppose that* $\phi : I \subseteq \mathbb{R} \to E$ *is a fuzzy-valued convex function, for every* $t_0 \in I^0$. *If there exists* $\phi'_{-g}(t_0)$ *and* $\phi'_{+g}(t_0) \in E$ *on* I^0, *then*

$$
\begin{cases}
\phi(t) \ominus_g \phi(t_0) \succeq \phi'_{+g}(t_0) \odot (t - t_0), \forall t \in I, \\
\phi(t) \ominus_g \phi(t_0) \succeq \phi'_{-g}(t_0) \odot (t - t_0), \forall t \in I.
\end{cases}
$$

Proposition 1 *Suppose that* $\phi : (-1, 1) \to E$ *is a fuzzy-valued convex function and* ϕ *is right and left uniformly* L_{gH}*-differentiable at* $t_0 \in (a, b)$. *So* ϕ *is g-subdifferential at each point* $t_0 \in (a, b)$ *and for every* $r \in [0, 1]$

$$
\partial_g \phi(t_0)_r = cl\left(conv \bigcup_{\beta \geq r} \left(\{\phi'_{-L_{gH}}(t_0)_\beta, \phi'_{+L_{gH}}(t_0)_\beta\}\right)\right).
$$

Example 2 Let $\phi : (-1, 1) \to E$, be defined by r-cuts for every $r \in [0, 1]$, $\phi_r(t) = [(r - 1)|t|, (1 - r)|t|]$ is described as below:

$$
\phi(t) = \langle -1, 0, 1\rangle \odot |t|,
$$

the functions $\phi_r^-(t)$ and $\phi_r^+(t)$ are not differentiable at $t_0 = 0$, so that ϕ is not L_{gH}-differentiable at t_0. But ϕ is g-subdifferential at $t_0 = 0$, indeed for every $r \in [0, 1]$, we have

$$
\partial_g \phi(0)_r = \left\{
\begin{aligned}
&m \in K_c : cl\left(conv \bigcup_{\beta > r} [\phi(t)]_\beta \ominus_{gH} [\phi(0)]_\beta\right) \\
&\geq cl\left(conv \bigcup_{\beta \geq r} m_\beta\right)(t - 0), \forall t \in \mathbb{R}
\end{aligned}
\right\}
$$

$$
= \left\{m \in K_c : cl\left(conv \bigcup_{\beta \geq r} [\beta - 1, 1 - \beta]\right) \geq cl\left(conv \bigcup_{\beta \geq r} m_\beta\right)\right\}.
$$

If $t > 0 : t > 0$:

$$\partial_g \phi(0)_r = \left\{ m \in K_c : cl\left(conv \bigcup_{\beta \geq r} [\beta - 1, 1 - \beta] \right) \geq cl\left(conv \bigcup_{\beta \geq r} m_\beta \right) \right\}.$$

If $t < 0 : t < 0$:

$$\partial_g \phi(0)_r = \left\{ \begin{array}{l} m \in K_c : cl\left(conv \bigcup_{\beta \geq r} [\beta - 1, 1 - \beta](-t) \right) \\ \geq cl\left(conv \bigcup_{\beta \geq r} m_\beta \right) t, \ \forall t \in \mathbb{R} \end{array} \right\}$$

$$= \left\{ m \in K_c : cl\left(conv \bigcup_{\beta \geq r} [\beta - 1, 1 - \beta] \right) \leq cl\left(conv \bigcup_{\beta \geq r} m_\beta \right) \right\}.$$

Hence

$$\partial_g \phi(0)_r = cl\left(conv \bigcup_{\beta \geq r} ([1 - \beta, \beta - 1], [\beta - 1, 1 - \beta]) \right), \ \forall r \in [0, 1].$$

For every $t \neq 0$ and for all $r \in [0, 1]$, both $\phi_r^-(t)$ and $\phi_r^+(t)$ are differentiable and

$$\partial \phi_r^-(t) = \{\phi_r^-(t)\} = \left\{ (r - 1)\frac{t}{|t|} \right\},$$

$$\partial \phi_r^+(t) = \{\phi_r^+(t)\} = \left\{ (1 - r)\frac{t}{|t|} \right\}.$$

5 Fuzzy Convexity and Fuzzy Convex Optimization

Now in this part, we consider applications of fuzzy convexity to some basic fuzzy optimization issues, such as the existence of optimal solutions. Let us consider an interval $I \subseteq \mathbb{R}$ and a fuzzy-valued function $\phi : I \subseteq \mathbb{R} \rightarrow E$. We focus on the problem of fuzzy minimizing $\phi(t)$ over $t \in I$.

Definition 9 Suppose that $\phi : I \subseteq \mathbb{R} \rightarrow E$. We say that $t_0 \in I$ is a global minimum point of ϕ over I, if $\phi(t_0) \preceq \phi(t)$, $\forall t \in I$, or equivalently $\phi(t_0) = \min_{t \in I} \phi(t)$.

Definition 10 Suppose that $\phi : I \subseteq \mathbb{R} \rightarrow E$. We say that $t_0 \in I$ is a local minimum point of ϕ over I, if there exists some such that $\phi(t_0) \preceq \phi(t)$, $\forall t \in I$, with $|t - t_0| \leq r$.

Theorem 11 *Suppose that $\phi : (a, b) \subseteq \mathbb{R} \rightarrow E$. Then*

(1) If ϕ is a fuzzy-valued convex function, then every local minimum point of ϕ is a global minimum point.

(2) If ϕ is a strictly fuzzy-valued convex function, then at most one global minimum point.

6 Conclusion

The important concept of g-difference was expressed by Bede and Stefanini [12], which we utilized to define g-differentiability for the one-dimensional fuzzy-valued convex functions. Moreover, the implications of g-subdifferential in terms of L_{gH}-differentiability were extended for one-dimensional fuzzy-valued convex functions. As a good suggestion, researchers should move towards the g-subdifferential fuzzy mapping and its usage in convex fuzzy optimization.

References

1. Nanda, S., Car, K.: Convex fuzzy mappings. Fuzzy Sets Syst. **48**, 129–132 (1992)
2. Amma, E.E.: On convex fuzzy mapping. J. Fuzzy Math. **14**(3), 501–512 (2006)
3. Nagata, F.: Convexity and local Lipschitz continuity of fuzzy valued mapping. Fuzzy Sets Syst. **93**, 113–119 (1998)
4. Shexiang, H., Gong, Z.: Generalized Differentiability for n-Dimensional Fuzzy-Number-Valued Functions and Fuzzy Optimization. Information Sciences (2016)
5. Yan, H., Xu, J.: A class convex fuzzy mappings. Fuzzy Sets Syst. **129**, 47–56 (2002)
6. Zhang, X.H., Yuan Lee, E.S.: Convex fuzzy mapping and operations of convex fuzzy mappings. Comput. Math. Appl. **51**, 143–152 (2006)
7. Syau, Y.R., Lee, E.S.: Fuzzy Weierstrass theorem and convex fuzzy mappings. Comput. Math. Appl. **51**, 1741–1750 (2006)
8. Furukawa, N.: Convexity and local Lipschitz continuity of fuzzy valued mapping. Fuzzy Sets Syst. **93**, 113–119 (1998)
9. Bao, Y.E., Wu, C.X.: Convexity and semicontinuity of fuzzy mappings. Comput. Math. Appl. **151**(12), 1809–1816 (2006)
10. Noor, M.A.: Fuzzy preinvex functions. Fuzzy Sets Syst. **64**, 95–104 (1994)
11. Stefanini, L.: A generalization of Hukuhara difference and division for interval and fuzzy arithmetic. Fuzzy Sets Syst. **161**, 1564–1584 (2010)
12. Bede, B., Stefanini, L.: Generalized differentiability of fuzzy-valued functions. Fuzzy Sets Syst. **230**, 119–141 (2013)
13. Gomes, L.T., Barros, L.C.: A note on the generalized difference and the generalized differentiability. Fuzzy Sets Syst. **280**, 142–145 (2015)
14. Wang, G., Wu, C.: Directional derivatives and subdifferential of convex fuzzy mappings and application in convex fuzzy programming. Fuzzy Sets Syst. **138**, 559–591 (2003)
15. Stefanini, L., Jimenez, M.A.: Karush-Kuhn-Tucker conditions for interval and fuzzy optimization in several variables under total and directional generalized differentiability. Fuzzy Sets Syst. **283** (2018)

Fuzzy Simulation

Applications of Autoregressive Process for Forecasting of Some Stock Indexes

Hilala Jafarova and Rovshan Aliyev

Abstract Autoregressive processes AR() are widely used in the effective solution of many management problems. In the presented article, AR(3) is applied in forecasting the Standard & Poor's 500 (S&P 500) and Nasdaq indices, which have a strong influence on the world economy in the international financial market. The main objective of the study is to investigate the results of a global pandemic of coronavirus disease 2019 (COVID-19) on the major stock market indices based on the index indicators used over the last ten years. Analysis ToolPak in MS Excel is used for application of statistical methods and for visualization.

Keywords Autoregressive models · Markov walk · Finance · Economic efficiency · Price of stock Indices · S&P 500 · Nasdaq · Analysis ToolPak in MS Excel

1   Introduction

Autocorrelation is usually a type of a strong positive correlation with past values influencing the current value separated by an interval. This in turn is a technique used to forecast time series. First-order autocorrelation characterizes the relationship between a series of magnitude values obtained at successive times, and pth-order autocorrelation characterizes the relationship between values with p periods from each other. Autocorrelation in data can be taken into account using autoregressive modeling methods (see, for example, Shiryaev [8]).

H. Jafarova (✉)
Department of Digital Technologies and Applied Informatics, UNEC, Baku, Azerbaijan
e-mail: hilala-jafarova@unec.edu.az

R. Aliyev
Department of Operations Research and Probability Theory, BSU, Baku, Azerbaijan
e-mail: rovshanaliyev@bsu.edu.az

Institute of Control System of ANAS, Baku, Azerbaijan

In Chakravarthy et al. [6] using AR model residual errors and AR model parameters as features, high specificity of coding DNA sequences is shown by AR residual error analysis, while AR feature-based analysis helps to distinguish between coding and non-coding DNA sequences.

In study Jafarova et al. (2019) focuses on the connection between USDX and Gold prices from 2010.01.01 to 2019.01.01. The aim of this paper is to analyze and determine the character of the movement between price levels.

Unlike the above studies, AR(3) with application to the S&P 500 and Nasdaq indices for 01.01.2012–01.01.2022.

One of the classic statistical methods that has been used for a long time and successfully in engineering practice is regression analysis. Autoregressive models are widely used as another approach to describe and predict the underlying trend of time series. Before building such models, the presence of autocorrelation in the studied series is evaluated.

2 Formulation of Problem

In economic forecasting, autoregressive models relate a time series variable to its past values.

The recent past of a variable should have the power to predict its near future. In the case of an AR(1) process, the distribution of the first term is taken equal to the unconditional distribution of the time series.

Equation (1) are used to forecast j years into the future from the current nth time period. Pth—order Autoregressive Forecasting equation [6]

$$\hat{Y}_{n+j} = a_0 + a_1 \hat{Y}_{n+j-1} + a_2 \hat{Y}_{n+j-2} + \cdots + a_p \hat{Y}_{n+j-p} \tag{1}$$

where $a_0, a_1, a_2, \ldots, a_p$—regression estimates of the parameters $A_0, A_1, \ldots, A_p$.

$\hat{Y}_{n+j-p}$ —forecast of from the current time period for $j - p > 0$, $\hat{Y}_{n+j-p}$ —observed value for Y_{n+j-p} for $j - p \leq 0$.

3 Mathematical Model and Solution Algorithm of the Problem

As it is known, S&P500 and Nasdaq, the world's leading indices, also influenced the change in the prices of other currency pairs. This paper presents these indices based on empirical data describing the AR(3) autoregressive process.

Table 1 Descriptive analyses of S&P 500 indices price

Indices	Max:	Min:	Difference:	Mean:	Change %:
S&P500	4818	1259	3559	2495	260

The S&P 500 is a stock index that consists of 505 stocks of 500 selected US listed public companies with the largest capitalization. It is one of the most commonly followed equity indices [10].

Initially, descriptive analysis was performed for S&P500 Price (Table 1).

As a result of the descriptive analysis, it was determined that the average price of the S&P 500 indices was 2495 USD, minimum price was 1259 USD in January 2012, and the maximum price was 4818 USD in January 2022. Percentage of change was 260% for considered the time interval.

Results of summary constructs the equation of the model AR(3) with application S&P 500 Price for 01/01/2012–01/01/2022:

Regression statistics						
Multiple R	0.991915855					
R square	0.983897062					
Adjusted R square	0.983473301					
Standard error	107.6526863					
Observations	118					
ANOVA						
	df	SS	MS	F	Significance F	
Regression	3	80,723,345.04	26,907,781.68	2331.817886	5.2888E-102	
Residual	114	1,321,157.499	11,589.10087			
Total	117	82,044,502.54				
	Coefficients	Standard error	T Stat	P-value	Lower 95%	Upper 95%
Intercept	− 13.19432234	32.57043608	− 0.405101188	0.686162316	− 77.71610535	51.32746066
Lag1	0.819973194	0.09551112	8.585107071	5.31485E-14	0.6307664	1.0099179987
Lag2	0.059587798	0.12635923	0.471574555	0.63813178	− 0.190728864	0.30990446
Lag3	0.140179522	0.098243518	1.42685773	0.156354064	− 0.054440129	0.334799174

$$Y = 0.821\text{Lag}1 + 0.06\text{Lag}2 + 0.14\text{Lag}3 - 13.19.$$

The adequacy of the model corresponds to the null hypothesis, the truth of which cannot be asserted, but it can be refuted in favor of an alternative, i.e. estimate statistical significance of inadequacy. The fact that the indicators indicating the adequacy (*Multiple R = 0.98, R-squared* = 0.99, *Adjusted R-squared* = 0.98), of the AR(3) model although high enough and the *P-values* = 0.87 are very large indicates a violation of the autoregressive process.

If the test resulted in a P < 0.05, then the null hypothesis is rejected, and the corresponding results are considered statistically significant.

Up to 5% is the probability that we erroneously concluded that the differences are significant, while they are unreliable in fact. In another way, we are only 95% sure that the differences are really significant, the two-tail t test with 117 degrees of freedom has critical values of ± 1.664. Because $-1.664 < t_{stat} = 1.42 < +1.664$ or because the *P-value* = 0.098 > 0.05 we do not reject H_0 ($H_0 = A_j = 0$) Conclude that the parameters of the autoregressive model is not significant and can be deleted.

The main reason for the inadequacy of the autoregressive model in the considered period may be the Coronavirus Pandemic. To confirm this view, let us examine the time series before the pandemic period.

The results obtained from the Autoregressive model AR(3) for 01.01.2012–01.01.2019 show that the model has a high degree of adequacy (*Multiple R = 0.99, R-squared* = 0.98, *Adjusted R-squared* = 0.97), using a 0.05 level of significance, the two-tail t test with 78 degrees of freedom has critical values of ± 1.664, $t_{stat} = 1.82 < +1.664$ or because the *P-value* = 0.007 < 0.05 we reject H_0 ($H_0 = A_j = 0$). In result the second-order parameter of the autoregressive model is significant and should remain in the model:

$$Y = 0.73\text{Lag1} + 0.29\text{Lag2} + 48.5$$

Let us consider the S&P500 stock prices from 2012 to 2022. Below we have plotted the stock prices from 01/01/2012 to 01/01/2022 and forecast line for 01/01/2022 - 01/12/2022 time period (Fig. 1).

The next economic indices to which we apply the autoregressive model are the Nasdaq indices. The NASDAQ-100 is calculated from 100 high-tech companies, including hardware, software, telecommunications, and biotech companies listed on the NASDAQ exchange [10] (Table 2).

As a result of the descriptive analysis, it was determined that the average price of the Nasdaq indices was 6798 USD, minimum price was 2814 USD in January 2012, and the maximum price was 15,645 USD in December 2021. Percentage of change was 460% for 01/01/2012–01/01/2022 time interval (Fig. 2).

Results of summary constructs the equation of the model AR(3) with application Nasdaq for 01/01/2012–01/01/2022:

Fig. 1 Price of S&P500 index between 01/01/2019–01/01/2022 and forecast line for 01/01/2022–01/12/2022 time period

Table 2 Descriptive analyses of Nasdaq indices price

Indices	Max:	Min:	Difference:	Mean:	Change %:
Nasdaq	15,645	2814	12,831	6798	460

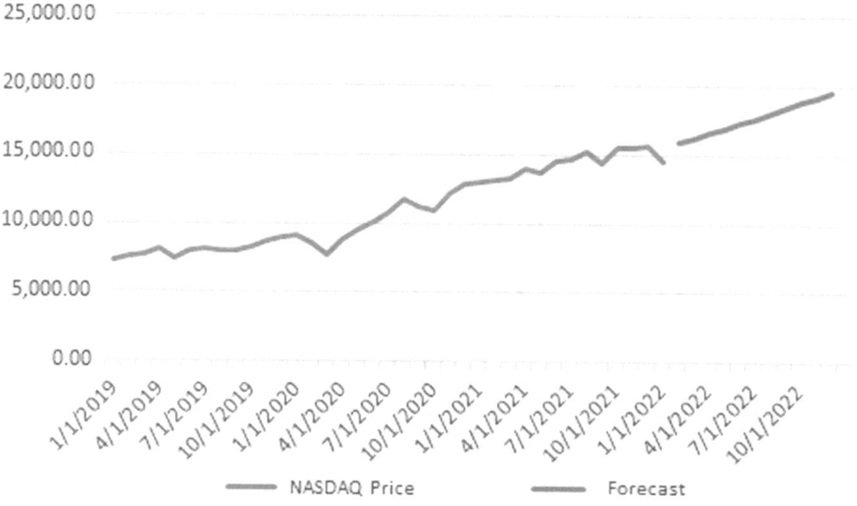

Fig. 2 Price of Nasdaq indices between 01/01/2019–01/01/2022 with forecast line to 01/01/2022

Regression statistics

Multiple R	0.994126119
R square	0.98828674
Adjusted R square	0.987978496
Standard error	375.3533555
Observations	118

ANOVA

	df	SS	MS	F	Significance F
Regression	3	13,555,160.223	451,720,074.2	3206.186533	7.0041E-110
Residual	114	16,061,476.13	140,890.1415		
Total	117	1,371,221,699			

	Coefficients	Standard error	T Stat	P-value	Lower 95%	Upper 95%
Intercept	7.637215448	79.43687199	0.096141946	0.923576645	− 149.7266161	165.001047
Lag1	0.873536927	0.101356829	8.618431849	4.45676E-14	0.672749828	1.074324025
Lag2	0.045311902	0.33405974	0.339654218	0.734741901	− 0.218964306	0.30958811
Lag3	0.097865721	0.104116903	0.939959971	0.349225502	− 0.108389062	0.304120504

$$Y = 0.87\text{Lag}1 + 0.04\text{Lag}2 + 0.09\text{Lag}3 + 7.64.$$

The two-tail t test with 114 degrees of freedom has critical values of ± 1.664. Because $-1.664 < t_{stat} = 1.42 < +1.664$ or because the $P\text{-value} = 0.33 > 0.05$ we do not reject H_0 $\left(H_0 = A_j = 0\right)$ Conclude that the parameters of the autoregressive model is not significant and can be deleted.

As in the S&P 500 index, the AR(3) model was applied to the Nasdaq indices to investigate the effects of the Covid-19 pandemic for 01/01/2012–01/01/2022.

The results obtained from the AR(3) for ten years show that the model has a high degree of adequacy (*Multiple R = 0.99, R-squared* = 0.98, *Adjusted R-squared* = 0.98), with 0.05 level of significance, the two-tail t test with 78 degrees of freedom has critical values of ± 1.664, $t_{stat} = 1.71 < +1.664$ or because the *P-value* = 0.002 < 0.05 we reject H_0 $\left(H_0 = A_j = 0\right)$. In result the second-order parameter of the autoregressive model is significant and should remain in the model:

$$Y = 0.77\text{Lag}1 + 0.21\text{Lag}2 + 64.3.$$

Although the adequacy ratios of the AR(3) model for the S&P 500 and Nasdaq indices are high, other conditions of the autoregressive model unsatisfied. This is due to significant changes in prices during the COVID-19 pandemic. Compared to the S&P 500 indices, the percentage change in the price of the Nasdaq indices is about twice as high.

4 Conclusion

In this paper, investigate AR(3) with application to the S&P 500 and Nasdaq indices which are the main indices in the international financial market for 01/01/2012–01/01/2022. The results from the model show that the COVID-19 pandemic has had a strong impact on price changes in stock market indices. Non-fulfilment of the necessary conditions, despite the high adequacy coefficients of the AR(3) model for the S&P500 and Nasdaq indices, the autoregressive model for the last ten years is not considered satisfactory. This is explained by the fact that the AR(3) model was created for both indices in the pre-pandemic period (01/01/2012–01/01/2019) and it was found that the AR(2) model is more adequate. For the period 01/01/2012–01/01/2022, the percentage of change of NASDAQ indices was 460%. That's almost twice as much as the S&P500 indexes. In descriptive analysis, visualization, Analysis ToolPak in MS Excel program was used to build the autoregressive AR model of stock indices.

Acknowledgements This work supported by the Science Development Foundation under the President of the Republic of Azerbaijan—Grant No EIF-ETL-2020-2(36)-16/05/1-M-05.

References

1. Aliyev, R., Ardic, O., Khaniyev, T.: Asymptotic approach for a renewal-reward process with a general interference of chance. Commun. Stat. Theory Methods **45**(14), 4237–4248 (2015)
2. Aliyev, R., Bayramov, V.: On the asymptotic behaviour of the covariance function of the rewards of a multivariate renewal–reward process. Stat. Probab. Lett. **127**, 138–149 (2017)
3. Aliyev, R.T., Khaniyev, T.A., Gever, B.: Weak convergence theorem for ergodic distribution of stochastic processes with discrete interference of chance and generalized reflecting barrier. Theory Probab. Appl. **60**(3), 502–513 (2016). https://doi.org/10.1137/S0040585X97T987806
4. Aliyev, R.T., Rahimov, F., Farhadova, A.: On the first passage time of the parabolic boundary by the Markov random walk. Commun. Stat. Theory Methods **1**, 1–10 (2022). https://doi.org/10.1080/03610926.2021.2024852
5. Berenson, M.L., et al.: Basic business statistics. Pearson Edu. **M11**, 21–912 (2011)
6. Chakravarthy, N., et al.: Autoregressive modeling and feature analysis of DNA sequences. EURASIP J. Adv. Signal Process. 952689 (2004). https://doi.org/10.1155/S111086570430925X
7. Saadatmand, A., Nematollahi, A.R., Sadooghi-Alvandi, S.M.: On the estimation of missing values in AR(1) model with exponential innovations. Commun. Stat. Theory Methods **46**(7), 3393–3400 (2016)
8. Shiryaev, A.N.: Essentials of Stochastic Finance: Facts, Models, Theory, p. 843. Word Scientific (2003)
9. Zhang, Y., Yang, X.: Limit theory for random coefficient first-order autoregressive process. Commun. Stat. Theory Methods **39**(11), 1922–1931 (2010)
10. Jafarova, H.A., Aliyeva, R.T.: The use of autoregressive process AR(1) in the analysis of some finance indices. Inform. Control Prob. **42**(1), 19–25 (2022)

Improving Intersection Traffic Management Solutions by Means of Simulation: Case Study

Fuad Dashdamirov⬤, Allaz Aliyev⬤, Turan Verdiyev⬤, and Ulvi Javadli⬤

Abstract Improving the organization of traffic is one of the important and complex decisions in the road network. To check the proposed solutions, implementation of simulation methods is of great importance. The article proposes the use of Anylogic software. One of the most difficult intersections in the city of Baku was chosen for the application of the program. According to the values of traffic intensity, the operating modes of traffic lights for the existing and planned option are determined. The disadvantages of applying the intersection at the same level with the help of the constructed models are shown. The proposed version of the organization of movement is justified with the visualization of the model. Instead of two-phase regulation, a model of three-phase regulation is constructed. Simulating this option, it turned out that this method is also not able to fully ensure the free passage of cars. The main stages of building a logical model of traffic organization and used basic tools of traffic library are shown.

Keywords Intersection · Traffic management · Optimization · Simulation · Anylogic

1 Introduction

One of the main problems about the organization of traffic in large cities is the regulation of traffic at intersections. Two or more roads intersect at an intersection where vehicles and pedestrians congregate. Incorrectly selected control modes sometimes lead to difficult situations, including traffic jams. To improve movement and to ensure the safety of vehicles and pedestrians, it is necessary to design intersections correctly and find the right solutions for organizing traffic at an intersection. To do this, you can use various tools, including simulation methods.

F. Dashdamirov (✉) · A. Aliyev · T. Verdiyev · U. Javadli
Azerbaijan Technical University, 25 H. Javid, Baku, Azerbaijan
e-mail: fuad.dashdamirov@aztu.edu.az

Traffic lights are mainly used to regulate traffic at the intersections. But the use of traffic light regulation does not always lead to the desired results. When the capacity of an intersection does not match the intensity of the traffic flow, alternative solutions must be taken.

Intersections are divided into intersections on one and at different levels. Intersections can also be classified according to the relative position of the intersected roads [1]. Depending on the mode of operation of a traffic signal, the organization of pedestrian traffic, vehicle delays, and the impact of delays on the state of the environment were studied in various works [2–5].

It should be noted that each intersection can have a unique configuration and properties. When organizing and improving the movement, many factors must be taken into account. Taking into account and analyzing the consequences of rationalization in an analytical way is a time-consuming job. Computer simulation can facilitate the optimization process. The use of computer simulation makes it possible to visualize the proposed variant of activities.

Currently, many simulation programs are used to simulate traffic at a road intersection.

As you know, simulation methods have three types: system dynamics, discrete-event modeling and agent-based modeling [6–8]. To build discrete-event models, many programs are used, such as GPSS World, Arena, FlexSim, Simul8, Anylogic etc. [9].

The use of existing simulation programs makes it possible to get a fairly clear idea of the expected situation as a result of the design. Simulation of logistics and transport processes has found wide application. Particularly effective are some simulation programs designed specifically for various areas of production and services. For the simulation of urban transport, there is software created by the PTV Group [10]. The Anylogic software is a versatile simulation tool. The program has a road library and with the help of the program you can build agent-based models. The program also makes it possible to enter the parameters of the movement of traffic and pedestrian flows and analyze the change in these parameters over time [11, 12].

The most suitable of the methods for building simulation models for road intersections is agent-based modeling. Such modeling can be carried out to a fairly good level using the Anylogic libraries. With this program, you can enter the parameters of road users: cars, pedestrians, passengers.

2   Analysis of the State of Organization of Traffic at the Intersection

As you know, the mode of operation of traffic lights is determined by the intensity of vehicles in the directions of movement. One of the problematic road junctions of the city of Baku is the intersection of the Binagadi highway with the Baksol road. Especially during peak hours there are serious congestion.

At present, two-phase traffic signal regulation is used at the intersection. During peak hours, there is a buildup of vehicles at the intersection, sometimes resulting in long queues and congestion. The main reason for delays at the intersection is left-turn traffic. The intensity of the movement of vehicles in these directions is quite high. Therefore, it is advisable to test alternative traffic control options by simulation at a given node.

Movement directions at the intersection is shown in Fig. 1.

The vehicle intensities at the considered intersection are shown in Table 1.

An analytical method for determining the mode of operation of traffic lights was created in the middle of the twentieth century. The formula proposed by Webster is as follows [13]:

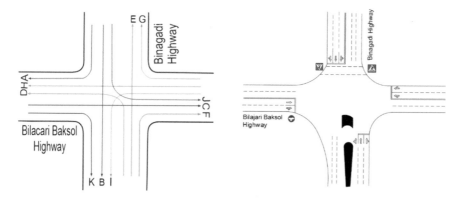

Fig. 1 Movement direction at the intersection of Binagadi and Baksol

Table 1 Movement intensity of vehicles at peak hours

Movement direction	Movement intensity of vehicles at peak hours (vehicle per hour)								Average intensity
	I	II	III	IV	V	VI	VII	VIII	
A	198	195	205	219	228	232	217	210	219
B	1114	1205	1253	1196	1185	1238	1141	1277	1206
C	173	195	185	198	167	188	110	183	174
D	388	396	412	370	427	408	385	401	398
E	1087	1160	1056	1155	1187	1132	1023	1210	1126
F	120	123	95	98	105	113	117	81	107
J	687	714	741	768	697	705	720	738	721
K	510	585	541	535	598	475	490	563	537
G	283	315	206	241	295	277	322	230	271
H	287	313	306	241	274	196	321	290	278
İ	133	105	176	87	91	112	122	60	110

Table 2 Traffic light mission timeline for existing option

Direction	C_{hcm}= 134 sec	t_g	t_y	t_r	t_y
A,B,C		81	4	49	0
D,E,F		81	4	49	0
G,H,I		45	0	85	4
J		45	0	85	4
K		134	0	0	0

t_g, t_y, t_r—respectively duration of green, yellow and red light

$$C_{hcm} = \frac{1.5 * I + 5}{1 - Y} \tag{1}$$

where C_{hcm}—denotes the desirable cycle length; I—is the total intermediate cycle duration per cycle.

The Higway Capacity Manual 2010 [14] proposed the desirable cycle length for signal intersections as follows.

$$C_{des} = \frac{L}{2 - \left[\frac{V_c}{s_i \cdot PHF \cdot \left(\frac{V}{c} \right)} \right]} \tag{2}$$

where C_{des} denotes the desirable cycle length; L is the total time per cycle; V_c denotes the sum of the critical lane volumes; s_i denotes the saturation flow rate of the approach; PHF denotes the peak hour factor; and $\frac{V}{c}$ denotes the target $\frac{V}{c}$ ratio for the critical movements in the intersection.

Table 2 shows the mode of operation of the traffic light object at the intersection for the existing option.

3 Improving the Organization and Regulation of Traffic in a Road Junction Using Anylogic

To build simulation models of the intersection, we use the tools of the road library and pedestrian library. The commands and their assignments for building a road intersection model in Anylogic 8.7.7 are shown in Table 3.

According to the existing mode of operation of traffic lights during peak hours, problems arise as mentioned above. Entering the data of the work of traffic lights for the existing variant of the organization of traffic, you can see the current situation at the intersection (Fig. 2). As can be seen from the figure, a traffic jam is formed at the intersection.

Let's determine the mode of operation of a traffic light object when applying three-phase regulation. By introducing this Formula (1), we determine the duration of the cycle of traffic lights. Now let's build a three-phase mode of operation of traffic

Table 3 The commands and their assignments for building a road intersection

Name of tools used	Function of tools
Road	Used to build roads. Using the properties of this tool, you can enter the main parameters of the road, such as the number and width of lanes
CarSource	Used to create and direct a group of cars. In the parameters section, you can enter the intensity of vehicles, speed, acceleration and deceleration
CarMoveTo	Used to indicate the section of the road on which cars will move
SelectOutput	Used to direct vehicles to different exits depending on conditions (probabilistic and deterministic). The condition can be determined depending on the state of the agent and external factors
CarDispose	Used to bring cars out of space and model
PedSource	Creates pedestrians as agents. This block creates the start point of the pedestrian flow. Pedestrian movement parameters are entered using properties. The field for pedestrians to cross the intersection is carried out using the **Restangular area** tool
PedGoTo	Specifies the target direction of pedestrians. The goal is set using the **TargetLine** tool

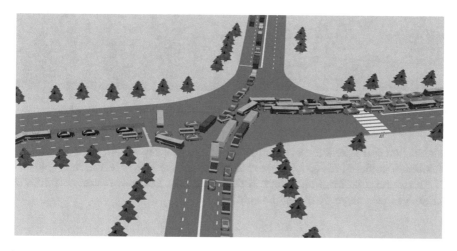

Fig. 2 Traffic management simulation for an existing intersection traffic control option

lights, which provides for the passage of left-turning cars in a separate phase. For this variant of traffic control, we construct a time-table. The results obtained for the existing option are entered in Table 4.

Based on the data obtained, we simulate the proposed option (Fig. 3). As can be seen from the figure, in this variant, accumulations of vehicles and congestion are also observed.

Calculations of the traffic light operation mode and the intersection simulation show that the optimization of the traffic light control variant does not lead to a complete reduction of delays at a given node. Therefore, it is proposed to create an

Table 4 Traffic light mission timeline for three-phase regulation

Direction	C_{hcm}= 125 sec		t_g	t_y	t_r	t_y
AB			60	4	61	0
EF			60	4	61	0
GHI			37	0	84	4
C			37	0	84	4
K			125	0	0	0
JD			20	0	0	0

Fig. 3 Simulation of traffic control for a three-phase variant of traffic control at the intersection

intersection at different levels with the use of a road overpass. An episode from the simulation model of the proposed variant of the intersection is shown in Fig. 4.

As can be seen from the figure in the proposed version of the organization of traffic, vehicles move freely and no conflicts arise.

4 Conclusion

The key points of the city road network are the intersections of city streets. These intersections have different configurations. As the number of points of conflict increases, the situation becomes more complicated. During peak hours, traffic jams are formed, which in turn contribute to delays in traffic flows.

It is advisable to use simulation models to suppress traffic control. With the use of the program Anylogic you can simulate the movement of vehicles and pedestrians, taking into account their basic traffic parameters. Building a simulation model for three-phase regulation at the intersection of Binagadi and Baksol streets does not improve traffic organization. The constructed simulation model shows that the

Fig. 4 An episode from the simulation model of the proposed variant of the intersection

creation of an intersection at different levels at this intersection allows you to avoid delays in the movement of vehicles and pedestrians.

This approach can be used to improve the management of similar and more complex transport nodes.

References

1. NamGung, H., Kim, C.S., Choe, K.C., Ri, C.U., Kim, Y.M., Ri, M.P.: Research progress of road intersection design analysis. Int. J. Sci. Res. Sci. Eng. Technol. **7**(6), 245–256 (2020)
2. Oskarbski, J., Guminska, L., Miszewski, M., Oskarbska, I.: Analysis of signalized intersections in the context of pedestrian traffic. Transp. Res. Proc. **14**, 2138–2147 (2016)
3. Shafii, M.A., Qin, C.S., Mei Ling, E.C., Shaffie, E.: Determination of Delay at Signalized Intersection: A Case Study, pp. 321–332. Technology Reports of Kansai University (2020)
4. Vajeeran, A., De Silva, G.L.D.I.: Delay analysis at a signalized T intersection. In: 3rd International Moratuwa engineering research conference, MERCon, pp. 325–330 (2017)
5. Zheng, Y., Hua, X., Wang, W., Xiao, J., Li, D.: Analysis of a signalized intersection with dynamic use of the left-turn lane for opposite through traffic. Sustainability **12** (2020)
6. Dashdamirov, F.S.: Creation of a simulation model of bus traffic in urban routes. Bull. Azov State Tech. Univ. Ser. Tech. Sci. **41**, 205–211 (2020)
7. Dashdamirov, F., Javadli, U.: Development of a methodology for creating an agent based model of transport hubs in suburban area. In: International conference on problems of logistics, management and operation in the East-West transport corridor (PLMO), pp. 153–157 (2021)
8. Ivanov, D.: Operations and Supply Chain Simulation with AnyLogic, 2nd edn. E-Textbook, Berlin School of Economics and Law (2017)
9. Dias, L.M.S., Vieira, A.A.C., Pereira, G.A.B., Oliveira, J.A.: Discrete simulation software ranking—a top list of the worldwide most popular and used tools. In: Proceedings of the 2016 Winter simulation conference, pp. 1060–1071 (2016)
10. https://www.ptvgroup.com/. Last accessed 10 Jan 2022
11. Nevers, B.L., Nguyen, K.M., Quayle, S.M., Zhou, X., Taylor, J.: The Effective Integration of Analysis, Modeling, and Simulation Tools. Technical Report (2013)

12. Gunesh, G.: Agent-based simulation and an example in anylogic. Yildiz technical university, industrial engineering department, İstanbul (2014)
13. Wagner, P., Gartner, N.H., Lu, T., Oertel, R.: Webster's Delay Formula—revisited. 93rd Annual Meeting Transportation Research Board. Washington, DC (2014)
14. Transportation Research Board of the National Academies: Chapter 18 Signalized Intersections, Highway Capacity Manual 2010, vol. 3, pp. 18–31. Transportation Research Board: Washington, DC, USA (2010)

Necessary Optimality Condition in Linear Fuzzy Optimal Control Problem with Delay

R. O. Mastaliyev and K. B. Mansimov

Abstract One linear fuzzy optimal control problem with delaying argument is considered and necessary optimality condition of Pontryagin maximum principle type is obtained. When obtaining the result, a modified version of the increment method was applied. Here, the control domain is an arbitrary no-empty and bounded set.

Keywords Fuzzy optimal control · Principle of optimality · Increment method · Fuzzy Hamiltonian function · Pontryagin maximum principle

1 Introduction

As know, Zadeh, in his first work [1], introduced the general concept of fuzzy sets as "continuum degrees of membership", defining the basic relations and operations on them. Subsequently, in [2–9] and others, introducing the definition of the derivative of a fuzzy function and other mathematical operations, fuzzy optimal control problems were studied and a number of first order necessary optimality conditions were obtained.

In this paper, we consider a Bolza control problem described by linear fuzzy differential equations with lagging argument and establish a first order necessary optimality condition.

2 Definitions and Preliminaries

Let us denote by F the class of fuzzy numbers, i.e., normal, convex, upper semicontinuous and compactly supported fuzzy subsets of the real numbers. For any given

R. O. Mastaliyev (✉) · K. B. Mansimov
Institute of Control Systems of Azerbaijan NAS, st. B. Vahabzade 68., Baku, Azerbaijan
e-mail: mastaliyevrashad@gmail.com

© The Author(s), under exclusive license to Springer Nature Switzerland AG 2023
Sh. N. Shahbazova et al. (eds.), *Recent Developments and the New Directions of Research, Foundations, and Applications*, Studies in Fuzziness and Soft Computing 423,
https://doi.org/10.1007/978-3-031-23476-7_26

$a \in F$, we denote as a^{α} its α—level of fuzzy number a and defined as the interval $a^{\alpha} = [L_a(\alpha), R_a(\alpha)] = \{x \in R, \mu(x) \geq \alpha\}$ for any $\alpha \in (0, 1]$. Let $a, b \in F, k \geq 0$, $a^{\alpha} = [L_a(\alpha), R_a(\alpha)]$, $b^{\alpha} = [L_b(\alpha), R_b(\alpha)]$.

Then $a^{\alpha} + b^{\alpha} = [L_a(\alpha) + L_b(\alpha), R_a(\alpha) + R_b(\alpha)]$, $ka^{\alpha} = [kL_a(\alpha), kR_a(\alpha)]$.

Definition 1 ([3, 8]). Let $x = (a_1, a_2) \in F \times F, y = (b_1, b_2) \in F \times F$. Then for any $x, y \in F \times F$ scalar product define as

$$x \circ y = \frac{1}{2} \int_0^1 [(L_{a_1}(\alpha) - L_{a_2}(\alpha))(L_{b_1}(\alpha) - L_{b_2}(\alpha))$$

$$+ (R_{a_1}(\alpha) - R_{a_2}(\alpha))(R_{b_1}(\alpha) - R_{b_2}(\alpha))]d\alpha.$$

Now, consider fuzzy function $f(t) \in F, t \in [t_0, t_1]$.

For any $\alpha \in [0, 1]$, $f_{\alpha_1}(t) = [L_{f(t)}(\alpha), R_{f(t)}(\alpha)]$ is called α—level of the function $f(t)$.

Definition 2 (Derivative of fuzzy function) Let such $\omega(t) \in F, \psi(t) \in F, t \in [t_0, t_1]$ exist, that

$$\lim_{\Delta t \to 0} \frac{(f(t + \Delta t), 0) - (f(t), 0)}{\Delta t} = (\omega(t), \psi(t)).$$

Then the pair $(\omega(t), \psi(t)) \in F \times F$ is a derivative of the function $f(t)$ at the point $t \in (t_0, t_1)$. This definition may be written in the following form

$$\lim_{\Delta t \to 0} \frac{(f_{\alpha}(t + \Delta t), 0) - (f_{\alpha}(t), 0)}{\Delta t} = (\omega_{\alpha}(t), \psi_{\alpha}(t)).$$

In addition, for any given $\eta(t) \in F \times F, f(t) \in F \times F$ the following identity is hold:

$$\int_{t_0}^T f'(t) \circ \eta(t)dt = f(t) \circ \eta(t)\big|_{t_0}^T - \int_{t_0}^T f(t) \circ \eta'(t)dt.$$

Note that this preliminary information can found in details in the works [3–9].

3   Problem Statement

Let us consider a fuzzy optimal control problem with delay. Suppose that behavior of an object described by a linear system

$$\dot{x}(t) = A(t)x(t) + B(t)x(t - h) + C(t)u(t), t \in (0, t_1], \tag{1}$$

$$x(t) = \varphi(t), t \in E_0 = [-h, 0), \tag{2}$$

$$x(0) = x_0, \tag{3}$$

where $x(t) \in F - n$—vector trajectory, $A(t), B(t) \in R^{n \times n}, C(t) \in R^{n \times r}$—given continuous matrix, $h = \text{const} > 0, x_0 \in F$—fuzzy number, $\varphi(t) \in F$—given continuous on E_0 initial fuzzy vector function.

The function $u(t) \in R^r$—satisfying the constraints

$$u(t) \in U \subset F, t \in (0, t_1], \tag{4}$$

is called the admissible control.

The task is to minimize functional (5)

$$S(u) = \int_0^{t_1} f(t, x(t), x(t - h), u(t)) dt + \phi(x(t_1)) \tag{5}$$

with restrictions (1)–(4).

Where $f(t, x(t), y(t), u(t)), \phi(x) \in R$—given continuously differentiable on LF [9] functions.

Using the increment method, a necessary optimality condition of Pontryagin maximum principle type was established [10].

4 Formula Increment Quality Criterion and Main Result

This section is aimed at deriving the necessary condition for the fuzzy optimal control problem of delay arguments.

Let $u(t)$ and $\bar{u}(t) = u(t) + \Delta u(t)$—be two admissible controls, $x(t)$ and $\bar{x}(t) = x(t) + \Delta x(t)$—solutions of system (1)–(2) corresponding to them.

Then from (1)–(3) it is clear that $\Delta x(t)$ satisfy the following system

$$\Delta \dot{x}(t) = A(t)\Delta x(t) + B(t)\Delta x(t - h) + C(t)\Delta u(t), t \in (0, t_1], \tag{6}$$

$$\Delta x(t) = 0, = [-h, 0]. \tag{7}$$

Multiplying both sides of the equality (6) by the yet unknown vector function $\psi(t) \in F$ and integrating by $[0, t_1]$ then we have

$$\int\limits_0^{t_1} \psi(t) \circ \Delta \dot{x}(t) dt = \int\limits_0^{t_1} \psi(t) \circ A(t) \Delta x(t) dt + \int\limits_0^{t_1} \psi(t) \circ B(t) \Delta x(t - h) dt$$

$$+ \int\limits_0^t \psi(t) \circ C(t) \Delta u(t) dt.$$

Hence, with the help of simple transformations, we have

$$\psi(t_1) \circ \Delta x(t_1) - \int\limits_0^{t_1} \dot{\psi}(t) \circ \Delta x(t) dt = \int\limits_0^{t_1} \psi(t) \circ A(t) \Delta x(t) dt$$

$$+ \int\limits_0^{t_1 - h} \psi(t + h) \circ B(t + h) \Delta x(t) dt$$

$$+ \int\limits_0^{t_1} \psi(t) \circ C(t) \Delta u(t) dt.$$

Taking into account the last identities, the formula for the increment of the quality functional (5) can be write as

$$\Delta S(u) = \phi(\overline{x}(t_1)) - \phi(x(t_1)) + \int\limits_0^{t_1} f(t, \overline{x}(t), \overline{x}(t - h), \overline{u}(t), \psi(t)) dt$$

$$- \int\limits_0^{t_1} f(t, x(t), x(t - h), u(t), \psi(t)) dt + \psi(t_1) \circ \Delta x(t_1)$$

$$+ \int\limits_0^{t_1} \dot{\psi}(t) \circ \Delta x(t) dt - \int\limits_0^{t_1} \psi(t) \circ A(t) \Delta x(t) dt$$

$$- \int\limits_0^{t_1 - h} \psi(t + h) \circ B(t + h) \Delta x(t) dt - \int\limits_0^{t_1} \psi(t) \circ C(t) \Delta u(t) dt. \qquad (8)$$

Assuming a fuzzy Hamilton-Pontryagin function

$$H(t, x, y, u, \psi) = \psi(t) \circ C(t) u(t) - f(t, x, y, u),$$

formula (8) be rewritten in the form

$$\Delta S(u) = \phi(\overline{x}(t_1)) - \phi(x(t_1)) + \int_0^{t_1} H(t, \overline{x}(t), \overline{x}(t-h), \overline{u}(t), \psi(t)) dt$$

$$- \int_0^{t_1} H(t, x(t), x(t-h), u(t), \psi(t)) dt + \psi(t_1) \circ \Delta x(t_1)$$

$$+ \int_0^{t_1} \dot{\psi}(t) \circ \Delta x(t) dt - \int_0^{t_1} \psi(t) \circ A(t) \Delta x(t) dt$$

$$- \int_0^{t_1-h} \psi(t+h) \circ B(t+h) \Delta x(t) dt - \int_0^{t_1} \psi(t) \circ C(t) \Delta u(t) dt$$

Hence using the Taylor formula, we have

$$H(t, \overline{x}(t), \overline{x}(t-h), \overline{u}(t), \psi(t)) - H(t, x(t), x(t-h), u(t), \psi(t))$$
$$= H_x(t, x(t), x(t-h), u(t), \psi(t)) \Delta x(t)$$
$$+ H_y(t, x(t), x(t-h), u(t), \psi(t)) \Delta x(t-h)$$
$$+ H(t, x(t), x(t-h), \overline{u}(t), \psi(t))$$
$$- H(t, x(t), x(t-h), u(t), \psi(t)) + o_1(\|\Delta(x) + \Delta(y)\|),$$
$$\phi(\overline{x}(t_1)) - \phi(x(t_1)) = \phi_x(x(t_1)) \Delta x(t_1) + o_2(\|\Delta x(t_1)\|).$$

Let $\psi(t)$ be solution following adjoint problem

$$\dot{\psi}(t) = H_x(t, x, y, u, \psi), t \in [0, T],$$
$$\dot{\psi}(t) = H_y(t+h, x(t+h), x(t), u(t+h), \psi(t+h))$$
$$+ \psi(t+h) \circ B(t+h), t \in [0, T-h],$$
$$\psi(t_1) = -\phi_x(x(t_1)).$$

Then formula increment (8) transforms in the form

$$\Delta S(u) = -\int_0^{t_1} H(t, x(t), x(t-h), \overline{u}(t), \psi(t)) dt$$

$$- \int_0^{t_1} H(t, x(t), x(t-h), u(t), \psi(t)) dt + \eta(\Delta(u)).$$

Hence, using traditional method [7–10], we verify the validity of the following theorem.

Theorem. Let $(x(t), u(t))$ be an optimal pair for the problem (1)–(9). Then for each $\dot{\forall} t \in (0, T)$ the following relation is fulfilled

$$H(t, x(t), x(t - h), u(t), \psi(t)) = \max_{v \in U} H(t, x(t), x(t - h), v, \psi(t)). \qquad (9)$$

It is clear that, the identities (9) represent as an analogue of the Pontryagin maximum principle for the considered fuzzy optimal control problem.

References

1. Zadeh, L.A.: Fuzzy sets. Inf. Control **8**(3), 338–353 (1965)
2. Chang, S.S.I., Zadeh, L.A.: On Fuzzy Mapping and Control, pp. 30–34 (1972)
3. Sakawa, M., Inuiguchi, M., Kato, K., Ikeda, T.: A fuzzy satisfying method for multiobjective linear optimal control problems. Fuzzy Sets Syst. **102**, 237–246 (1999)
4. Facchinetti, G., Giove, S., Pacchiarotti, N.: Optimization of a fuzzy non linear function. Soft Comput. J. **6**(6), 476–480 (2002)
5. Fard, O.S., Soolaki, J., Torres, D.F.M.: A necessary condition of Pontryagin type for fuzzy fractional optimal control problems. Discrete Cont. Dyn. Syst. **11**(1), 59–76 (2018)
6. Dubois, D., Prade, H.: Operations on fuzzy numbers. Int. J. Syst. Sci. **9**(6), 613–626 (1978)
7. Niftiyev, A.A., Zeynalov, C.I., Poormanuchehri, M.: Fuzzy optimal control problem with non-linear functional. Fuzzy Inform. Eng. **3**(3), 311–320 (2011)
8. Niftiyev, A.A., Zeynalov, C.I., Pur, M.: Fuzzy optimal control prosecution problem. Rep. ANA Sci. **4**, 18–28 (2010)
9. Niftiyev, A.A., Poormanoocheri, M., Zeynalov, C.I.: Fazzy optimal control problem with non-linear functional. News BSU 29–34 (2013)
10. Gabasov, R., Kirillova, F.M.: The Maximum Principle in Optimal Control Theory, p. 272. URSS, Moscow (2011)

Modeling the Movement of Mudflows in River Basins

A. B. Hasanov⃝ and **S. Yu. Guliyeva**⃝

Abstract In the present work, the movement of mudflows is investigated using the methods of hydraulics of multicomponent media with a variable flow rate and with a free boundary. The simple calculated dependencies obtained in this way, when applied to practical engineering problems, often give satisfactory results. The maximum velocity in any cross-section of the flow, as a rule, differs little from the average velocity, especially this applies to hyper concentrated sediment mudflows (Natishvili and Tevzadze, in Fundamentals of mudflow dynamics. Publishing House "Metsniereba", Tbilisi (2007) [1]; Natishvili and Tevzadze, in Waves in mudflows. Tbilisi (2011) [2]; Vabishevich, in Numerical methods for solving problems with a free boundary. Moscow University Publishing House (1987) [3]; Hasanov, in Mathematical modeling of floods in rivers. Baku (2014) [4]). In such cases, the movement can be considered as one-dimensional with some average cross-section velocity.

Keywords Mudflows · River basins · The variable expense

1 Introduction

For forecasting and operational assessment of the impact of mudflows in river basins on hydraulic and other economic facilities, the use of intelligent expert systems based on mathematical methods with the use of modern information and communication technologies is required. This requires taking into account the geomorphological and climatic factors of the territory, as well as changes in characteristic fractal parameters of the region, under which the destructive processes of the earth's surface are

A. B. Hasanov
Laboratory of Computer Modeling of Dynamic Systems, Institute of Control Systems of ANAS, Az 1141, Baku, Azerbaijan
e-mail: hesenli_ab@mail.ru

S. Yu. Guliyeva (✉)
Department of Education, Institute of Control Systems of ANAS, Az 1141, Baku, Azerbaijan
e-mail: quliyevasevinc2012@mail.ru

© The Author(s), under exclusive license to Springer Nature Switzerland AG 2023
Sh. N. Shahbazova et al. (eds.), *Recent Developments and the New Directions of Research, Foundations, and Applications*, Studies in Fuzziness and Soft Computing 423,
https://doi.org/10.1007/978-3-031-23476-7_27

calculated. To assess the ecological situation in the river basin, it is required to study changes in the degree of damage to fractal parts, to establish the current value of the erosion coefficient. Within the one-dimensional approach, the velocity, pressure, density, and other flow parameters depend on one coordinate. Depending on the speed of mudflows in each section of the natural watercourse, the process of channel formation occurs. When a solid phase settles, a part of its mass is separated from the main flow, and when washed out, it joins it. These issues can be investigated taking into account the movement of coherent mudflows, taking into account the variability of mudflows in a channel with an uneven bottom [1, 2, 4–6].

The movement of siel mudflows consisting of a mixture of liquids and solid particles is very complex. The data of the scientific and technical literature [1, 2, 5, 7, 8] allow us to identify the basis for the development of a suitable hydrodynamic theory of the motion of nano-containing flows. Despite a large number of works devoted to solving this problem, the general theory of studying the dynamics of channel flows with variable flow and with a free boundary has not yet been developed. The existing mathematical models are very closely related to the hydrogeological data of a particular riverbed. Therefore, for a qualitative study of such movements, a fundamentally new mathematical approach is required, taking into account the unevenness of the bottom of the basin and the variability of the flow rate. To calculate the capacity of a mudflow on active tributaries of the river, it is necessary to combine the analytical and modeling capabilities of geoinformation systems (GIS) to create predictive modeling complexes to prevent, minimize and eliminate the consequences of emergencies. Methods and algorithms for calculating the hydraulic characteristics of flows in different parts of the basin are given in many works, in particular in [1–9]. Therefore, we will not repeat these approaches. Our goal is to develop a fundamentally new approach suitable for qualitative research of this phenomenon, allowing the calculation of quantitative values of parameters in subsequent stages.

The main part: The movement of flows consisting of a mixture of liquid and solid particles (including mudflows) in a general setting is investigated using the methods of hydrodynamics of multicomponent media with variable flow rates. Depending on the speed, during the time t, in each section of the natural watercourse, the process of changing the channel occurs. During the deposition of the solid phase, part of its mass is separated from the main stream, and when washed out it joins it. These issues can be investigated based on the theory of motion of multiphase media, taking into account the variability of flow in a channel with an uneven bottom (Fig. 1).

There Ω_1 is the flow domain, Ω_2 is the boundary of the solid bottom, which in general can be defined as a continuous or discrete, even random function, Ω_0 is the free surface of the flow.

Similar tasks in different aspects are set and investigated in [3]. The analysis of the works given in the direction under consideration reveals the need to build the most advanced models that take into account the forces acting in the flow, the unevenness of the bottom and the variability of the flow rate.

We will write the equations of motion of real fluid in the form:

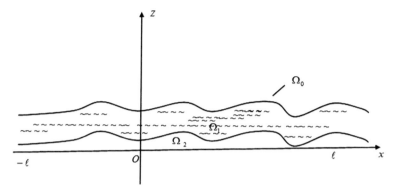

Fig. 1 Flow in a flat channel with an uneven bottom

$$(V\nabla)\, V = -\nabla P, \ \nabla = \frac{\partial}{\partial x_1} + \frac{\partial}{\partial x_2}, \ x = (x_1, x_2) \in \Omega_1 \tag{1}$$

$V = (V_1; V_2)$ is speed, P is pressure. We introduce the current function ϕ by the relations:

$$V_1 = \frac{\partial \phi}{\partial x_2}; \quad V_2 = -\frac{\partial \phi}{\partial x_1}. \tag{2}$$

We satisfy the incompressibility condition and denote by ω the velocity vortex which is defined by the expression:

$$\omega = \frac{\partial V_2}{\partial x_1} - \frac{\partial V_1}{\partial x_2}, \tag{3}$$

$$\Delta\phi = \frac{\partial^2 \phi}{\partial x_1^2} + \frac{\partial^2 \phi}{\partial x_2^2} = -\omega, \ x \in \Omega_2. \tag{4}$$

Let's write (1) in coordinate wise form:

$$V_1 \frac{\partial V_1}{\partial x_1} + V_2 \frac{\partial V_2}{\partial x_2} = -\frac{\partial P}{\partial x_1}, \tag{5}$$

$$V_1 \frac{\partial V_2}{\partial x_1} + V_2 \frac{\partial V_2}{\partial x_2} = -\frac{\partial P}{\partial x_2}. \tag{6}$$

Differentiating (5) by x_2 and (6) by x_1 subtracting them from each other, after some simple algebraic transformations we obtain:

$$\frac{\partial \phi}{\partial x_1} \cdot \frac{\partial \omega}{\partial x_2} - \frac{\partial \phi}{\partial x_2} \cdot \frac{\partial \omega}{\partial x_1} = 0. \tag{7}$$

Hence it follows that ϕ and ω are connected by the functional dependence

$$\omega = \omega(\phi). \tag{8}$$

Substituting (8) into (4) we obtain the following nonlinear Poisson equation for the current function ϕ in the flow domain Ω_2:

$$\Delta\phi = -\omega(\phi), \tag{9}$$

with some right part. In the simplest case of a vortex—free potential flow $\omega(\phi) = 0$ we obtain

$$\Delta\phi = 0, \tag{10}$$

whose solution is $\phi = \phi_0$. The general solution of Eq. (9) is obtained by the method of successive approximations, i.e. we assume that at the initial moments of time when the flow enters the channel, the movement is vortex-free, and at subsequent moments of time, some "fictitious forces" appear in the flow due to the occurrence of the vortex, calculated using the solutions of previous iterations. Then the general solution of Eq. (9) will be in the form:

$$\phi = \phi_0 + \phi_1 + \phi_2 + \cdots = \sum_{k=0}^{\infty} \phi_k \tag{11}$$

For finding ϕ_1, substituting the resulting expression ϕ_0 to the right side of (9), we obtain an inhomogeneous Poisson equation with the known right-hand side and solving it we find ϕ_1 under zero boundary conditions, etc.

Let us now consider the boundary conditions for Eqs. (9) and (10) that must satisfy the solution ϕ_0.

Suppose that at the inlet and outlet of the channel the height of which is equal to h, the bottom is flat, and the fluid velocity is constant and set. In this case

$$V(x) = V_0, \ x_1 = \pm l, \ x_2 \in (0; h) \tag{12}$$

It follows from (2) and (12) that

$$\phi(x) = U_0 \cdot x_2, \ \text{by} \ x_2 = \pm l, \ x_2 \in (0; h) \tag{13}$$

If the bottom Ω_2 is solid, we apply impermeability conditions: $(V \cdot \vec{n}) = 0 \, x \in \Omega_2$, which, taking into account (13), give the ratio

$$\phi(x) = 0, x \in \Omega_2 \tag{14}$$

The condition on the unknown boundary, i.e. on the free surface is obtained in the form of (non-flow condition):

$$\phi(x) = V_0 \cdot h, \quad x \in \Omega_0 \tag{15}$$

In order to derive the equation of one-dimensional motion of a connected mudflow with a variable flow rate along the path, we use the well-known equation of hydraulics, which is based on the energy principle.

In addition, on Ω_0, as on a current line, the resulting expression has a constant value (similar to Formula 3.1 in [2]). Denoting by P_0 the atmospheric pressure we get:

$$P_0 + \frac{1}{2}\left(V_1^2 + V_2^2\right) + \rho\, g x_2 = P_0 + \frac{1}{2} V_0^2 + \rho\, g h \tag{16}$$

where ρ is the density, g is the acceleration of gravity. From (15) and (16) follows the condition

$$\left(\frac{\partial \phi}{\partial \vec{N}}\right)^2 = |\nabla \phi|^2 = V_0^2 + 2\rho\, g(h - x_2), \quad x \in \Omega_0, \tag{17}$$

where $\vec{N}$ is a normal to Ω_0. As a result, we come to a problem with a free boundary, which is solved using numerical methods.

When solving a theoretical problem to illustrate the results, the profile of the bed bottom (area Ω_2) can be represented, for example, in the form $z_0 = z_1 \cos^2 \frac{k\pi}{2L} x$, or in another form, as a continuous, discrete or random function. The following initial data can be used for numerical calculations

$$L = 1000\,\text{m}, \quad z_1 = 0, 1\,\text{m}, \quad k \in Z, \quad h = 20\,\text{m},$$

and the corresponding hydraulic characteristics of the muddy flow mixture, given for example, in [7]. As a result, we obtain the values of the force characteristics of the flow, the time of the flow travel depending on the various initial values of the hydraulic parameters, the values of other force effects on the associated obstacles when the mixture moves in the channel according to the well-known methodology given in [1, 2], etc. The main difference of the proposed methodology is a more advanced approach to the study of the problem, in which the calculation modules are an integral intermediate part of the forecasting system. The knowledge database has a dynamic structure and is constantly being filled with new information obtained from the continuous monitoring system (GIS), etc. The functional scheme of the proposed intelligent expert system has the form (Fig. 2).

The main purpose of the proposed system will be operational monitoring of the hydrological situation in the basins of mudflow-dangerous rivers, constant monitoring of environmental safety, forecasting of mudflow phenomena and theoretical

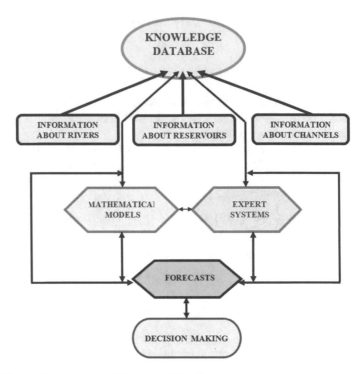

Fig. 2 The functional scheme of the proposed intelligent expert system

calculation of the hydraulic characteristics of the expected flow and decision-making to prevent dangers.

References

1. Natishvili, O.G., Tevzadze, V.I.: Fundamentals of Mudflow Dynamics. Publishing House "Metsniereba", Tbilisi (2007), 214 pp
2. Natishvili, O.G., Tevzadze, V.I.: Waves in Mudflows, 162 pp. Tbilisi (2011)
3. Vabishevich, P.N. Numerical Methods for Solving Problems with a Free Boundary, 164 pp. Moscow University Publishing House (1987)
4. Hasanov, A.B.: Mathematical Modeling of Floods in Rivers, 144 pp. Baku (2014)
5. Takahashi, T.: Debris flow. In: IAHR Monograph Series, Balkema Publishers, Netherlands (1991)
6. Imanov, F.A.: Restoration and Ecological Flow of Rivers, 96 pp. Baku (2019)
7. Shayusupov, M.: Movement of Multiphase Flows with Variable Flow Rate in the Channels. Tashkent (1981)
8. Zhu, P., He, Z., Wang, Y., et al.: A Study of Typical Mountain Hazards Along Sichuan-Tibet Highway. Chengdu Science and Technology University Publishing House (1999)
9. Rickenmann, D., Chen, C. (eds.): Debris Flow Hazards Mitigation: Mechanics, Prediction, and Assessment, vols. 1, 2, 1325 pp. MillPress, Rotterdam (2003)

Fuzzy Applications and Measurement Systems

Development of a Combined Test Algorithm to Increase the Intelligence of Measurement Systems

Mazahir Isayev, Mahmudbeyli Leyla, and Khasayeva Natavan

Abstract The combined test algorithms increase the accuracy of measurement by using the simple additive and multiplicative tests. This enables one to determine the measured values of non-electrical quantities by the results of additional test measurements used for the identification of nonlinear transformation functions (TF) of information measurement systems (IMS).

Keywords Test method · Increasing accuracy of measurement · Nonlinearity of transformation function

1 Introduction

The main problem in designing contemporary measuring systems is the increase in quality indices of their metrological characteristics (MC). It is known that this is performed by the following two ways: making renewal in the construction of the initial transformers-technological method or by not intervening in their construction using the algorithmic-test method [1, p. 7]. In this paper, by using the second method, at the expense of additional information obtained by performing additional measuring

---

The original version of this chapter was revised: The chapter author name "K. Natavan" has been changed to "Khasayeva Natavan" in the Chapter opening page. The correction to this chapter can be available at https://doi.org/10.1007/978-3-031-23476-7_31

M. Isayev (✉)
Institute of Control Systems, St. B. Vahabzadeh 9, AZ 1141 Baku, Azerbaijan
e-mail: mezahir@bk.ru

M. Leyla
Azerbaijan University of Architecture and Construction, St. A. Sultanova, 5, AZ 1073 Baku, Azerbaijan

Khasayeva Natavan
Azerbaijan Technical University, H. Javid 25, AZ 1073 Baku, Azerbaijan
e-mail: n.xasayeva1@gmail.com

© The Author(s), under exclusive license to Springer Nature Switzerland AG 2023, corrected publication 2023
Sh. N. Shahbazova et al. (eds.), *Recent Developments and the New Directions of Research, Foundations, and Applications*, Studies in Fuzziness and Soft Computing 423, https://doi.org/10.1007/978-3-031-23476-7_28

operations, special test equations are set up and the joint solution of these equations increases the accuracy of measurement of measuring systems.

The article aims to decrease the number N of test equations as far as possible, to minimize the order of each equation, and to determine the coefficients of polynomials with high accuracy by using algorithmic-test methods. For that, the capabilities of the algorithmic-test method were analyzed, and the possibility of making combined simple test equations of different kinds was investigated.

2 Problem Statement

In the general form, the transformation function of the measurement system, its mathematical model of *n-th* order, is written in the form of the following polynomial [2, pp. 61–70]:

$$y_i = \sum_{i=0}^{n} a_i x^i, \tag{1}$$

is the measured value contained in the coefficients of the transformation function (TF) of MS, and their real values are defined by the test algorithms, increasing the accuracy of measurement at each periodical measurement. Therefore, a problem in the definition of a_i quantities of TF with high accuracy is stated, and the change in the values of these coefficients with respect to time $A_j(x)$, $j = 1, ...n.$ is investigated. Changes caused by external factors are determined, as are the correlation dependence of these changes, and the errors of measurement results and their constituents are studied and estimated. As a result, a generalized structure of an information-measurement system (IMS) is given.

3 Problem Solution

Let us consider the elaboration of the algorithm for increasing the accuracy of measurement based on simple additive tests. As by measuring with, tests the TF of initial MS depends on the definition of n-th order unknown quantities x, $a_1, ..., a_n$ with high accuracy, the system of equations consisting of $(n + 1)$ number equations is composed [1, p. 23]:

$$
\begin{cases}
y_0 = \displaystyle\sum_{i=1}^{n} a_i x^{i-1} \\[2mm]
y_1 = \displaystyle\sum_{i=1}^{n} a_i [A_1(x)]^{i-1} \\[2mm]
y_2 = \displaystyle\sum_{i=1}^{n} a_i [A_2(x)]^{i-1} \\[2mm]
\quad\cdots \\[2mm]
y_n = \displaystyle\sum_{i=1}^{n} a_i [A_n(x)]^{i-1}
\end{cases}
\tag{2}
$$

The expression (2) is a system of linear equations with respect to quantities $a_1, ..., a_n$ and is solved by the Cramer law.

Writing the obtained values of the quantity a_i in (1), we obtain the following expression for the main test equation:

$$
y_0 = \sum_{i=1}^{n} (-1)^{i+1} \sum_{j=1}^{n} y_j x^{j-1} \underset{B\neq j,\, 1\leq g<B\leq n,\, g\leq j}{\Pi\left[A_b(x) - A_g(x)\right]} \cdot
$$

$$
\sum_{d\neq y}^{C_n^{n-i}} \frac{A_{d_1}(x) A_{d_2}(x) \dots A_{dn-1}(x)}{\Pi\left[A(x) - A_g(x)\right] 1 \leq g < B \leq n}.
\tag{3}
$$

From the known values of n and from the set of tests determined from (3), we can get the processing algorithm of MR. But as the TF of initial MS is nonlinear, depending on the number of tests used in IMS and their quality ratio (the ratio of the number of additive tests to the number of multiplicative tests), the obtained algorithm makes the determination of the sought-for quantity difficult for this or other reasons.

Therefore, one of the main properties of the optimality of the test set used simultaneously for the identification of nonlinear TF in IMS is that the order of the main test equation obtained at their realization is minimum. Simultaneously, the use of tests in this or other form is connected with the accuracy received by θ additive and k multiplicative constant components that they create. This feature is explained by the fact that the accuracy of MN obtained by the tested IMS in the first turn is determined by the accuracy of tests performed in the system.

In this connection, we should note that the simple additive test realized by the given accuracy in comparatively simple measurements in both electric and nonelectric quantities is in this form. This θ is an additive constant component and has the same physical nature as the quantity x.

From the carried-out investigations, we get those any initial measuring sensors designed on the basis of up-to-date technologies have no ideal linear MC and the mathematical model of their TF in the majority of cases is in the form of a square, and very rarely in the form of a cubic polynomial. Even the curves in the complicated

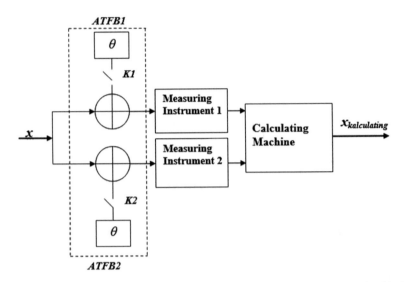

Fig. 1 The structural scheme of differential TIMS ATFB. ATFB-additive test formation block

form may be reduced to this form using the algorithmic-structural approximation method [3, pp. 32–39].

In Fig. 1, the structural scheme of TIMS equipped with a joint test method and differential scheme for increasing the accuracy of measurement of the initial IS whose TF is in the form of square trinomial $y = a_1 + a_2x + a_3x^2$ is given as an example.

The measuring process consists of four stages. In the first two stages, the quantities x and $-x$ in IS1 and IS2, in the next two stages the quantities $x + \theta$ and $-(x + \theta)$ are measured.

The processing algorithm of the measurement result realized with respect to measurement quantity x will be in the following form:

$$x = \frac{\Delta_{(y_1 - y_2)}}{\Delta_{(y_3 - y_4)} - \Delta_{(y_1 - y_2)}}\theta \qquad (4)$$

As seen in formula (4) the accuracy of MR is independent of the quantities a_1, a_2, a_3 and is determined only by the absolute similarity and constancy of the additive test formation blocks (ATFB1 and ATFB2).

This result has been obtained on the conditional identity of MS1 and MS2. But in reality, only such IMS-s possess the indicated properties, that the IT-s contained in their structure have differentiation properties [4, pp. 65–68].

The mentioned factors rather restrict the usage possibilities of the considered method. But using the method suggested by us, we can simplify the solution of the problem by piecewise approximation [5].

Note that only by using the additive tests, the differential scheme of the initial MS is not required.

In some cases, in MS, the joint use of the test methods for increasing the accuracy of measurements based on the realization of additive tests and the measurement method based on inverse-transformation (IT) is possible. This time, the influence of the error in IT on MR is excluded.

The structural scheme of IMS constructed by the initial MS possessing linear TF is given in Fig. 2. IMS consists of two identical MS (MS1 and MS2), IT and a calculating device (CD).

ATFB I and ATFB 2 are linked based on IT and MS2 inputs.

The measurement process is constructed by the following algorithmic succession.

At the output of MS1, the measurement quantity x is transformed into the output quantity y_1:

$$y_1 = x(a_{2N} + \Delta_{a_2}) + a_1$$

Here $\Delta_{a_2} x$ and a_1 are multiplicative and additive errors of MS1 and MS2, respectively.

The output quantity y_1 in IT with the β transformation quantity is transformed to the homogeneous quantity x_1:

$$x_1 = y_1(\beta_N + \Delta_\beta)$$

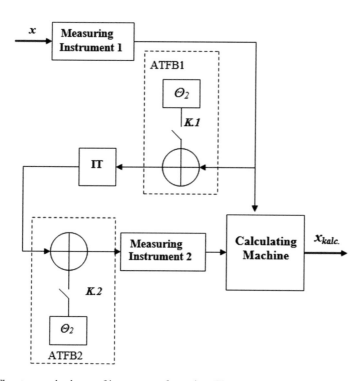

Fig. 2 The structural scheme of inverse transformation (IT) tested IMS

Here Δ_β is the current error of IT.

The quantity x_1 is transformed into y_2 in MS2:

$$y_2 = y_1[a_{2N}(\beta_N + \Delta_\beta) + \Delta_{a_2}(\beta_N + \Delta_\beta)] + a_1$$

Then the additive test $y_1 + \theta_1$ adjoins the input of IT and at the output of MS2 we get the signal y_3:

$$y_3 = y_2 + \theta_1[a_{2N}(\beta_1 + \Delta_\beta) + \Delta_{a_2}(\beta_N + \Delta_\beta)].$$

At the next step, the additive test $\beta_N(y_1 + \theta_1) + \Delta_\beta(y_1 + \theta_1) + \theta_2$ adjoins the input of MS2 and the result of the transformation becomes:

$$y_4 = y_3 + \theta_2(a_{2N} + \Delta_{a_2}).$$

In the calculating device, by processing the results of measurements y_1, y_2, y_3, y_4 with respect to the measuring quality x we get the following expression:

$$x = \frac{2y_1 - y_2}{y_4 - y_3}\theta_2 + \frac{y_3 - y_2 - \theta_1}{\theta_1(y_4 - y_3)}y_1\theta_2. \tag{5}$$

The realization of the algorithm (5) enables us to decrease simultaneously the additive and multiplicative components of the resulting measurement error using one additive test. This time, the accuracy of IT doesn't influence MR and the identity of ATFB1 and ATFB2 is not required.

But this method has some restrictions in its practical realization:

- the identity of MS (MS1 and MS2) is required:
- the initial MS possesses nonlinear TF, the use tests make the processing algorithm rather difficult. But, as was mentioned above, by joint use of approximation by parts and differential measurement methods, these restrictions are eliminated, and as a result, we obtain a perfect TIMS.

As another approach, we can consider the practical realization of the method for increasing the accuracy of measurement by using the additional measurements of additive tests. This time, taking into account the restrictions and requirements imposed on MS, we can note that if it is possible to form a multiplicative test in the system, then a test algorithm based on the realization of the set of simple additive and multiplicative tests would be more effective and universal.

4 The Combined Test Method Increases the Accuracy of Measurement

The test algorithms increase the accuracy of measurement using the combined test method, which usually consists of a set of simple additive and multiplicative tests [1, p. 32]. By using the joint use of the test equations, the efficiency of accuracy of measurement will be expressed in the system by the set of these or other kinds of tests:

$$A^* + M^* = N,$$

where A^* is the number of additive tests in the system, M^* is the number of the multiplicative tests, N is the number of the quantities a_i of the transformation function (TF) of the measurement system.

From the investigations, it becomes clear that the use of $A^* = (n - 1)$ tests of one kind, $M^* = 1$ number of test equations of another kind is considered the minimal limit and the identification of the nonlinearity of the transformation function of MS is obtained. But by using separately the additive and multiplicative tests, the identity of the main test equations is disturbed for the general case.

By applying a simple additive test in the form $A_B(x) = x + \theta_B$ $(B = 1, ..., n)$ the main test equations take the form:

$$y_0 = \sum_{j=1}^{n} y_j \frac{\prod\limits_{1 \le B \le n,\, B \ne j}[\theta_B]}{\prod\limits_{\substack{1 \le B \le n,\, \ell \ne j \\ B \ne j}}[\theta - \theta_l]}. \tag{6}$$

By applying a simple multiplicative test in the form $K_g x$ $(g = 1, ..., n)$, the main test equations take the form:

$$y_0 = \sum_{j=1}^{n} y_j \frac{\prod\limits_{1 \le g \le n,\, g \ne j}\left[K_g - 1\right]}{\prod\limits_{\substack{1 \le g \le n \\ g \ne j}}\left[K_g - K_\ell \atop \ell \ne j \right]}. \tag{7}$$

As it is seen from Eqs. (6) and (7), in this case, the main test equations are independent of the measured quantity x.

It should be noted that as the nonlinearity of the transformation function of the initial measuring system increases, it must write its mathematical model with rather high accuracy. When increasing the accuracy of the measuring using the algorithmic method, the general number of tests contained in MS increases in proportion to the order of the approximating polynomial, and simultaneously, according to the accuracy of the composition of tests, the error components of the measurement results of the tested MS will increase. Furthermore, by measuring the non-electrical quantities, the introduction of excessive tests complicates its construction or it is not always possible. Therefore, the development of an optimal structure is a very urgent problem.

Considering the above deficiencies and restrictions, the tested information-measurement system obtained from the joint use of additive and multiplicative tests in rather nonlinear TF will have high accuracy and stability.

For that, the non-linear TF of MS is studied along the entire range and successively separated into optimal intervals (beginning with the zero value of x). Unlike the known methods, for attaining a higher accuracy, the greatest curvature, minimum and maximum points are determined by means of special algorithms. The curve between extremity points is approximated in two steps and is iterated in other intervals as well. Such an algorithmic solution of the problem is said to be an approximation method by the algorithmic-structural way and as a result, a set of simple curves along the characteristic is obtained. The system of primary equations describing these curves is jointly solved and we get exact values of the coefficients of the curve at each interval. This method eliminates the iterative measurement and reduces the number of unknown quantities (for each interval) of each approximating intermediate polynomial to three.

It is known that in the field Δx_s (where $s = 1, ..., \ell^*$ is the number of the approximated fields), the mathematical dependence between the value of the measured quantity x, the tests $A_i(x)$ $(i = 1, ..., e$, here e is the number of unknown coefficients in the approximating polynomial) and their measurement results y_{is} will be as follows:

$$y_{se} = \sum_{i=1}^{e} \mathcal{B}_{si}^{**}[A_e(x)]^{i-1} \tag{8}$$

(here $\mathcal{B}_{si}^{**}$ are the parameters of the TF of MS in s—approximation field). This test algorithm enables us to determine the real result of the measured quantity with high accuracy at each measurement step using the constant parameters $\mathcal{B}_{si}^{**}$. For each quantity Δx_s some contradictory requirements are imposed. On one hand, the quantity Δx_s should be so rather small in each field that the inequality $\Delta_{ap\cdot s} < \Delta_{\max .ver.}$, be satisfied, on the other hand, according to the composition conditions of test algorithms, the measured quantity at the same time should be located in the s approximation field.

Here $\Delta_{\max .ver.}$ is the maximum measurement error given by the specifications; $\Delta_{ap\cdot s}$ is the approximation error of the real TF of the initial MS in the given field. At this time, the current real value of $\mathcal{B}_{si}^{**}$ may be determined.

It should be noted that in the general cases, the quantity Δx_s takes insignificant value, $x_{\max_s} / x_{\min_s} \rightarrow 1$; here $x_{\max_s}$ and $x_{\min_s}$ are maximum and minimum values of the measured quantity x, more specifically, all the intermediate values of the quantities x and $A_i(x)$ will be located in $\Delta x_{\max}$.

As $x_{\max_s}/x_{\min_s} \to 1$, the y_{is} results of all additional measurements of the tests $A_i(x)$ according to the measurement period s given in the approximating field Δx_s, the quantity x and the measurement result y_{s0} will get closer values. Thus, their difference, in the general case will get small values comparable with the value of the statistical random error $\overset{\circ}{\Delta} y(t)$. Therefore, during the realization process of the processing algorithm, when the initial MS has a rather nonlinear TF, final measurement results may have rather great error.

The contradictory requirements imposed on the value of Δx make necessary functioning of IMS in such a working range of initial MS that it could enable to satisfy them in this or other degree. This condition restricts the use possibilities of all working range of the initial MS.

It should be noted that only the decrease in orders of the main test equations of IMS equipped with additive test costs excessive time to the system. To realize the identification of TF of the initial MS with high accuracy, in addition to these tests there appears a necessity for additional tests and as a final result, the ordinary test algorithms providing the full functioning of the system increase from the provided $(n + 1)$ number steps to $2n$ number measurement step.

By the similar mathematical transformations by using the joint use of both tests, we can get a general expression for the measurement algorithm. This algorithm represents $n - 1$ number additive $x + \theta$ test and one multiplicative (k, x) test.

At it is shown, the necessary condition for the existence of the main test equation (MTE) is the joint realization of simple additive and multiplicative tests in TIMS. In this time, to obtain minimum order of MTE with respect to it is necessary to realize $n - 1$ number one kind and one another kind test.

We can write formula as follows:

$$y_0 = \sum_{j=1}^{n} y_j \frac{\prod\limits_{1 \le \mathcal{B} \le n, \mathcal{B} \ne j} \left| A_g(x) - x \right|}{\prod\limits_{1 \le \mathcal{B} \le n, \mathcal{B} \ne \ell, \ell = j} \left| A_g(x) - A_\ell(x) \right|} \tag{9}$$

If we find the values of the parameters $a_1, a_2, \ldots, a_n$ of TF of TIMS from (9) and write them in (2.3), we can get a mathematical model of TF of MS.

As it is seen from (9), for $a_i \ne 0$ by using the total tests of MTE in TIMS, we can get an identity with respect to x.

Indeed, assume that for the identification of TF of IMS in the form (1); in the system the n number additional tests are realized using the following combined test:

$$A_\mathcal{B}(x) = k_\mathcal{B} x + \theta_\mathcal{B}, \quad (\mathcal{B} = 1, \ldots n).$$

It we write the values of the y_j results of the combined test and their measurements in the expression (9), for the given TIMS we get the following mathematical expression:

$$y_0 = \sum_{j=1}^{n} y_j \frac{\Pi\left[\begin{array}{c} x(k_B - 1) + \theta_B \\ {\scriptstyle 1 \leq B \leq n,} \quad {\scriptstyle B \neq j} \end{array}\right]}{\Pi\left[\begin{array}{c} x(k_B - k_l) + (\theta_B - \theta_l) \\ {\scriptstyle 1 \leq B \leq n,} \quad {\scriptstyle B \neq j,} \quad {\scriptstyle \ell = j} \end{array}\right]} \tag{10}$$

It is seen from formula (10) that for $\theta_B \neq 0$, $k_B \neq 0$, $k_B \neq 1$ the MTE of TIMS realized only on the basis of combined tests has at least one solution with respect to the measured quantity x.

Note that, as a rule, it is practically difficult to realize the exact values of the quantities k_B with respect to the given values of the constant θ_B components of the additive tests. Therefore, by forming in TIMS the combined tests in the form $A_B(x) = k_B x + \theta_B$, $(B = 1, ..., n)$ the parameter $k = const$ is saved and the parameter θ_B is changed. With respect to this, we can describe the expression (10) as follows:

$$\begin{aligned}
y_0 = {} & y_1 \frac{[x(k-1)+\theta_n][x(k-1)+\theta_{n-1}]\ldots[x(k-1)+\theta_2]}{(\theta_n - \theta_1)(\theta_{n-1} - \theta_1)\ldots(\theta_2 - \theta_1)} \\
& - y_2 \frac{[x(k-1)+\theta_n][x(k-1)+\theta_{n-1}]\ldots[x(k-1)+\theta_3][x(k-1)+\theta_1]}{(\theta_n - \theta_2)(\theta_{n-1} - \theta_2)\ldots(\theta_3 - \theta_2)(\theta_2 - \theta_1)} \\
& + y_j \frac{[x(k-1)+\theta_n]\ldots[x(k-1)+\theta_{j-1}][x(k-1)+\theta_{j-1}]\ldots[x(k-1)+\theta_1]}{(\theta_n - \theta_1)\ldots(\theta_{j-1} - \theta_j)(\theta_{j-1} - \theta_j)\ldots(\theta_1 - \theta_j)} \\
& + \cdots + y_n \frac{[x(k-1)+\theta_{n-1}]\ldots[x(k-1)+\theta_1]}{(\theta_{n-1} - \theta_n)(\theta_{n-2} - \theta_n)\ldots(\theta_1 - \theta_n)} \tag{11}
\end{aligned}$$

For nominators of all summands of the expression (11) are $(n - 1)$ order polynomials with respect to quantity x and in the denominator, the coefficients $\mu_j = F(\theta_1, \ldots, \theta_n)$, are independent of x. Here, for all the values $a_i \neq 0$ and by using only the combined tests in the form $A_B(x)$, the MTE will be $(n - 1)$ ordered equation with respect to the quantity x.

So, as in the resulting expression, consisting of a combined test, there are no $n - 1$ number the same multipliers with the quantity x, the θ_B and k don't become identities in the domains where the parameters participate.

5   Result

It should be noted that it is possible to increase the accuracy of MR to some extent by means of the combined test algorithm on the basis of simple additive and multiplicative tests, and that the application of simple tests in the form $kx + \theta$ and $kx - \theta$ making no difficulties in practical realization of TIOS are more appropriate.

References

1. Bromberg, E.M., Kulikovskii, K.L.: Test Methods for Increasing Accuracy of Measurements, 176 p. Energia (1980) (Russian)
2. Kulikovskii, K.L., Kuper, V.YA.: Methods and Means of Measurements, 448 p. Energoatomizdat (1986) (Russian)
3. Gadjiev, Ch.M., Isayev, M.M.: Methods of Processing of Measurements in IMM for Determining the Mass of Oilproducts in Reservoirs. ELM, Baku (2000), 94 p. (Russian)
4. Isayev, M.M.: Metrological supply of control and measurement of the quantity of oil products in commercial units. Izvestia Vyshshikh Tekhnicheskikh uchebnikh zavedeniy Azerbaydzhana **1**(53), 65–68 (2008) (Russian)
5. Melikov, Ch.M., Isayev, M.M.: Pressure measuring equipment. Patent No 99 (001607, class G 01 L 19/00, Industrial property. Offic. Bull. **1** (2001)

Simulation of Statistical and Spectral Characteristics of Surficial Sea-Ways in Reservoirs

Allahverdi Hasanov and Ramin Gasanov

Abstract A fundamentally new method for determining statistical and spectral characteristics of real sea-ways based on the information of wave recorders or other self-recording devices, is offered. Wave loads acting on multi-support offshore deep water hydraulic structures are determined.

As a result, the formulas are obtained for calculating the values of vertical and horizontal dynamical loads on the construction.

Keywords Wind waves · Disturbance spectrum sea-ways · Statistical characteristics · Wave record · Of real sea-way · Security function

1 Introduction

Theory of harmonic and trochodial waves on the surface of heavy liquid does not describe the properties of real sea-way. Because at any degree of sea-way on the surface of a reservoir, the waves of various forms and sizes can be observed [1, 2]. They have height, length, speed and propagation direction whose important feature is randomness. The exceptions are sea waves in the form of a swell. Variety of waves observed on the water surface are analyzed well by the probability theory and mathematical statistics methods that allow to characterize the given sea-way by the mean values of the elements of the observed waves and distribution functions.

A fully developed wave d sea-way just like a steady state wave regime are random processes [3, 4] that posses two important properties: stationarity and ergodicity.

This work is partially supported by NSF Grant #2003168 and CNSF Grant #9972988

A. Hasanov · R. Gasanov (✉)
Institute of Control System of Azerbaijan National Academy of Sciences, Baku, Azerbaijan
e-mail: ramingasanov@mail.ru

A. Hasanov
e-mail: hesenli_ab@mail.ru

Experiments affirm that implementation of processes derived from sea way, the loads, reaction of structures, also will possess these properties. The stationarity and ergodicity properties greatly facilitates theoretical analysis of random processes and improve modern calculation methods.

In this paper we determine frequency energy spectrum of a random two-dimensional wave characterizing energy distribution between spectral components.

2 Problem Statement

Statistical characteristics of sea-way describe its outward manifestation without affecting the physics of the process. The development of spectral theory of waves based on general provisions of spectral theory of centered stationary ergodic processes made it possible to start studying the physical and energy [5] essence of the process of random wind waves:

Energy spectrum is an intrinsic physical characteristic of a random surface sea-way.

The spectral theory of random processes is used to study random wave actions on the structure and to determine reactions of such structures. Dynamical calculation of general reactions of structures (of displacements of the surface part of the stationary platform, support forces, etc.) is made taking into account all external influences of deterministic and random nature.

In an experimental study of influences of such random natural factors as wind and storm waves on a structure, the implementations are continuous readings of sensors or recorders (self-recorders, oscillographs, wave recorders, etc.) recorded on magnetic film or a special paper. In this case, firstly the oscillations of the surface, velocity and acceleration of wave particles at any point in depth, secondly external loads from these factors [6, 7] acting on the structure (point pressures, specific loads in different depth sections, resultant loads, overturning moments) and thirdly, the reactions arising in the elements of structures from external loads (bending moments, normal and cutting forces, stresses, strains) can be fixed in the form of implementations.

In the paper we study statistical and spectral characteristics of a real sea-way based on a sufficiently long implementation, the records of wave oscillations at any point of the sea surface performed by a wave recorder. The main parameters of this implementation and also its probability characteristics as of a random function were considered.

A part of the graph of implementations of random wave oscillations over time T_x is shown in Fig. 1.

The ridge top

Sole of the hollow

Similarly, elastic deformations of the structure caused by the action of waves in the form of random oscillations occur near its average position and therefore also are random sign-alternating functions oscillating around zero.

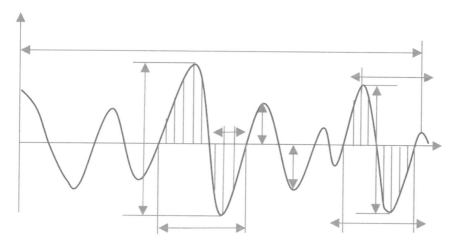

Fig. 1 Centered implementation of waves

A similar type of random functions refers to the so-called centered random process with zero mean value, since the mean value of positive and negative coordinates is zero.

3   Problem Solutions

We will analyze the implementation of random processes by a numerical method using a specially developed [8, 9] version of the DRAKULA software package.

For that, we divide continuous implementations into N time inserwals (time discrete) $\Delta t = T_x/N$ and automatically, with the help of the special program we determine $N + 1$ values of ordinates of Xi implementation at appropriate times $t_1 = 0$, $t_2 = \Delta t$, $t_3 = 2\Delta t$,...., $t_n + 1 = T_x = N \cdot \Delta t$.

For marine data processing with a realization length equal to $150 \div 200$ waves, the value of time step is accepted as

$$\Delta t = (0.05 \div 0.1) \times \tau_{cp} \tag{1}$$

where τ_{cp} is average period of waves and is determined by the formula

$$\tau_{cp} = \sum_{i=1}^{M} \tau_i / M \tag{2}$$

M is the number of waves.

Thus, the number of ordinates for all implementation from $100 \div 200$ waves equals: $N = 1000 \div 4000$.

When analyzing the centered implementation of random wave oscillations and reaction of structures we differ the following main parameters:

- duration of the implementation T_x, or the number of ordinates, $N + 1$, where $N = T_x/\Delta t$;
- signalternating values of centered ordinates $x\ i$, or ordinates of excited surface;
- amplitude values of the centered random function a_{x_i}, corresponding to its maximum and minimum values counted from zero value, that are widely used in the next calculations of stresses and strains;
- amplitudes of the centered random function η_{x_i} of the quantity characterizing the distance between the maximums of the random function located above the zero line and the closest minimums lying below this line.

The amplitudes will equal the wave height h_i.
We determine the average height by the formula:

$$h_{cp} = \sum_{i=1}^{M} h_i / M \tag{3}$$

Average height of waves and average period (τ_{cp}) are the most important characteristics of the storm sea-way. Multiyear experimental studies showed that for deep-water sea waves the security function for the wave height [10] is determined by the Rayleigh law. For dimensionless wave heights $h_0 = h/hcp$ the Rayleigh law is expressed by the formula

$$F_1(h_0) = \exp\left(-\frac{\pi}{4} h_0^2\right) \tag{4}$$

Distribution of displacement of the free surface of the sea $\varphi(x, y, t)$ is described better by the Gramme-Charlet distribution. The slope of this surface in the direction of wind $\vec{a}$ is determined by the following formula:

$$\frac{\partial \varphi}{\partial \vec{a}} = \frac{\partial \varphi}{\partial x} \cdot \cos(\vec{a}\,\vec{x}) + \frac{\partial \varphi}{\partial y} \cdot \cos(\vec{a}\,\vec{y}) \tag{5}$$

The distribution function of slopes $f(\varphi x, \varphi y)$ in the [11] first approximation satisfies the normal two-dimensional low:
We have obtained a formula that approximates the frequency spectrum divided into separate domains.
We divide the spectrum into 5 intervals: gravitational, Kitaygorodsky interval, Leikin-Rosenberg interval, capillary and viscous [2]. Each interval offers its own type of dependence of the spectrum.
Taking into account three-dimensionality of the process, the distribution of directional spectral components can be described in the linear approximation by means of the frequency-angular spectrum C (ω, θ).

The estimates $C(\omega, \theta)$ can be obtained through the analysis of synchronous measurements of elevations of the surface and slopes at several points.

$$C(\omega, \theta) = C(\omega) \times R(\omega, \theta)$$

where $C(\omega)$ is a frequency spectrum, $R(\omega, \theta)$ is a angular distribution function satisfying the condition

$$\int_{-\pi/2}^{\pi/2} R(\omega, \theta)d\theta = 1$$

This function determines distribution of wave energy in the given directional interval frequency has a pronounced maximum coinciding with general direction of wind θ_0.

We can approximate this function in the form

$$C(\omega) = A \times \omega{-}m \times \exp(-B \times \omega{-}n)$$
$$R(\omega, \theta) = A(n) \times \cos n(\theta{-}\theta_0)$$

where m, n dimensionless, A, B are dimensional parameters with various values for each region of the spectrum.

For example, for three point studies

$$R(\omega, \theta) = 2^k \cdot \frac{\Gamma(2 + 2\kappa)}{\Gamma^2(1 + \kappa)} \cdot \cos^k(\theta - \theta_0)$$

here $\kappa = 1, 8\omega 0/\omega$, Γ is a gamma function.

The value $D_x = \int_0^{\infty} C(\omega)d\omega$ equals the random process variance and characterizes total energy of the random process, while $C(\omega)$ is distribution of this energy over spectral components.

Allowing for this, we can write

$$h_{cp}^2 = 2\pi\sigma_x^2 = 2\pi D_x$$

where $\sigma_x = \pm\sqrt{D_x}$ in a standard (mean square deviation of a random stationary function).

Considering what has been said above, after some calculations, for calculating the [12, 13] values of the specific maximum horizontal wave load on pile foundations

$$q_{x\,max}(y) = \gamma\pi \frac{2h_{max}}{\pi_{max}} \cdot \frac{D\delta^2}{4} \frac{Chky}{ChkH} \cdot C_m\left(\frac{\pi D\delta}{\lambda}\right)$$

$$P_{z\,\max}(x) = \gamma \frac{h_{\max}}{2} \cdot \frac{ChkH\delta}{ChkH} \cdot \cos\frac{2\pi}{\lambda}x$$

the values of the function $C_m\left(\frac{\pi D\delta}{\lambda}\right)$ and corresponding designations are detailed in [13]. Vertical wave pressures are calculated by these formulas.

4 Conclusions

Thus, a fundamentally new method was developed to determine statistical and spectral characteristics of a real sea way based on the information of wave recorders or other self-recording devices.

The obtained relations allow to determine the values of horizontal and vertical wave load on cylindrical elements of multi-support offshore deep-water structures with a random three-dimensional sea-way.

References

1. Moiseev, N.N.: Numerical Methods in the Theory of Optimal Systems, p. 424. Moscow, Nauka (1971) (in Russian)
2. Register of the USSR, Wind and waves in the oceans and seas. L.Transport, p. 1225 (1974)
3. Moiseev, N.N.: Elements of the Theory of Optimal Systems, p. 526. Moscow, Nauka (1975) (in Russian)
4. Caspian Sea.: Hydrology and Hydrochemistry. In: Kosarev, A.N. p. 262, MGU Publishing House (1986)
5. Hasanov, A.B.: Reaction of Mechanical Systems to Nonstatinary External Influences, p. 248. "Elm" Publishing House, Baku (2004)
6. Sadigov, A.B.: Models and Technologies for Solving Problems of Management in Emergency Situations, p. 372. "Elm", Baku (2017) (in Russian)
7. Aida-zade, K., Asadova, J., Suleymanov, N.: Numerical analysis of methods of solving optimal control problems with non-fixed time. In: Proceedings of the 2nd International Conferences "Problems of Cybernetics and Informatics", pp. 94–97, Baku, Azerbaijan. Section 5, "Control and Optimization", Sept 10–12, 2008
8. Riedinger, P., Morarescu, I.-C.: A numerical framework for optimal control of switched affine systems with state constraint. In: Proceedings of the 4th IFAC Conference on Analysis and Design of Hybrid Systems (ADHS 12), pp. 141–146. Eindhoven, The Netherlands, 6–8 June 2012
9. Zhang, W., Xu, H., Wang, H., Lin, Z.: Stochastic systems and control: Theory and applications. Math. Prob. Eng. p. 4, Article ID 4063015 (2017)
10. Menshikh, V., Samorokovskiy, A., Avsentev, O.: Models of resource allocation optimization when solving the control problems in organizational systems. J. Phys. Conf. Ser. **973**, 012040 (2018)
11. Sprocka, T., Bocka, C., McGinnis, F.: Survey and classification of operational control problems in discrete event logistics systems (DELS). Inter J Prod Res (2018). https://doi.org/10.1080/00207543.2018.1553314
12. Levine, W.S.: Control System Applications, p. 360. Published by CRC Press, ISBN 9780367399061 (2019)
13. Sadigov, A.B., Zeynalov, R.M.: Optimal control in the problems of calculating the benefit/cost ratio in emergency response. Inform. Control Problems **40**(1), 47–56 (2020)

The Main Directions of the Use of Innovations in Industrial Production in Azerbaijan

Aygun A. Aliyeva⊙

Abstract Innovation as a process that changes the type of reproduction, the quality of socio-economic relations, creates a new innovative directed benefit, offers a unique form, nature of accumulation and fund of natural wealth. In modern conditions in time usage of innovations in industrial enterprises of Azerbaijan is guaranteed of successful functioning of whole industry of the country. The fundamental prerequisite for the formation the innovative sector of the economy is to achieve high rates of economic development, which is expressed in the constant growth of the main macroeconomic indicators. The globalization of the world economy, which has been actively developing in recent decades, has opened various economic opportunities and diversified the methods of competition. Innovation today plays a decisive role and is the main factor in the competitive struggle for both domestic and foreign markets.

Keywords Innovation · Industrial · Modern technologies

The main purpose of the study is to identify promising areas for the development of the industrial sector of the Republic of Azerbaijan, incl. local businesses in today's economic environment. The main task in this direction is to improve the activities of industrial enterprises, which contributes to the implementation of the following tasks: improving the efficiency of the activities and management of industrial enterprises; reducing costs and expanding the scale of sales; maintaining the competitive position of the enterprise; stimulation of investors, etc. [1].

Systematic and large-scale reforms carried out in the Republic of Azerbaijan at the level of state leadership in recent years in the national economic spheres have ensured high socio-economic development in the country. It should be noted that all this has allowed our country to constantly expand cooperation with various international

A. A. Aliyeva (✉)
The Ministry of Science and Education of the Republic of Azerbaijan, Institute of Control Systems, B. Vahabzade Street, 68, Baku, Azerbaijan
e-mail: aliyevaaygun1978@gmail.com

© The Author(s), under exclusive license to Springer Nature Switzerland AG 2023
Sh. N. Shahbazova et al. (eds.), *Recent Developments and the New Directions of Research, Foundations, and Applications*, Studies in Fuzziness and Soft Computing 423,
https://doi.org/10.1007/978-3-031-23476-7_30

financial institutions in the direction of integration into the world economy. It should be noted that the realization of the national economic interests of Azerbaijan is based on the sustainable development of the national economy. In recent years, during the political and financial crises in all countries of the world, sustainable development has been observed in Azerbaijan.

The formation of the national economy and its effective integration into the world economic system determines the competitiveness of the country's economy as a whole and the industry that forms its basis. The industrial policy formed in our country is in line with the concept of ensuring the competitiveness of the national economy.

The diversification of the national economy, increasing its efficiency, increasing product quality and competitiveness, modernizing all spheres of public life, raising the living standards of the population and reaching the level of developed countries in all these areas are developing rapidly in our country. Azerbaijan's economic achievements are highly valued by the world's leading international organizations.

At present, the socio-economic development strategy set in our country is being successfully implemented. The concept of socio-economic development developed on the basis of the order of the country's leadership "Azerbaijan 2020: Vision for the Future" identifies the creation of a diversified, innovation-oriented, efficient and sustainable economy in the coming years as the main goals of our country's development [2]. Paragraph 4 of the relevant concept is entitled towards a highly competitive economy and includes the formation of an economic model based on effective government regulation, mature market relations, improving the structure of the economy, developing the non-oil sector, supporting scientific potential and innovation.

Industry practice of developed countries shows that in economic development projects (for industrial enterprises) there is a newer methodological approach that corresponds to modern conditions. That includes: program and project management; reengineering of business processes; ABC/AVM methodology; correct and timely method; general management approach to the quality of knowledge management methods; information technologies and methodologies, etc. With a deeper consideration of these methods, we will be able to determine their essence.

Project management. The restructuring of industrial enterprises, within a well-organized sequence, covers problems that are solved through specific studies. The initial stages of the processes consist of studying the current state of enterprises, preparing for restructuring, and developing its concept.

The processed data is converted into information intended for decision-making at the next stage of the development of economic development programs, and after that the direct execution of programs is carried out. It should be borne in mind that each project, from its creation and execution, goes through successive stages. These stages also include the following: idea, concept, planning, execution and completion. The restructuring program determines the effectiveness of ongoing projects and the investment attractiveness of enterprises [3].

Business process reengineering is a set of techniques planned to improve the economic performance of a business by modeling, analyzing and redesigning business processes.

Reengineering is the creation of modern business processes that increase the efficiency of enterprises. Business process is a set of business operations designed to create products for a consumer. In general, the purpose of business processes is to offer the consumer quality goods and services.

Business processes are also considered as operations to restore intangible assets and financial flows. It should be noted that business reengineering projects cover four stages: increasing the prospective reengineering of the enterprise; business activity research; drawing up new business projects; implementation of new business projects. Reengineering ensures the achievement of effective goals through the use of information technology.

The main features of business process analysis and reengineering are as follows: several projects are concentrated in the work and the performers make decisions individually; the process itself continues sequentially; the work process is carried out at the appropriate places; processes have individual actuators, etc.

The results of business process reengineering include the following: in the functional structures of enterprises, there is a transition to project structures; the functions of managers are completely changed; the criterion of efficiency in the work process is taken into account more; administrative duties are being improved; the process of preparing works is changing in the direction of professionalism [4].

1 Statement of the Main Material of the Research

The cost management methodology for industrial enterprises ABC/ABM as a single system is created on the basis of an accounting base in accordance with international standards. The methodology drawn up on the basis of cost restructuring is considered to be the main part of these processes. The ABC methodology is a method for determining the cost of information and services. The basis of this process is the determination of the costs of performing work and functions in accordance with the provided resources. It should be noted that a group of three objects is established: resources, functions, and value objects (products, goods, and services). Cumulative cost of functions reflects the total of individual cost elements and can be grouped in a hierarchical structure and function centers. This mutual coordination forms the ABC model.

The ABC methodology makes it possible to accurately analyze the activities of industrial enterprises, determine the process, create value using the exact time method. This term implies the process of ensuring the release of products with all the necessary resources, which as a result ensures the intensity of output in accordance with production programs. In this case, 2 important conditions of the production stages must be met: the first is the timely provision of production with all the necessary resources; the second is the activity of all the constituent parts of production in close coordination.

As a result, it can be seen that "just-in-time production is only possible with quality assurance. Quality is the initial stage of organizing a system". Therefore, the organization of the system under consideration contributes to the effective solution of the main problems of quality management [5].

Quality management is one of the initial priorities in the implementation of the restructuring of industrial enterprises, it is considered the formation and certification of a quality management system that guarantees the stable quality of all developed and received products. It is precisely as such a guarantor of stability that a high-level quality system is required in a manufacturing company. Quality management is the basis of the industrial enterprise management system. The use of quality management methods during restructuring guarantees the effective operation of the enterprise, as well as maintaining the quality of products and business processes.

Knowledge management. Recently, this concept has gained importance. In this aspect, to improve the efficiency of industrial enterprises, the experience of relevant sources, knowledge gained in practice and accumulated intellectual capabilities are used.

In this process, in order to form the accumulated and used sphere of knowledge (experience useful for business, skills, information), technologies, organizational structure, business relationships, etc. are coordinated. It should be noted that in order to increase the efficiency (profitability) of knowledge management, it is necessary to properly use the intellectual capital of the enterprise.

To determine the acquired knowledge, it is prominent to model the activities of the firm. It should be considered that the use of information technologies (computer software, the Internet, conceptual modeling, semantic networks, etc.) is a necessary factor in knowledge management.

Cooperative information technologies. The use of information technology is the basis of the company's activities. The use of information systems constitutes a certain restructuring process of the management system. At the same time, working methods are changing, modern technological and information sources are being formed. It should be noted that the effective restructuring of industrial enterprises is not possible without the use of information technology, information support for business processes, development of communications, etc.

Innovation as a process that changes the type of reproduction, the quality of socio-economic relations, creates a new innovative directed benefit, offers a unique form, nature of accumulation and fund of natural wealth. In modern conditions in time usage of innovations in industrial enterprises of Azerbaijan is guarantee of successful functioning of whole industry of the country. The fundamental prerequisite for the formation the innovative sector of the economy is to achieve high rates of economic development, which is expressed in the constant growth of the main macroeconomic indicators.

In this respect, modern technologies, and their scaled usage in enterprises of the field is necessary for boost of national economy to high level of development. It should be noted that implementation of new achievements of scientific-technical progress ensure base for dynamic social-economic development for enterprises of industry, anticipating followings: establishment of necessary size of manufacture incurring from demands of the market; enhancing competitiveness of manufactured production; improvement of conditions of labor; lightening solution of social problems; smoothing seasonality in separate branches of industry; creation of opportunities for reprocessing in place of wastes and garbage of industry. It should be emphasized that, cooperate manufacture plays important role in provision of incessant work of industrial enterprises and today it is developed in the fields manufacturing constructive complex production and requiring its special technical characteristic and parameters [6].

Recently, the role of the Information and Communication Technologies (ICT) sector in the development of the country's economy, which is the main priority of modern innovative technologies in the Azerbaijani economy, and the state pays special attention to the widespread application of ICT achievements in the country. In this direction, very serious measures have been taken in recent years to establish the State Fund for ICT Development, the High Technology Park, the Information Security Center and the State Agency for Information Security [8].

The launch of Azerbaijan's first telecommunications satellite is a very important success in technical sciences and indicates that ICT will develop more rapidly in the country in the future. In accordance with the strategy defined by the President, the revenues from ICT in Azerbaijan in 2025 are expected to reach revenues from oil exports.

The economic experience of developed countries demonstrates that one of the most effective options for solving the problem of financing the innovation process is the activity of venture capital funds. Their application in the national economy through the direct use of international experience sometimes does not meet the expectations, because the specificity of the country's economic development is not taken into account.

Creating favorable conditions for the development of technological innovations in economic areas, along with the modernization of the technological base of the Azerbaijani economy, can also increase the competitiveness and export orientation of national industrial products. The above makes it necessary to develop a state innovation policy and a mechanism for its implementation, as well as the creation of a national innovation system. In the context of globalization of the world economy, only the innovative activity of industrial enterprises will give impetus to increase their competitiveness [7].

In recent years, serious measures have been taken in all countries of the world to apply modern innovations in national economic spheres. Thus, according to the International Competitiveness Report for 2019, even though some post-Soviet countries have different positions on innovation opportunities among 141 countries, economic measures are being taken in these countries to move in the right direction. Thus, despite the fact that according to the latest report, Germany ranks first among all

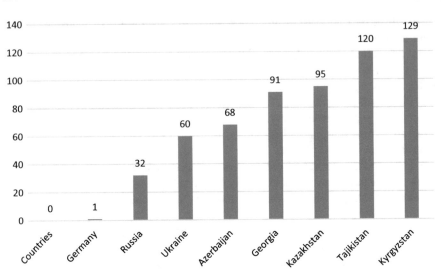

Fig. 1 Positions of some countries on innovation opportunities among 141 countries in the International Competitiveness Report (2019) [8]

countries in the world in terms of innovation opportunities, Russia is in the best position among the countries of the post-Soviet space—32nd, Ukraine—60th. Azerbaijan's position was 68th. In this list, Azerbaijan ranks Georgia (91st place), Kazakhstan (95th place), Tajikistan (120th place) for innovation opportunities. It is ahead of countries such as Kyrgyzstan (129th place) (Fig. 1). The application of modern innovative technologies in industrial production processes in Azerbaijan, the increase in the volume of export-oriented production has played an essential role in increasing the country's reputation as an innovative country [8].

Analysis of statistical indicators covering recent years shows that the share of science-based products in the structure of GDP in Azerbaijan is about 0.2%, oil and gas, oil refining products are more than 95% on average, the main production in the refining industry high depreciation of funds (AMF), etc. Such issues show that the organizational and economic environment, system of incentives and rules that are important for innovation in our country are not yet fully formed. Due to the low level of many scientific developments in the field of industry, it is impossible to sell them as a market product and apply them in production. The growing pace of imports in the domestic market in Azerbaijan also highlights the importance of comprehensive application of innovation in industries. An improved national innovation strategy should be formed to prevent these negative trends and increase innovation activity in the country, and it should be considered as one of the most important global directions of the state's important scientific, technical, and socio-economic policy.

Economic policy based on international experience in our country should be organized based on long-term forecasts based on a comprehensive assessment of the country's intellectual and production potential, taking into consideration the prospects for the development of the market of intellectual products. The experience of developed countries shows that development of the National Innovation System (NIS) is primarily related to its integration into regional and global systems. At present, developed countries pursue a sustainable policy on the overall innovation development strategy and mechanisms for the implementation of these processes in areas, such as innovation systems, human resource development, information and communication technologies, business environment [9].

In recent years, Azerbaijan has witnessed significant changes in total expenditures on product innovations and process innovations in various industries. For example, in 2013, expenditures on product innovations amounted to 11,899.5 thousand manat, while in 2016 this figure increased to 20,313.8 thousand manat. In 2017, compared to the previous year, this figure decreased to 10,439.6 thousand manat. Total expenditures on process innovations in industrial areas also decreased in 2017 and amounted to 5696.1 thousand manat [10]. As can be seen from the analysis, in 2017, compared to 2016, there was a decrease in expenditures on product and process innovations in our country (Fig. 2).

If we determine the trend line between process innovations in industry and the cost of product innovation in our country, we see that there is a linear correlation between them expressed by the regression equation $y = 0.249x + 5471$ (Fig. 3). The fact that the corresponding correlation coefficient $R = 0.126$ indicates that the relationship between the y-dependent variable and the free variables representing the factors included in the model, i.e. between process innovations and product innovation costs, is very weak. According to the Cheddock scale (0.1–0.3), this connection is very weak. The coefficient of determination is $R_2 = 0.016$. This means that the corresponding regression equation is explained by 1.6% of the variance result, and 98.4% by the influence of other factors. It is important to note that the

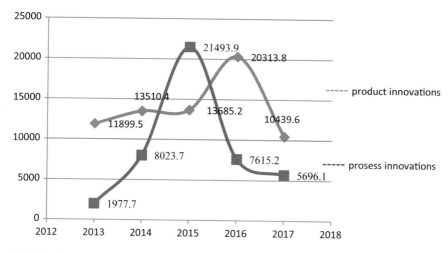

Fig. 2 Total expenditures on technological innovations in industry by types of innovations in Azerbaijan (in thousands of manats) [10]

low coefficient of determination indicates that the regression equation is weaker than the initial data and that a smaller portion of the outcome factor (1.6%) is explained by the factors included in the model. As can be seen from the analysis, the annual increase in the cost of process innovations and product innovations in the field of industry in our country can ensure an increase in production in this area [10] (Fig. 4).

If we analyze the number of industrial enterprises specializing in the application of innovations in all types of property in the Republic of Azerbaijan on the basis of annual statistics, we can see that recently significant changes have been recorded in this area (Fig. 1). For example, in 2015, the number of industrial enterprises operating in our country on all types of property was 2583 units, while in 2018 and 2019 these indicators increased to 2837 and 3169 units, respectively [9]. All this growth process is associated with the sustainable development of enterprises in the non-oil sector in our country, the increase in investment in industry, the expansion of export opportunities of industrial enterprises and private innovative entrepreneurship.

In general, the analysis shows that in the coming years it is expedient to implement the following measures to apply modern innovations in the field of industry in Azerbaijan:

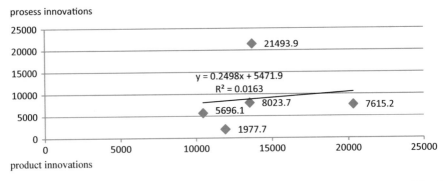

Fig. 3 Correlation between process innovations and product innovation costs in the Republic of Azerbaijan (in thousands of manats) [10]

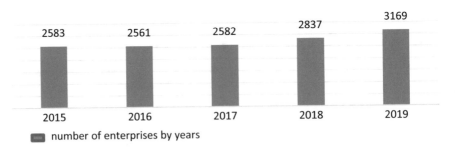

Fig. 4 Number of industrial enterprises operating in the Republic of Azerbaijan on all types of property (by types of property), unit [10]

- In order to ensure the innovation orientation of the industrial sector, it is necessary to develop a comprehensive and balanced development of innovation infrastructure, innovation capital market formation (innovation and venture funds) and information and advisory services in the field of innovation, improvement of legal framework for intellectual property protection.
- It is important to create an organizational and legal framework for the protection of property rights in industrial areas, information transparency and the improvement of methods of regulating the prices of natural monopolies [11].
- It is expedient to carry out new reforms in the tax system related to industry. It should be based on measures aimed at stimulating production, including the development of innovative small and medium enterprises, as well as attracting local and foreign investors [12].
- The role of the state should be increased to create a mechanism and concrete measures to ensure the formation of a national innovation system in our country and to ensure the development of innovative entrepreneurship in industry, etc.

References

1. Shakaraliyev, A.Sh.: Ekonomiceskaya politika gosudarstva: torjestva ustoycivoqo istabilnoqo razvitiya [Economic policy of state: triumph of sustainable and stable development]. Baku, Azerbaijan (2011)
2. www.prezident.az
3. Aliyeva, A.A.: The main promising directions of modern innovative economic development in the republic of Azerbaijan. Int. Econ. Bull. Ukraine **34**(6), 15 (2020)
4. Kolesov, V.P., Kulakov, M.V.: International Economy, p. 345. Moscow: M. (2009)
5. Shirai, V.I.: World Economy and International Economic Relations, p. 528. Dashkov and K. Publishing House, Moscow (2003)
6. Huseynova, A.A.: Innovations in industrial manufacture of Azerbaijan. In: 2012 IV International Conference "Problems of Cybernetics and Informatics" (PCI), pp. 1–2 (2012)
7. Gasimov, F.H.: The state of innovation in Azerbaijan and its development prospects. News of ANAS; Science and Innovation Series, №. 1; Baku, pp. 4–10 2009
8. www.weforum.org/gcr_KlausSchwab/TheGlobalCompetitivenessReport-2019/WorldEconomicForum/Geneva_(Switzerland)
9. Babashkina, A.M.: State regulation of the national economy. Financ. Stat. p. 254 (2017)
10. www.stat.gov.az
11. Kuzmin, D.V.: National competitiveness, global instability and macroeconomic equilibrium. Science, p. 245 (2018)
12. Ivanova, N.: Formation and evolution of national innovation systems. M: IMEMO, p. 355 (2018)

Correction to: Development of a Combined Test Algorithm to Increase the Intelligence of Measurement Systems

Mazahir Isayev, Mahmudbeyli Leyla, and Khasayeva Natavan

Correction to:
Chapter "Development of a Combined Test Algorithm to Increase the Intelligence of Measurement Systems" in: Sh.N. Shahbazova et al. (eds.), *Recent Developments and the New Directions of Research, Foundations, and Applications*, Studies in Fuzziness and Soft Computing 423, https://doi.org/10.1007/978-3-031-23476-7_28

This book was inadvertently published with the incorrect chapter author name, the following belated corrections have been incorporated. The chapter author name "K. Natavan" has been changed to "Khasayeva Natavan" in the Chapter opening page.

The correction chapter and the book have been updated with the changes.

The updated version of this chapter can be found at
https://doi.org/10.1007/978-3-031-23476-7_28

Sh. N. Shahbazova et al. (eds.), *Recent Developments and the New Directions of Research, Foundations, and Applications*, Studies in Fuzziness and Soft Computing 423, https://doi.org/10.1007/978-3-031-23476-7_31

Author Index

Printed in the United States
by Baker & Taylor Publisher Services